TAKING SIDES

Clashing Views on Controversial

Environmental Issues

ELEVENTH EDITION, EXPANDED

W9-ANY-603

TAKING SIDES

Clashing Views on Controversial

Environmental Issues

ELEVENTH EDITION, EXPANDED

Selected, Edited, and with Introductions by

Thomas A. Easton
Thomas College

Photo Acknowledgment
Cover image: PhotoLink/Getty Images

Cover Acknowledgment
Maggie Lytle

Copyright © 2006 by McGraw-Hill Contemporary Learning Series,
A Division of The McGraw-Hill Companies, Inc., Dubuque, IA 52001

Manufactured in the United States of America

Eleventh Edition, Expanded

123456789DOCDOC098765

Library of Congress Cataloging-in-Publication Data
Main entry under title:
Taking sides: clashing views on controversial issues in environmental issues/selected, edited, and with introductions by Thomas A. Easton—11th ed.
Includes bibliographical references and index.
1. Environmental policy. 2. Environmental protection. I. Easton, Thomas A. *comp.*

363.7

0-07-351441-1
1091-8825

Printed on Recycled Paper

Preface

Most fields of academic study evolve over time. Some evolve in turmoil, for they deal with issues of political, social, and economic concern. That is, they involve controversy.

It is the mission of the *Taking Sides* series to capture current, ongoing controversies and make the opposing sides available to students. This book focuses on environmental issues, from the philosophical to the practical. It does not pretend to cover all such issues, for not all provoke controversy or provoke it in suitable fashion. But there is never any shortage of issues that can be expressed as pairs of opposing essays that make their positions clearly and understandably.

The basic technique—presenting an issue as a pair of opposing essays—has risks. Students often display a tendency to remember best those essays that agree with the attitudes they bring to the discussion. They also want to know what the "right" answers are, and it can be difficult for teachers to refrain from taking a side or from revealing their own attitudes. Should teachers so refrain? Some do, though rarely so successfully that students cannot see through the attempt. Some do not, but of course they must still cover the spectrum of opinion if they wish to do justice to the scientific method and the complexity of an issue.

For any *Taking Sides* volume, the issues are always phrased as yes/no questions. Which answer—yes or no—is the correct answer? Perhaps neither. Perhaps both. Perhaps we will not be able to tell for another century. Students should read, think about, and discuss the readings and then come to their own conclusions without letting my or their instructor's opinions dictate theirs. The additional readings mentioned in the introductions and postscripts should prove helpful.

For each issue in this book, an *introduction* provides historical background and a brief description of the debate. The *postscript* that follows each pair of readings offers recent contributions to the debate, additional references, and sometimes a hint of future directions. *On the Internet* page that accompanies each part opener provides Internet site addresses (URLs) that should prove useful as starting points for further research.

Changes to this edition This eleventh edition of *Taking Sides: Clashing Views on Controversial Environmental Issues* contains 38 sections arranged in pro and con pairs to form 19 issues. About half of this book consists of new material. Two issues, *Will Hydrogen Replace Fossil Fuels for Cars?* (Issue 10), and *Should Existing Power Plants Be Required to Install State-of-the-Art Pollution Controls?* (Issue 11), were added for the 2004 partial revision. There are two completely new issues: *Is It Time to Revive Nuclear Power?* (Issue 12) and *Are Marine Reserves Needed to Protect Global Fisheries?* (Issue 15). In addition, for eleven of the issues retained from the previous edition, one or both of the selections have been replaced. In all, 22 of the selections in this edition were not in the tenth edition.

A word to the instructor   An *Instructor's Manual With Test Questions* (multiple-choice and essay) is available through the publisher for the instructor using *Taking Sides* in the classroom. Also available is a general guidebook, *Using Taking Sides in the Classroom*, which offers suggestions for adapting the pro-con approach in any classroom setting. An online version of *Using Taking Sides in the Classroom* and a correspondence service for Taking Sides adopters can be found at http://www.mhcls.com/usingts/.

Taking Sides: Clashing Views on Controversial Environmental Issues is only one title in the Taking Sides series. If you are interested in seeing the table of contents for any of the other titles, please visit the Taking Sides Web site at http://www.mhcls.com/takingsides/.

Thomas A. Easton
Thomas College

Contents In Brief

PART 1   Environmental Philosophy   1

Issue 1.   Is the Precautionary Principle a Sound Basis for International Policy?   2
Issue 2.   Is Sustainable Development Compatible With Human Welfare?   22
Issue 3.   Should a Price Be Put on the Goods and Services Provided by the World's Ecosystems?   39

PART 2   Principles versus Politics   65

Issue 4.   Is Biodiversity Overprotected?   66
Issue 5   Should Environmental Policy Attempt to Cure Environmental Racism?   81
Issue 6.   Can Pollution Rights Trading Effectively Control Environmental Problems?   94
Issue 7.   Do Environmentalists Overstate Their Case?   110

PART 3   Energy Issues   125

Issue 8.   Should the Arctic National Wildlife Refuge Be Opened to Oil Drilling?   126
Issue 9.   Should Society Act Now to Forestall Global Warming?   143
Issue 10. Will Hydrogen End Our Fossil-Fuel Addiction?   156
Issue 11. Should Existing Power Plants Be Required to Install State-of-the-Art Pollution Controls?   168
Issue 12. Is It Time to Revive Nuclear Power?   194

PART 4   Food and Population   213

Issue 13. Is Limiting Population Growth a Key Factor in Protecting the Global Environment?   214
Issue 14. Is Genetic Engineering an Environmentally Sound Way to Increase Food Production?   232
Issue 15. Are Marine Reserves Needed to Protect Global Fisheries?   262

PART 5   Toxic Chemicals   279

Issue 16. Should DDT Be Banned Worldwide?   280
Issue 17. Do Environmental Hormone Mimics Pose a Potentially Serious Health Threat?   300
Issue 18. Is the Superfund Program Successfully Protecting the Environment from Hazardous Wastes?   317
Issue 19. Should the United States Continue to Focus Plans for Permanent Nuclear Waste Disposal Exclusively at Yucca Mountain?   329
Issue 20. Should the Military Be Exempt from Environmental Regulations?   348
Issue 21. Is Additional Federal Oversight Needed for the Construction of LNG Import Facilities?   372

Contents

Preface   v

Introduction   xv

PART 1   ENVIRONMENTAL PHILOSOPHY   1

Issue 1.   Is the Precautionary Principle a Sound Basis for International Policy?   2

YES: Marco Martuzzi and Roberto Bertollini, "The Precautionary Principle, Science and Human Health Protection," *International Journal of Occupational Medicine and Environmental Health* (January 2004)   4

NO: Henry I. Miller and Gregory Conko, from "The Perils of Precaution," *Policy Review* (June & July 2001)   9

Marco Martuzzi and Roberto Bertollini, researchers with the World Health Organization (WHO), argue that although the Precautionary Principle, which demands preventive action in the face of credible threats of harm (even lacking full scientific certainty), may be difficult to apply, it can be valuable in the effort to protect human health and the environment. Henry I. Miller, a research fellow at Stanford University's Hoover Institution, and policy analyst Gregory Conko argue that the precautionary principle leads "regulators to abandon the careful balancing of risks and benefits," blocks progress, limits the freedom of scientific researchers, and restricts consumer choice.

Issue 2.   Is Sustainable Development Compatible With Human Welfare?   22

YES: Dinah M. Payne and Cecily A. Raiborn, from "Sustainable Development: The Ethics Support the Economics," *Journal of Business Ethics* (July 2001)   24

NO: Ronald Bailey, from "Wilting Greens," *Reason* (December 2002)   34

Professor of management Dinah M. Payne and professor of accounting Cecily A. Raiborn argue that environmental responsibility and sustainable development are essential parts of modern business ethics and that only through them can both business and humans thrive. Environmental journalist Ronald Bailey states that sustainable development results in economic stagnation and threatens both the environment and the world's poor.

Issue 3.   Should a Price Be Put on the Goods and Services Provided by the World's Ecosystems?   39

YES: Janet N. Abramovitz, from "Putting a Value on Nature's 'Free' Services," *World Watch* (January/February 1998)   41

NO: Marino Gatto and Giulio A. De Leo, from "Pricing Biodiversity and Ecosystem Services: The Never-Ending Story," *BioScience* (April 2000)   53

Janet N. Abramovitz, a senior researcher at the Worldwatch Institute, argues that if we fail to attach economic value to supposedly free services provided by nature, we are more likely to misuse and destroy the ecosystems that provide those services. Professors of applied ecology Marino Gatto and Giulio A. De Leo contend that the pricing approach to valuing nature's services is misleading because it falsely implies that only economic values matter.

PART 2   PRINCIPLES VERSUS POLITICS   65

Issue 4.   Is Biodiversity Overprotected?   66

YES: David N. Laband, from "Regulating Biodiversity: Tragedy in the Political Commons," *Ideas on Liberty* (September 2001)   68

NO: Howard Youth, from "Silenced Springs: Disappearing Birds," *Futurist* (July/August 2003)   72

Professor of economics David N. Laband argues that the public demands excessive amounts of biodiversity largely because decision makers and voters do not have to bear the costs of producing it. Wildlife conservation researcher and writer Howard Youth argues that the actions needed to protect biodiversity not only have economic benefits but also are the same actions needed to ensure a sustainable future for humanity.

Issue 5.   Should Environmental Policy Attempt to Cure Environmental Racism?   81

YES: Robert D. Bullard, from "Environmental Justice for All," *Crisis (The New)* (January/February 2003)   83

NO: David Friedman, from "The 'Environmental Racism' Hoax," *The American Enterprise* (November/December 1998)   87

Professor of sociology Robert D. Bullard argues that even though environmental racism has been recognized for a quarter of a century, it remains a problem, and Bush administration policies threaten to undo what progress has been achieved. Writer and social analyst David Friedman denies the existence of environmental racism. He argues that the environmental justice movement is a government-sanctioned political ploy that will hurt urban minorities by driving away industrial jobs.

Issue 6.   Can Pollution Rights Trading Effectively Control Environmental Problems?   94

YES: Charles W. Schmidt, from "The Market for Pollution," *Environmental Health Perspectives* (August 2001)   96

NO: Brian Tokar, from "Trading Away the Earth: Pollution Credits and the Perils of 'Free Market Environmentalism,'" *Dollars & Sense* (March/April 1996)   102

Freelance science writer Charles W. Schmidt argues that economic incentives such as emissions rights trading offer the most useful approaches to reducing pollution. Author, college teacher, and environmental activist Brian Tokar maintains that pollution credits and other market-oriented environmental protection policies do nothing to reduce pollution while transferring the power to protect the environment from the public to large corporate polluters.

Issue 7. Do Environmentalists Overstate Their Case? 110

YES: **Ronald Bailey**, from "Debunking Green Myths," *Reason* (February 2002) 112

NO: **David Pimentel**, from "Skeptical of the Skeptical Environmentalist," *Skeptic* (vol. 9, no. 2, 2002) 116

Environmental journalist Ronald Bailey argues that the natural environment is not in trouble, despite the arguments of many environmentalists that it is. He holds that the greatest danger facing the environment is not human activity but "ideological environmentalism, with its hostility to economic growth and technological progress." David Pimentel, a professor of insect ecology and agricultural sciences, argues that those who contend that the environment is not threatened are using data selectively and that the supply of basic resources to support human life is declining rapidly.

PART 3 ENERGY ISSUES 125

Issue 8. Should the Arctic National Wildlife Refuge Be Opened to Oil Drilling? 126

YES: **Dwight R. Lee**, from "To Drill or Not to Drill?" *Independent Review* (Fall 2001) 128

NO: **Katherine Balpataky**, from "Protectors of the Herd," *Canadian Wildlife* (Fall 2003) 137

Professor of economics Dwight R. Lee argues that the economic and other benefits of Arctic National Wildlife Refuge (ANWR) oil are so great that even environmentalists should agree to permit drilling—and they probably would if they stood to benefit directly. Katherine Balpataky argues that cost-benefit analyses do not support the case for drilling in the ANWR and that the damage done by drilling both to the environment and to the traditional values of the indigenous people, the Gwich'in, cannot be tolerated.

Issue 9. Should Society Act Now to Forestall Global Warming? 143

YES: **George Marshall and Mark Lynas**, from "Why We Don't Give a Damn," *New Statesman* (December 2003) 145

NO: **Stephen Goode**, from "Singer Cool on Global Warming," *Insight on the News* (April 27, 2004) 150

George Marshall and Mark Lynas argue that despite a remarkable level of agreement that the threat of global warming is real, human psychology keeps us "in denial." But survival demands that we escape denial and seek more positive action. Long-time anti-global warming spokesman Fred Singer argues in an interview by Stephen Goode that global warming just is not happening in any significant way and if it were, it would—judging from the past—be good for humanity.

Issue 10. Will Hydrogen End Our Fossil-Fuel Addiction? 156

YES: **Jeremy Rifkin**, from "Hydrogen: Empowering the People," *The Nation* (December 23, 2002)   158

NO: **Henry Payne and Diane Katz**, from "Gas and Gasbags... Or, the Open Road and Its Enemies," *National Review* (March 25, 2002)   162

Social activist Jeremy Rifkin maintains that fossil fuels are approaching the end of their usefulness and that hydrogen fuel holds the potential not only to replace them but also to reshape society. Writer Henry Payne and director of science, environment, and technology policy at the Mackinac Center for Public Policy Diane Katz argue that hydrogen can only be made widely available if society invests heavily in nuclear power. Market mechanisms will keep fossil fuels in play for years to come.

Issue 11. Should Existing Power Plants Be Required to Install State-of-the-Art Pollution Controls? 168

YES: **Eliot Spitzer**, from Testimony before the United States Senate Committee on Environment and Public Works and the Committee on the Judiciary (July 16, 2002)   170

NO: **Jeffrey Holmstead**, from Testimony before the United States Senate Committee on Environment and Public Works and the Committee on the Judiciary (July 16, 2002)   186

New York attorney general Eliot Spitzer states that removing regulatory requirements for power plant pollution controls will prevent needed improvements in air quality. Environmental Protection Agency assistant administrator Jeffrey Holmstead argues that removing regulatory requirements for power plant pollution controls in favor of a markets-based approach will improve air quality.

Issue 12. Is It Time to Revive Nuclear Power? 194

YES: **Stephen Ansolabehere, et al.**, from "The Future of Nuclear Power," *An Interdisciplinary MIT Study* (MIT 2003)   196

NO: **Karl Grossman**, from "The Push to Revive Nuclear Power," *Synthesis/Regeneration* 28 (http://www.greens.org/s-r/28/28-21.html) (Spring 2002)   205

Professor Stephen Ansolabehere, et al. argue that greatly expanded use of nuclear power should not be excluded as a way to meet future energy needs and reduce the carbon emissions that contribute to global warming. Professor of journalism Karl Grossman argues that to encourage the use of nuclear power is reckless. He concludes that it would be wiser to promote renewable energy and energy efficiency.

PART 4   FOOD AND POPULATION   213

Issue 13. Is Limiting Population Growth a Key Factor in Protecting the Global Environment? 214

YES: **Lester R. Brown**, from "Rescuing a Planet under Stress," *The Humanist* (November/December 2003)   216

NO: **Stephen Moore**, from "Body Count," *National Review* (October 25, 1999)   224

Lester R. Brown, president of the Earth Policy Institute, argues that stabilizing world population is central to preventing overconsumption of environmental resources. Stephen Moore, director of the Cato Institute, argues that human numbers pose no threat to human survival or the environment but that efforts to control population do threaten human freedom and worth.

Issue 14.   Is Genetic Engineering an Environmentally Sound Way to Increase Food Production?   232

YES: **Royal Society of London et al.**, from "Transgenic Plants and World Agriculture," A Report Prepared Under the Auspices of the Royal Society of London, the U.S. National Academy of Sciences, the Brazilian Academy of Sciences, the Chinese Academy of Sciences, the Indian National Science Academy, the Mexican Academy of Sciences, and the Third World Academy of Sciences (July 2000)   234

NO: **Brian Halweil**, from "The Emperor's New Crops," *World Watch* (July/August 1999)   249

The national academies of science of the United Kingdom, the United States, Brazil, China, India, Mexico, and the Third World argue that genetically modified crops hold the potential to feed the world during the twenty-first century while also protecting the environment. Brian Halweil, a researcher at the Worldwatch Institute, argues that the genetic modification of crops threatens to produce pesticide-resistant insect pests and herbicide-resistant weeds, will victimize poor farmers, and is unlikely to feed the world.

Issue 15.   Are Marine Reserves Needed to Protect Global Fisheries?   262

YES: **Robert R. Warner**, from "Marine Protected Areas," Statement Before the Subcommittee on Fisheries Conservation, Wildlife and Oceans Committee on House Resources, United States House of Representatives (May 23, 2002)   264

NO: **Sean Paige**, from "Zoned to Extinction," *Reason* (October 2001)   269

Professor of marine ecology Robert R. Warner argues that marine reserves, areas of the ocean completely protected from all extractive activities such as fishing, can be a useful tool for preserving ecosystems and restoring productive fisheries. Sean Paige, a fellow at the market-oriented Competitive Enterprise Institute, argues that marine reserves are based on immature and uncertain science and that they will have a direct and detrimental effect on commercial fishermen.

PART 5   TOXIC CHEMICALS   279

Issue 16.   Should DDT Be Banned Worldwide?   280

YES: **Anne Platt McGinn**, from "Malaria, Mosquitoes, and DDT," *World Watch* (May/June 2002)   282

NO: **Alexander Gourevitch**, "Better Living through Chemistry," *Washington Monthly* (March 2003)   291

Anne Platt McGinn, a senior researcher at the Worldwatch Institute, argues that although DDT is still used to fight malaria, there are other, more effective and less environmentally harmful methods. She maintains that DDT should be banned or reserved for emergency use. Alexander Gourevitch, an *American Prospect* writing fellow, argues that, properly used, DDT is not as dangerous as its reputation insists and that it remains the cheapest and most effective way to combat malaria.

Issue 17.   Do Environmental Hormone Mimics Pose a Potentially Serious Health Threat?   300

YES: **Michele L. Trankina**, from "The Hazards of Environmental Estrogens," *The World & I* (October 2001)   302

NO: **Michael Gough**, from "Endocrine Disrupters, Politics, Pesticides, the Cost of Food and Health," *Cato Institute* (December 15, 1997)   308

Professor of biological sciences Michele L. Trankina argues that a great many synthetic chemicals behave like estrogen, alter the reproductive functioning of wildlife, and may have serious health effects—including cancer—on humans. Michael Gough, a biologist and expert on risk assessment and environmental policy, argues that only "junk science" supports the hazards of environmental estrogens.

Issue 18.   Is the Superfund Program Successfully Protecting the Environment from Hazardous Wastes?   317

YES: **Robert H. Harris, Jay Vandeven, and Mike Tilchin**, from "Superfund Matures Gracefully," *Issues in Science & Technology* (Summer 2003)   319

NO: **Margot Roosevelt**, from "The Tragedy of Tar Creek," *Time* (April 26, 2004)   323

Environmental consultants Robert H. Harris, Jay Vandeven, and Mike Tilchin argue that although the Superfund program still has room for improvement, it has made great progress in risk assessment and treatment technologies. Journalist Margot Roosevelt argues that because one-quarter of Americans live near Superfund sites, and sites like Tar Creek, Oklahoma, remain hazardous, Superfund's work is clearly not getting done.

Issue 19.   Should the United States Continue to Focus Plans for Permanent Nuclear Waste Disposal Exclusively at Yucca Mountain?   329

YES: **Spencer Abraham**, from *Recommendation by the Secretary of Energy Regarding the Suitability of the Yucca Mountain Site for a Repository Under the Nuclear Waste Policy Act of 1982* (February 2002)   331

NO: **Gar Smith**, from "A Gift to Terrorists?" *Earth Island Journal* (Winter 2002–2003)   342

Secretary of Energy Spencer Abraham argues that the Yucca Mountain, Nevada, nuclear waste disposal site is suitable technically and scientifically and that its development serves the U.S. national interest in numerous ways. Environmentalist writer Gar Smith argues that transporting nuclear waste to Yucca Mountain will expose millions of Americans to risks from accidents and terrorists.

Issue 20.   Should the Military Be Exempt from Environmental Regulations?   348

YES: **Benedict S. Cohen**, from "Impact of Military Training on the Environment," Testimony before the Senate Committee on Enviornment and Public Works (April 2, 2003)   350

NO: **Jamie Clark**, from "Impact of Military Training on the Environment," Testimony before the Senate Committee on Environment and Public Works (April 2, 2003)   361

Benedict S. Choen argues that environmental regulations interfere with military training and other "readiness" activities, and that though the U.S. Department of Defense will continue "to provide exemplary stewardship of the lands and natural resources in our trust," those regulations must be revised to permit the military to do its job without interference. Jamie Clark argues that reducing the Department of Defense's environmental obligations is dangerous because both people and wildlife would be threatened with serious, irreversible, and unnecessary harm.

Issue 21.   Is Additional Federal Oversight Needed for the Construction of LNG Import Facilities?   372

YES: **Edward J. Markey**, from "LNG Import Terminal and Deepwater Port Siting: Federal and State Roles," Testimony before House Committee on Government Reform, Subcommittee on Energy Policy, Natural Resources, and Regulatory Affairs (June 22, 2004)   374

NO: **Donald F. Santa, Jr.**, from "LNG Import Terminal and Deepwater Port Siting: Federal and State Roles," Testimony before House Committee on Government Reform, Subcommittee on Energy Policy, Natural Resources, and Regulatory Affairs (June 22, 2004)   382

Edward J. Markey argues that the risks—including those associated with terrorist attack—associated with LNG (liquified natural gas) tankers and terminals are so great that additional federal regulation is essential in order to protect the public. Donald F. Santa, Jr., argues that meeting demand for energy requires public policies that "do not unreasonably limit resource and infrastructure development." The permitting process for LNG import facilities should be governed by existing Federal Regulatory Commission procedures without additional regulatory impediments.

Contributors 390

Index 395

Introduction

Environmental Issues:
The Never-Ending Debate

Thomas A. Easton

One of the courses I teach is "Environmentalism: Philosophy, Ethics, and History." I begin by explaining the roots of the word "ecology," from the Greek *oikos* (house or household), and assigning the students to write a brief paper about their own household. How much, I ask them, do you need to know about the place where you live? And why?

The answers vary. Some of the resulting papers focus on people—roommates if the "household" is a dorm room, spouses and children if the students are older, parents and siblings if they live at home—and the needs to cooperate and get along, and perhaps the need not to overcrowd. Some pay attention to houseplants and pets, and occasionally even bugs and mice. Some focus on economics—possessions, services, and their costs, where the checkbook is kept, where the bills accumulate, the importance of paying those bills, and of course the importance of earning money to pay those bills. Some focus on maintenance—cleaning, cleaning supplies, repairs, whom to call if something major breaks. For some the emphasis is operation—garbage disposal, grocery shopping, how to work the lights, stove, fridge, and so on. A very few recognize the presence of toxic chemicals under the sink and in the medicine cabinet and the need for precautions in their handling. Sadly, a few seem to be oblivious to anything that does not have something to do with entertainment.

Not surprisingly, some students object initially that the exercise seems trivial. "What does this have to do with environmentalism?" they ask. Yet the course is rarely very old before most are saying, "Ah! I get it!" That nice, homey microcosm has a great many of the features of the macrocosmic environment, and the multiple ways people can look at the microcosm mirror the ways people look at the macrocosm. It's all there, as is the question of priorities: What is important? People, fellow creatures, economics, maintenance, operation, waste disposal, food supply, toxics control, or entertainment? Or are they all equally important?

And how does one decide? I try to illuminate this question by describing a parent trying to teach a teenager not to sit on a woodstove. In July, the kid asks, "Why?" and continues to perch. In August, likewise. And still in September. But in October or November, the kid yells "Ouch!" and jumps off in a hurry.

That is, people seem to learn best when they get burned.

This is surely true in our homely *oikos*, where we may not realize our fellow creatures deserve attention until the houseplants die of neglect or cockroaches invade the cupboards. Similarly, economics comes to the fore when the phone gets cut off, repairs when a pipe ruptures, air quality when the air conditioner breaks or strange fumes rise from the basement, and garbage disposal when the bags pile up and begin to stink. Toxics control suddenly matters when a child or pet gets into the rat poison.

In the larger *oikos* of environmentalism, such events are analogous to the loss of a species or an infestation by another, to floods and droughts, to lakes turned into cesspits by raw sewage, to air turned foul by industrial smokestacks, to the contamination of groundwater by toxic chemicals, to the death of industries and the loss of jobs, and to famine and plague and even war.

If nothing is wrong in our households, we are not very likely to realize there is something we should be paying attention to. And this too has its parallel in the larger world. Indeed, the history of environmentalism is in important part a history of people carrying on with business as usual until something goes obviously awry. Then, if they can agree on the nature of the problem (Did the floor cave in because the joists were rotten or because there were too many people at the party?), they might learn something about how to prevent its recurrence.

The Question of Priorities

There is of course a crucial "if" in that last sentence: If people can agree ... It is a truism to say that agreement is difficult. In environmental matters, people argue endlessly over whether anything is actually wrong, what its eventual impact will be, what (if anything) can or should be done to repair the damage, and how to prevent recurrence. Not to mention who is to blame and who should take responsibility for fixing the problem! Part of the reason is simple: Different things matter to different people. For example, individual citizens might want clean air and water, or cheap food, or a convenient commute. Politicians might favor sovereignty over international cooperation. Economists and industrialists might think a few coughs (or worse) a cheap price to pay for wealth or jobs.

No one now seems to think that protecting the environment is not important. But different groups—even different environmentalists—have different ideas of what "environmental responsibility" means. To a paper company that cuts trees for pulp, it might mean leaving a screen of trees (a "beauty strip") beside the road and minimizing erosion. To hikers following trails through or within view of the same tract of land, that might not be enough; they might want the trees left alone. The hikers might also object to people using trail bikes and all-terrain-vehicles on the trails. They might even object to hunters and anglers, whose activities they see as diminishing the wilderness experience. They might push for protecting the land as limited-access wilderness. The hunters and anglers would object to that, of course, because they want to be able to use their vehicles to bring their game home, or to bring their boats to their favorite rivers and lakes. They could also argue, with some justification, that their license fees support a great deal of environmental protection work.

To a corporation, dumping industrial waste into a river might make perfect sense, for alternative ways of disposing of waste are likely to cost more and diminish profits. Of course, the waste renders the water less useful to wildlife or downstream humans, who might well object. Yet preventing the corporation from dumping might be seen as depriving it of property. A similar problem arises when regulations prevent people and corporations from using land—and making money—as they had planned. Conservatives have argued that environmental regulations thus violate the Fifth Amendment to the U.S. Constitution, which says "No person shall ... be deprived of ... property, without due process of law; nor shall private property be taken for public use, without just compensation."

One might think that the dangers of such things as dumping industrial waste into rivers are obvious. But scientists can and do disagree about the consequences of such activities, even given the same evidence. For instance, a chemical in waste may clearly cause cancer in laboratory animals. Is it therefore a danger to humans? A scientist working for a company that is dumping that chemical in a river might maintain that no such danger has been proven. Yet a scientist working for an environmental group such as Greenpeace might insist that the danger is obvious because carcinogens generally affect more than one species.

Scientists are human. They have values rooted in political ideology and religion. They might feel that the individual matters more than corporations or society, or vice versa. They might also favor short-term benefits over long-term benefits, or vice versa. And scientists, citizens, corporations, and government all reflect prevailing social attitudes. When America was expanding westward, the focus was on building industries, farms, and towns. When problems arose, there was vacant land waiting to be moved to. But when the expansion was done, problems became more visible and less avoidable. People could see that there were trade-offs involved in human activity: more industry meant more jobs and more wealth, but there was a price in air and water pollution and adverse effects on human health (among other things).

Nowhere, perhaps, are these trade-offs more obvious than in the former Soviet Union, which was infamous for refusing to admit that industrial activity was anything but desirable. Any citizen who spoke up about environmental problems risked being jailed. The result, which became visible to Western nations after the fall of the Iron Curtain in 1990, was industrial zones in which rivers had no fish, children were sickly, and life expectancies were reduced. The fate of the Aral Sea, a vast inland body of water once home to a thriving fishery and a major regional transportation route, is emblematic: Because the Soviet Union wanted to increase its cotton production, it diverted the rivers that delivered most of the Aral Sea's fresh water supply for irrigation. The Aral Sea then began to lose more water to evaporation than it gained, and it rapidly shrank, exposing a sea-bottom that is so contaminated by industrial wastes and pesticides that wind-borne dust is now responsible for a great deal of human illness. The fisheries are dead, and freighters lie rusting on bare ground where waves once lapped.

The Environmental Movement

The twentieth century saw immense changes in the conditions of human life and in the environment that surrounds and supports human life. According to historian J. R. McNeill, in *Something New Under the Sun: An Environmental History of the Twentieth-Century World* (W. W. Norton, 2000), the environmental impacts that resulted from the interactions of burgeoning population, technological development, shifts in energy use, politics, and economics during that period are unprecedented in both degree and kind. Yet a worse impact might be that we have come to accept as "normal" a very temporary situation that "is an extreme deviation from any of the durable, more 'normal,' states of the world over the span of human history, indeed over the span of earth history." Thus, people are not prepared for the inevitable and perhaps drastic changes ahead.

Environmental factors cannot be denied their role in human affairs. Nor can human affairs be denied their place in any effort to understand environmental change. As McNeill says, "Both history and ecology are, as fields of knowledge go, supremely integrative. They merely need to integrate with each other."

The environmental movement, which grew during the twentieth century in response to increasing awareness of human impacts, is a step in that direction. Yet environmental awareness was evident long before the modern environmental movement. When he was young, John James Audubon (1785–1851), famous for his bird paintings, was an enthusiastic slaughterer of birds (some of which he used as models for his paintings). Later in life, he came to appreciate that birds were diminishing in numbers, as was the American bison, and he called for conservation measures. His was a minority voice, however. It was not until later in the century that John Muir (1838–1914), founder of the Sierra Club, began to call for the preservation of natural wilderness, untouched by human activities. In 1890, Gifford Pinchot (1865–1946) found "the nation ... obsessed by a fury of development. The American Colossus was fiercely intent on appropriating and exploiting the riches of the richest of all continents." Under President Theodore Roosevelt, he became the first head of the U.S. Forest Service and a strong voice for conservation (not to be confused with preservation; Gifford's conservation meant using nature but in such a way that it was not destroyed; his aim was "the greatest good of the greatest number in the long run"). In the 1930s, Aldo Leopold (1887–1948), best known for his concept of the "land ethic" and his book, *A Sand County Almanac, and Sketches Here and There* (Oxford University Press, 1949), argued that people had a responsibility not only to maintain the environment but also to repair damage that was done in the past.

The modern environmental movement was kick-started by Rachel Carson's *Silent Spring* (Houghton Mifflin, 1962). In the 1950s, Carson realized that the use of pesticides was having unintended consequences—the death of non-pest insects, food-chain accumulation of poisons and the consequent loss of birds, and even human illness—and meticulously documented the case. When her book was published, she and it were immediately vilified by pesticide proponents in government, academia, and industry (most notably, the pesticides industry). There was no problem, the critics said; the negative effects, if any, were outweighed by the benefits; and she—a *woman* and a nonscientist—could not possibly know what she was talking about. But the facts won out. A decade later, DDT was banned and other pesticides were regulated in ways unheard of before Carson spoke out.

Other issues have followed or are following a similar course.

The situation before Rachel Carson and *Silent Spring* is nicely captured by Judge Richard Cudahy, who in "Coming of Age in the Environment," *Environmental Law* (Winter 2000), writes, "It doesn't seem possible that before 1960 there was no 'environment'—or at least no environmentalism. I can even remember the Thirties, when we all heedlessly threw our trash out of car windows, burned coal in the home furnace (if we could afford to buy any), and used a lot of lead for everything from fishing sinkers and paint to no-knock gasoline. Those were the days when belching black smoke meant a welcome end to the Depression and little else."

Historically, humans have felt that their own well-being mattered more than anything else. The environment existed to be used. Unused, it was only wilderness or wasteland, awaiting the human hand to "improve" it and make it valuable. This is not surprising, for the natural tendency of the human mind is to appraise all things in relation to the self, the family, and the tribe. An important aspect of human progress has lain in enlarging our sense of "tribe" to encompass nations and groups of nations. Some now take it as far as the human species. Some include other animals. Some embrace plants as well, and bacteria, and even landscapes, as well.

The more limited standard of value remains common. Add to that a sense that wealth is not just desirable but a sign of virtue (the Puritans brought an explicit version of this with them when they colonized North America; see Lynn White, Jr., "The Historical Roots of Our Ecological Crisis," *Science* [March 10, 1967]), and it is hardly surprising that humans have used and still use the environment intensely. People also tend to resist any suggestion that they should restrain their use out of regard for other living things. Human needs, many insist, come first.

The unfortunate consequences include the loss of other species. For example, lions vanished from Europe about 2000 years ago. The dodo of Mauritius was extinguished in the 1600s (see the American Museum of Natural History's account at http://www.amnh.org/exhibitions/expeditions/treasure_fossil/Treasures/Dodo/dodo.html?dinos). And the last of North America's passenger pigeons died in a Cincinnati zoo in 1914 (see http://www.amnh.org/exhibitions/expeditions/treasure_fossil/Treasures/Passenger_Pigeons/pigeons.html?dinos). Concern for species was at first limited to those of obvious value to humans. In 1871, the U.S. Commission on Fish and Fisheries was created and charged with finding solutions to the decline in food fishes and with promoting aquaculture; the first federal legislation designed to protect game animals was the Lacey Act of 1900. It was not until 1973 that the U.S. Endangered Species Act adopted to shield all species from human impacts.

Other unfortunate consequences of human activities include dramatic erosion, air and water pollution, oil spills, accumulations of hazardous (including nuclear) waste, famine, and disease. Among the many "hot stove" incidents that have caught public attention are the following:

- The Dust Bowl—in 1934 wind blew soil from drought-stricken farms in Oklahoma all the way to Washington, DC;
- Cleveland's Cuyahoga River caught fire in the 1960s;
- The Donora, Pennsylvania, smog crisis—in one week of October 1948, 20 died and over 7,000 were sickened;
- The London smog crisis in December 1952—4000 dead;
- The *Torrey Canyon* and *Exxon Valdez* oil spills, which fouled shores and killed seabirds, seals, and fish;
- Love Canal, where industrial wastes seeped from their burial site into homes and contaminated ground water;
- Union Carbide's toxics release at Bhopal, India—3,800 dead and up to 100,000 ill, according to Union Carbide; others claim a higher toll;
- The Three Mile Island and Chernobyl nuclear accidents;
- The decimation of elephants and rhinoceroses to satisfy a market for tusks and horns;
- The loss of forests—in 1997, fires set to clear Southeast Asian forest lands produced so much smoke that regional airports had to close;
- Ebola, a virus that kills nine tenths of those it infects, apparently first struck humans because growing populations reached into its native habitat;
- West Nile Fever, a mosquito-borne virus with a much less deadly record, was brought to North America by travelers or immigrants from Egypt;
- Acid rain, global climate change, and ozone depletion, all caused by substances released into the air by human activities.

The alarms have been raised by many people in addition to Rachel Carson. For instance, in 1968 (when world population was only a little over half of what it is today), Paul Ehrlich described the ecological threats of a rapidly growing population in *The Population Bomb* (Ballantine Books) and Garrett Hardin described the consequences of using self-interest alone to guide the exploitation of publicly-owned resources (such as air and water) in his influential essay, "The Tragedy of the Commons," *Science* (December 13, 1968). (In 1974, Hardin introduced the unpleasant concept of "lifeboat ethics," which says that if there are not enough resources to go around, some people must do without). In 1972, a group of economists, scientists, and business leaders calling themselves "The Club of Rome" published *The Limits to Growth* (Universe Books), an analysis of population, resource use, and pollution trends that predicted difficult times within a century; the study was redone in 1992 using more powerful computer models, and the researchers came to very similar conclusions (see *Beyond the Limits: Confronting Global Collapse, Envisioning a Sustainable Future* [Chelsea Green, 1992]).

The following list of selected U.S. and U.N. laws, treaties, conferences, and reports illustrates the national and international responses to these alarms:

1967   The U.S. Air Quality Act set standards for air pollution.

1968   The U.N. Biosphere Conference discussed global environmental problems.

1969   The U.S. Congress passed the National Environmental Policy Act, which (among other things) required federal agencies to prepare environmental impact statements for their projects.

1970   The first Earth Day demonstrated so much public concern that the Environmental Protection Agency (EPA) was created; the Endangered Species Act, Clean Air Act, and Safe Drinking Water Act soon followed.

1971   The U.S. Environmental Pesticide Control Act gave the EPA authority to regulate pesticides.

1972   The U.N. Conference on the Human Environment, held in Stockholm, Sweden, recommended government action and led to the U.N. Environment Programme.

1973   The Convention on International Trade in Endangered Species of Wild Fauna and Flora (CITES) restricted trade in threatened species; because enforcement was weak, however, a black market flourished.

1976   The U.S. Resource Conservation and Recovery Act and the Toxic Substances Control Act established control over hazardous wastes and other toxic substances.

1979   The Convention on Long-Range Transboundary Air Pollution addressed problems such as acid rain (recognized as crossing national borders in 1972).

1982   The Law of the Sea addressed marine pollution and conservation.

1982   The second U.N. Conference on the Human Environment (the Stockholm +10 Conference) renewed concerns and set up a commission to prepare a "global agenda for change," leading to the 1987 Brundtland Report (*Our Common Future*).

1983   The U.S. Environmental Protection Agency and the U.S. National Academy of Science issued reports calling attention to the prospect of global warming as a consequence of the release of greenhouse gases such as carbon dioxide.

1987   The Montreal Protocol (strengthened in 1992) required nations to phase out the use of chlorofluorocarbons (CFCs), the chemicals responsible for stratospheric ozone depletion.

1987   The Basel Convention controlled cross-border movement of hazardous wastes.

1988   The U.N. assembled the Intergovernmental Panel on Climate Change, which would report in 1995, 1998, and 2001 that the dangers of global warming were real, large, and increasingly ominous.

1992   The U.N. Convention on Biological Diversity required nations to act to protect species diversity.

1992   The U.N. Conference on Environment and Development (also known as the Earth Summit), held in Rio de Janeiro, Brazil, issued a broad call for environmental protections.

1992   The U.N. Convention on Climate Change urged restrictions on carbon dioxide release to avoid climate change

1994   The U.N. Conference on Population and Development, held in Cairo, Egypt, called for the stabilization and reduction of global population growth, largely by improving women's access to education and health care.

1997   The Kyoto Protocol attempted to strengthen the 1992 Convention on Climate Change by requiring reductions in carbon dioxide emissions, but U.S. resistance limited its success.

2001   The UN Stockholm Convention on Persistent Organic Pollutants required nations to phase out use of many pesticides and other chemicals. It took effect May 17, 2004, after ratification by over fifty nations (not including the United States and the European Union).

2002   The U.N. World Summit on Sustainable Development, held in Johannesburg, South Africa, brought together representatives of governments, nongovernmental organizations, businesses, and other groups to examine "difficult challenges, including improving people's lives and conserving our natural resources in a world that is growing in population, with ever-increasing demands for food, water, shelter, sanitation, energy, health services and economic security."

2003   The World Climate Change Conference, held in Moscow, Russia, concluded that global climate is changing, very possibly because of human activities, and the overall issue must be viewed as one of intergenerational justice. "Mitigating global climate change will be possible only with the coordinated actions of all sectors of society."

Rachel Carson would surely have been pleased by these responses, for they suggest both concern over the problems identified and determination to solve those problems. But she would just as surely have been frustrated, for a simple listing of laws, treaties, and reports does nothing to reveal the endless wrangling and the way political and business forces try to block progress whenever it is seen as interfering with their interests. Agreement on banning chlorofluorocarbons was relatively easy to achieve because CFCs were not seen as essential to civilization and because substitutes were available. Restraining greenhouse gas emissions is harder because fossil fuels are considered essential. Also, although fuel substitutes do exist, they are more expensive.

The Globalization of the Environment

Years ago, environmental problems were largely seen as local. A smokestack belched smoke and made the air foul. A city sulked beneath a layer of smog. Bison or passenger pigeons declined in numbers and even vanished. Rats flourished in a dump where burning garbage produced clouds of smoke and runoff contaminated streams and groundwater and made wells unusable. Sewage, chemical wastes, and oil killed the fish in streams, lakes, rivers, and harbors. Toxic chemicals such as lead and mercury entered the food chain and affected the health of both wildlife and people.

By the 1960s, it was becoming clear that environmental problems did not respect borders. Smoke blows with the wind, carrying one locality's contamination to others. Water flows to the sea, carrying sewage and other wastes with it. Birds migrate, carrying with them whatever toxins they have absorbed with their food. In 1972, researchers were able to report that most of the acid rain falling on Sweden came from other countries. Other researchers have shown that the rise and fall of the Roman Empire can be tracked in Greenland, where glaciers preserve lead-containing dust deposited over the millennia—the amount rises as Rome flourished, falls with the Dark Ages, and rises again with the Renaissance and Industrial Revolution. Today it is common knowledge that pesticides and other chemicals can show up in places where they have never been used (such as the Arctic), even years after their use has been discontinued. The 1979 Convention on Long-Range Transboundary Air Pollution has been strengthened several times with amendments to address persistent organic pollutants, heavy metals, and other pollutants.

There are also new environmental problems that exist only in a global sense. Ozone depletion, first identified in the stratosphere over Antarctica, threatens to increase the amount of ultraviolet light reaching the ground, and thereby increase the incidence of skin cancer and cataracts, among other things. The cause is the use of chlorofluorocarbons (CFCs) in refrigeration, air conditioning, aerosol cans, and electronics (for cleaning grease off circuit boards). The effect is global. Worse yet, the cause is rooted in northern lands such as the United States and Europe, but the worst effects may be felt where the sun shines brightest—in the tropics, which are dominated by developing nations. A serious issue of justice or equity is therefore involved.

A similar problem arises with global warming, which is also rooted in the industrialized world and its use of fossil fuels. The expected climate effects will hurt worst the poorer nations of the tropics, and perhaps worst of all those on low-lying South Pacific islands, which are expecting to be wholly inundated by rising seas.

Both the developed and the developing world are aware of the difficulties posed by environmental issues. In Europe, "green" political parties play a growing part in government. In Japan, some environmental regulations are more demanding than those of the United States. Developing nations understandably place dealing with their growing populations high on their list of priorities, but they play an important role in UN conferences on environmental issues, often demanding more responsible behavior from developed nations such as the United States (which often resists these demands; it has refused to ratify international agreements such as the Kyoto Protocol, for example).

Western scholars have been known to suggest that developing nations should forgo industrial development because if their huge populations ever attain the same per-capita environmental impact as the populations of wealthier lands, the world will be laid waste. It is not hard to understand why the developing nations object to such suggestions; they too want a better standard of living. Nor do they think it fair that they should suffer for the environmental sins of others.

Are global environmental problems so threatening that nations must surrender their sovereignty to international bodies? Should the US or Europe have to change energy supplies to protect South Pacific nations? Should developing nations be obliged to reduce birth rates or forgo development because their population growth is seen as exacerbating pollution or threatening biodiversity?

Questions such as these play an important part in global debates today. They are not easy to answer, but their very existence says something important about the general field of environmental studies. This field is based in the science of ecology, a word whose root is that same *oikos* with which I began. Ecology focuses on living things and their interactions with each other and their surroundings. It deals with resources, limits, and coexistence. It can see problems, their causes, and even potential solutions. And it can turn its attention to human beings as easily as it can to deer mice.

Yet human beings are not mice. We have economies and political systems, vested interests, and conflicting priorities and values. Ecology is only one part of environmental studies. Other sciences—chemistry, physics, climatology, epidemiology, geology, and more—are involved. So are economics, history, law, and politics. Even religion can play a part.

Unfortunately, no one field sees enough of the whole to predict problems (the chemists who developed CFCs could hardly have been expected to realize what would happen when these chemicals reached the stratosphere). Environmental studies is a field for teams. That is, it is a holistic, multidisciplinary field.

This gives us an important basic principle to use when evaluating arguments on either side of any environmental issue: Arguments that fail to recognize the complexity of the issue are necessarily suspect. On the other hand, arguments that endeavor to convey the full complexity of an issue may be impossible to understand. A middle ground is essential for clarity, but any reader or student must realize that something important might be left out.

Current Environmental Issues

In 2001, the National Research Council's Committee on Grand Challenges in Environmental Sciences published *Grand Challenges in Environmental Sciences* (National Academy Press) in an effort to reach "a judgment regarding the most important environmental research challenges of the next generation—the areas most likely to yield results of major scientific and practical importance if pursued vigorously now." These areas include the following:

- Biogeochemical cycles (the cycling of plant nutrients, the ways human activities affect them, and the consequences for ecosystem functioning, atmospheric chemistry, and human activities)
- Biological diversity
- Climate variability
- Hydrologic forecasting (groundwater, droughts, floods, etc.)
- Infectious diseases
- Resource use
- Land use
- Reinventing the use of materials (e.g., recycling)

Similar themes appeared when Issues in *Science and Technology* celebrated its twentieth anniversary with its Summer 2003 issue. The editors noted that over the life of the magazine to date, some problems have hardly changed, nor has our sense of what must be done to solve them. Others have been affected, sometimes drastically, by changes in scientific knowledge, technological capability, and political trends. In the environmental area, the magazine paid special attention to:

- Biodiversity
- Overfishing
- Climate change
- The Superfund program
- The potential revival of nuclear power
- Sustainability.

Many of the same basic themes were reiterated when *Science* magazine (published weekly by the American Association for the Advancement of Science) published in November and December 2003 a four-week series on the "State of the Planet," followed by a special issue on "The Tragedy of the Commons." In the introduction to the series, H. Jesse Smith began with these words: "Once in a while, in our headlong rush toward greater prosperity, it is wise to ask ourselves whether or not we can get there from here. As global population increases, and the demands we make on our natural resources grow even faster, it becomes ever more clear that the well-being we seek is imperiled by what we do."

Among the topics covered in the series were:
- Human population
- Biodiversity
- Tropical soils and food security
- The future of fisheries
- Freshwater resources
- Energy resources
- Air quality and pollution
- Climate change
- Sustainability
- The burden of chronic disease

Many of the topics on these lists are covered in this book. There are, of course, a great many other environmental issues—many more than can be covered in any one book such as this one. I have not tried to deal here with invasive species, the Endangered Species Act, the removal of dams to restore populations of anadromous fishes such as salmon, the depletion of aquifers, floodplain development, urban planning, nor many others. My sample of the variety of available issues begins with the more philosophical ones. For instance, there is considerable debate over the "precautionary principle," which says in essence that even if we are not sure that our actions will have unfortunate consequences, we should take precautions just in case (see Issue 1). This principle plays an important part in many environmental debates, from those over the value of preserving biodiversity (see Issue 4), to the wisdom of opening the Arctic National Wildlife Refuge, to oil drilling (see Issue 8), to the folly (or wisdom) of burying nuclear waste under Yucca Mountain in Nevada (see Issue 19).

I said above that many people believed (and still believe) that nature has value only when turned to human benefit. One consequence of this belief is that it may be easier to convince people that nature is worth protecting if one can somehow calculate a cash value for nature "in the raw." Some environmentalists object to attmept to do this, on the grounds that economic value is not the only value, or even the value that should matter (see Issue 3).

What other values might be considered? Previous editions of this book have considered whether nature has a value all its own or a right to exist unmolested, and whether or not human property rights should take precedence. Here we discuss whether or not we should strive for social justice (see Issue 5).

Should we be concerned about the environmental impacts of specific human actions or products? Here, too, we can consider the wisdom of opening the Arctic National Wildlife Refuge to oil drilling (see Issue 8), as well as the conflict between the value of DDT for preventing malaria and its impact on ecosystems (see Issue 16), the hormone-like effects of some pesticides and other chemicals on both wildlife and humans (see Issue 17), and the hazards of air pollutants (Issue 11) and global warming (see Issue 9). Genetic engineering promises to do wonders for food production, but some worry about effects on ecosystems (see Issue 14).

Waste disposal is a problem area all its own. It encompasses both nuclear waste (see Issue 19) and hazardous waste (see Issue 18). A new angle on hazardous waste comes from the popularity of the personal computer—or more specifically, from the huge numbers of computers that are discarded each year.

What solutions are available? Some are specific to particular issues, as the hydrogen car (see Issue 10) or a revival of nuclear power (see Issue 12) may be to the problems associated with fossil fuels (see Issue 10), or as marine reserves may be to declining fish stocks (see Issue 15). Some are more general, as evidenced by the issue of whether or not population growth is a primary cause of environmental problems (see Issue 13).

Some analysts argue that whatever solutions we need, does not need to impose them all. Private industry may be able to do the job if government can find a way to motivate industry, as with the idea of tradable pollution rights (see Issue 6).

The overall aim, of course, is to avoid disaster and enable human life and civilization to continue prosperously into the future. The term for this is *sustainable development* (see Issue 2), and it was the chief concern of the UN World Summit on Sustainable Development, held in Johannesburg, South Africa, in August 2002. On the other hand, there are people who deem this a non-issue, for today people are better off than ever before in history, and "environmentalism" might more honestly be called "exaggerationism" (see Issue 7).

Leadership for Environment and Development International

Leadership for Environment and Development (LEAD), set up in 1991 by The Rockefeller Foundation, is an international network of professionals committed to sustainable development. Among other things, it provides a Web site where people can calculate their "ecological footprint"—a rough estimate of how much of the Earth's land, water, and other resources must be used to meet the needs of their personal lifestyle.

http://www.lead.org/leadnet/footprint/

The Natural Resources Defense Council

The Natural Resources Defense Council (NRDC) is one of the most active environmental research and advocacy organizations. Its home page lists its concerns as clean air and water, energy, global warming, toxic chemicals, nuclear waste, and much more.

http://www.nrdc.org/

National Councils for Sustainable Development

Since the 1992 Rio do Janeiro Earth Summit, over 70 countries have established multistakeholder groups to promote and implement sustainable development (National Councils for Sustainable Development). This is the Earth Council's NCSD site, dedicated to facilitating and coordinating the work of the NCSDs.

http://www.ncsdnetwork.org/

The United Nations Environment Programme

The United Nations Environment Programme "works to encourage sustainable development through sound environmental practices everywhere." Its activities cover a wide range of issues, from atmosphere and terrestrial ecosystems, the promotion of environmental science and information, to an early warning and emergency response capacity to deal with environmental disasters and emergencies.

http://www.unep.org/

The International Institute for Sustainable Development

The International Institute for Sustainable Development advances sustainable development policy and research by providing information and engaging in partnerships worldwide. It promotes the transition toward a sustainable future and seeks to demonstrate how human ingenuity can be applied to improve the well-being of the environment, economy, and society.

http://www.iisd.org/default.asp

PART 1

Environmental Philosophy

*E*nvironmental debates are rooted in questions of values—what is right? what is just?—and inevitably are political in nature. People who consider themselves to be environmentalists can be found on both sides of most of the issues in this section. They differ in what they see as their own self-interests and even in what they see as humanity's long-term interest.

Understanding the general issues raised in this initial section is useful preparation for examining the more specific controversies that follow in later sections.

- Is the Precautionary Principle a Sound Basis for International Policy?

- Is Sustainable Development Compatible With Human Welfare?

- Should a Price Be Put on the Goods and Services Provided by the World's Ecosystems?

ISSUE 1

Is the Precautionary Principle a Sound Basis for International Policy?

YES: Marco Martuzzi and Roberto Bertollini, from "The Precautionary Principle, Science and Human Health Protection," *International Journal of Occupational Medicine and Environmental Health* (January 2004)

NO: Henry I. Miller and Gregory Conko, from "The Perils of Precaution," *Policy Review* (June & July 2001)

ISSUE SUMMARY

YES: Marco Martuzzi and Roberto Bertollini, researchers with the World Health Organization (WHO), argue that although the Precautionary Principle, which demands preventive action in the face of credible threats of harm (even lacking full scientific certainty), may be difficult to apply, it can be valuable in the effort to protect human health and the environment.

NO: Henry I. Miller, a research fellow at Stanford University's Hoover Institution, and policy analyst Gregory Conko argue that the precautionary principle leads "regulators to abandon the careful balancing of risks *and* benefits," blocks progress, limits the freedom of scientific researchers, and restricts consumer choice.

The traditional approach to environmental problems has been reactive. That is, first the problem becomes apparent—wildlife or people sicken and die, or drinking water or air tastes foul. Then researchers seek the cause of the problem, and regulators seek to eliminate or reduce that cause. The burden is on society to demonstrate that harm is being done and that a particular cause is to blame.

An alternative approach is to presume that *all* human activities—construction projects, new chemicals, new technologies, etc.—have the potential to cause environmental harm. Therefore, those responsible for these activities should prove in advance that they will not do harm and should take suitable steps to prevent any harm from happening. A middle ground is occupied by the "precautionary principle," which has played an increasingly important part in environmental law ever since it first appeared in Germany in the mid-1960s. On the international scene, it has been applied to climate change, hazardous waste management, ozone depletion, biodiversity, and fisheries management. In 1992 the Rio Declaration on Environment and Development, listing it as Principle 15, codified it thus:

> In order to protect the environment, the precautionary approach shall be widely applied by States according to their capabilities. When there are threats of serious or irreversible damage, lack of full scientific certainty shall not be used as a reason for postponing cost-effective measures to prevent environmental degradation.

Other versions of the principle also exist, but all agree that when there is reason to think—not absolute proof—that some human activity is or might be harming the environment, precautions should be taken. Furthermore, the burden of proof should be on those responsible for the activity, not on those who may be harmed. This has come to be broadly accepted as a basic tenet of ecologically or environmentally sustainable development. See Paul L. Stein, "Are Decision-Makers Too Cautious With the Precautionary Principle?" *Environmental and Planning Law Journal* (February 2000).

The precautionary principle also contributes to thinking in the areas of risk assessment and risk management in general. Human activities—the manufacture of chemicals and other products; the use of pesticides, drugs, and fossil fuels; the construction of airports and shopping malls; and even agriculture—can damage health and the environment. Some people insist that action need not be taken against any particular activity until and unless there is solid, scientific proof that it is doing harm, and even then risks must be weighed against each other. Others insist that mere suspicion should be grounds enough for action.

Since solid, scientific proof can be very difficult to obtain, the question of just how much proof is needed to justify action is vital. Not surprisingly, if action threatens an industry, that industry's advocates will argue against taking precautions, generally saying that more proof is needed. A good example can be found in Stuart Pape, "Watch Out for the Precautionary Principle," *Prepared Foods* (October 1999): "In recent months, U.S. food manufacturers have experienced a rude introduction to the 'Precautionary Principle.' ... European regulators have begun to adopt extreme definitions of the Principle in order to protect domestic industries and place severe restrictions on the use of both old and new materials without justifying their action upon sound science."

In the following selection, Marco Martuzzi and Roberto Bertollini, researchers with the World Health Organziation (WHO), argue that although the Precautionary Principle, which demands preventive action in the face of credible threats of harm (even lacking full scientific certainty), may be difficult to apply, it can be valuable in the effort to protect human health and the environment. It also enhances the interface between science and public policy. In the second selection, Henry I. Miller and Gregory Conko argue that the precautionary principle blocks progress, limits the freedom of scientific researchers, and restricts consumer choice.

3

Marco Martuzzi and
Roberto Bertollini

The Precautionary Principle, Science and Human Health Protection

The Precautionary Principle: Definitions and Interpretations

What Uncertainty?

The Precautionary Principle (PP) is a valuable tool for developing adequate course of action in situations where there is large uncertainty. Such uncertainty can derive from: patchy scientific evidence about the health effects of an agent; sporadic reports of episodical adverse effects, unconfirmed or not reproducible; or limited knowledge of the dynamics of complex systems, resulting in effective ignorance on a series of chain events. Uncertainty can be of different magnitude and degree, but essentially discussion around the PP has focused on elements and criteria that should be addressed when making decision under this kind of "undetermined" uncertainty, i.e., not easily measured or quantified. This contrasts with uncertainties linked to lack of accuracy, for example when a risk is established, but incomplete information about mechanisms of action, individual vulnerability, quality of exposure assessment and exportability of risk estimates from a population to another, to name a few, produce large confidence bounds, reflecting uncertainty on the relevant figures.

Notwithstanding the absence of an established definition, some consensus [has] emerged on key aspects of the PP, which can help describe its role and relevance for public health protection: 1) the PP prescribes that uncertainty cannot be used as a pretext to delay action; 2) it suggests that the burden of proof might be reversed, from "recipients" to prove that an agent or technology is harmful to "proponents," to prove that it is innocuous; 3) it underlines the importance of switching the debate from arguments of acceptable risks to considering alternatives, preferably at early stages of the process; and 4) it recommends that the decision making process should be as transparent and democratic as possible, throughout its development.

The Precautionary Principle in Action

Consensus around these distinctive tracts has promoted the debate and has produced much progress in the field. However, given the fact that the PP remains effectively undetermined, there are many open questions, both theoretical and practical. Some argue that the PP is not a principle, in that it does not enunciate an explicit "statement," but rather reiterate[s] consolidated ones, for example those in the Hippocratic oath. But for public health, the practical implications of how the PP translates into protective policies are of crucial importance.

In the debate aiming at clarifying these practical aspects, there seem to be two main parallel views, somewhat conflicting but not necessarily mutually exclusive, on how the PP can help deal with decision making under uncertainties. According to some, the PP consists of a set of rules, considerations, evaluations, procedures that are applied when faced with a concrete decision, often of a dichotomous nature: can we go ahead with trading of GMOs? Should we ban beef imports? With this view, the debate tends to address the question of whether or not the PP is applicable, which would result in more or less stringent regulatory responses. For example, a recent report of the Health Council of the Netherlands on mobile telephones finds "no reason [...] to apply the PP and lower the [...] limits for partial body exposure." This approach, in other words, considers the available information and establishes if there are grounds to take "special," perhaps additional, action for health protection. Under this formulation, of applicability versus non-applicability, it is thus appropriate to develop criteria of application: indeed the EC has proposed, in an influential communication, several such criteria, for deciding on the PP applicability. Emphasis is given to consistency across different areas of application, and substantial weight is put on "proportionality" and adequacy of any action in terms of its costs to the society.

A different view seems to have developed in the debate that is mainly taking place in North America. Some authors refer to the PP as an "overarching principle," i.e., a set of considerations, criteria, recommendations, guidelines to inform the whole process that goes from initial proposal of introduction of, for example, a new technology or industrial facility, through the decision on whether to proceed and even after implementation, to monitor potential consequences. In this view, emphasis is given to all steps of the process, and precaution must be applied throughout. Analysis of possible alternatives becomes prominent, clarification of all stakeholders' interests is essential, as well as openness in the way decisions are reached at all stages. In this framework, the PP is always applicable and applied, in that it should guide the entire course of action. Precaution should inspire all decisions that are to be made, and help identify the most pressing needs in terms of research.

Both views have advantages and disadvantages, and both can be valuable for human health protection. In the first of the two, the "European" approach, the PP is used to clarify the question of how much evidence is needed to take a certain action, typically involving large costs. Arguably, this is not a new question, and indeed the whole concept of prevention hinges around the same evaluation. However, the PP does provide an additional contribution: it suggests that, given the increasingly complex and far-reaching threats to health and the environment, we might need to "reset" such threshold, if one exists. On the other hand, there might be the danger that the PP is used, or rather misused, against technological development and scientific advancement. Although this is often an argument that is put forward by *a priori* critics of the PP, there are sometimes good reasons, because the PP is occasionally wielded as a definitive, irrevocable veto. These controversies might be explained by the difficulty of conducting a democratic and transparent debate on a dichotomous question, with the inevitable polarisation of opinions. This tendency is probably the reason why the European way to the PP does not lend itself too well to resolve questions other than a yes/no type of decisions, but when a line must be drawn at some point along a spectrum, i.e., when quantitative protection standards must be defined. But again on the "pro" side, it is possible to specify coherent criteria of application, and ensure some kind of consistency in the way the PP is applied in different instances, an attractive feature for decision makers.

The "American" view, the second one described above, holds the PP as an overarching principle, which, among other advantages, might help adopt precaution earlier on. For example, there would be little scope in discussing about the PP when a technology is being introduced following large investments on the part of the industry (and hence of society). Similarly, evaluating the applicability of the PP at specific stages may mislead the debate onto treacherous terrains of cost-benefit considerations, where excessive credit may be given to weak scientific evidence. In other words, this view or use of the PP can be beneficial to raise and promote the public debate around the issues of real relevance. There is a shortcoming, however, in the necessary implication that the PP always applies, because it is not obvious what provision does the PP make exactly. It is very desirable that all steps, from hazard identification to risk assessment, from policy making to implementation and monitoring, are adjusted to allow for the extra degree of complexity that we face nowadays; but when a compelling case calls for extraordinary protection measures it might be difficult to invoke the PP to support such decisions.

Although this classification is somewhat over-simplistic, it is difficult to deny that a variety of open questions are raised in the frequent controversies arising in the area of environment and health. It might be useful to explore how the differences between these two approaches (and certainly others) to precaution have developed, as a function of the legal framework, the cultural environment, the political and public opinion response. Also, it might be beneficial to clarify ways in which different approaches can be reconciled and harmonized.

PP and Evidence-Based Policy Making

As previously discussed, the PP has emerged as a tool for dealing with uncertainty of an "undetermined" type, i.e, where available knowledge indicate[s] the potential occurrence of harm, but little is known about its likelihood and its magnitude. Although this is probably the most unfavorable circumstance for decision making, experience suggests that decision making is fraught with difficulties also when evidence is more robust. It should not be forgotten that rational decisions, and especially the development of satisfactory policies, [are functions] of knowledge and science, as well as of societal and ethical values. Accommodating all these variables is a demanding exercise in complex democratic societies, even when information from science is exhaustive. In addition, the boundary between the region where scientific information is univocal and where it is equivocal or controversial is extremely blurry.

Thus, although the PP "specializes" in informing decision making under uncertainty, it appears advantageous to clarify its role and relevance in the wider context of the science-policy interface. Other lines of work endeavor to support and shed light to the mechanisms that underlie the translation of scientific evidence into policy making, and it seems likely that taking a wider perspective might help clarify the contribution of the PP, and of other approaches, to this complex issue.

The Ethical Framework

The ethical values and principles are a crucial component of the question, which is occasionally overlooked by exchanges of opinions on whether the PP applies or not. It has been described by several authors that several alternative ethical frameworks can be used in decision making: utilitarian, libertarian, distributive. The choice of one of these frameworks, often done implicitly, has implications on the value and relevance of available scientific information, and can determine different courses of action. It has been compellingly stated that it is crucial that these values are made explicit, however discussion around the PP is sometimes scanty of open acknowledgments in this respect. In particular, criticism is often made of decisions taken applying risk assessment and cost-benefit analysis, on the grounds that these methods fail to consider interactions in complex systems, or that instability in the available evidence prevents meaningful results. Although legitimate, such criticisms do not always address another important aspect, i.e., that most cost-benefit analyses are based on a utilitarian criterion, aiming at maximizing the average "common good," while controversies might stem from unequal distribution of exposures. This omission is potentially dangerous: by keeping the focus of the discussion on the issue of whether science is strong enough to warrant a cost-benefit analysis, divergence of opinions tends to be ascribed only to different assessment of the scientific literature. Thus, while the debate is seemingly about what level of protection is warranted by the weight of the evidence, given the practical constraints, a more fundamental difference in terms of what ethical framework is applied might explain the controversy.

Health Impact Assessment

Health impact assessment (HIA) is a set of tools and procedures for assessing the health consequences of policies, developments and plans, typically implemented outside the health sector. Since it has become apparent how such policies are key health determinants, HIA addresses "upstream" factors and characterizes their likely health implications. HIA is a special case of other kinds of impact assessments, and in particular it builds on environmental impact assessment. HIA has become a recognized tool for supporting policy making in some European countries over the last decade. Examples of application include the creation of local industrial facilities, urban development schemes within cities, regional transport policies and agricultural, food and nutrition policies at the national level. These exercises are normally very difficult, because attempts are made to identify and analyze all pathways through which health can be affected as a result of changes in policy or social organization that can have a very wide spectrum of consequences, very often mediated by socioeconomic mechanisms. In addition, HIA is based on the WHO's view of health, that is, it is concerned not only with disease occurrence, but also with well-being and quality of life. Thus HIA's mandate is to study detrimental health effects as well as beneficial consequences of policies.

Even from a summary description of HIA, it should not be surprising that HIA shares several features with the PP. Given HIA's interest in complex chains of events, from non-health policies to health effects in modem society, uncertainty is a recurrent theme. HIA, however, takes a proactive approach, where, despite all uncertainties, health is put at the center of the debate, in order to support the identification and development of policies that take health into high consideration. Among its distinctive characteristics, HIA is based on an open and transparent process, where all relevant stakeholders should be involved. HIA is most effective when several policy options are being analyzed, and is valuable in identifying possible mitigation strategies.

Thus, three of the four elements that characterize the main "definitions" of the PP, described above (i.e., taking action in the face of uncertainty, transparency of the process, analysis of alternatives), are also central to HIA. It seems therefore of interest to explore better how these close connections can be mutually beneficial and ultimately enhance the science-policy interface.

NO

Henry I. Miller and
Gregory Conko

The Perils of Precaution

Environmental and public health activists have clashed with scholars and risk-analysis professionals for decades over the appropriate regulation of various risks, including those from consumer products and manufacturing processes. Underlying the controversies about various specific issues—such as clorinated water, pesticides, gene-spliced foods, and hormones in beef—has been a fundamental, almost philosophical question: How should regulators, acting as society's surrogate, approach risk in the absence of certainty about the likelihood or magnitude of potential harm?

Proponents of a more risk-averse approach have advocated a "precautionary principle" to reduce risks and make our lives safer. There is no widely accepted! definition of the principle, but in its most common formulation, governments should implement regulatory measures to prevent or restrict actions that raise even conjectural threats of harm to human health or the environment, even though there may be incomplete scientific evidence as to the potential significance of these dangers. Use of the precautionary principle is sometimes represented as "erring on the side of safety," or "better safe than sorry"—the idea being that the failure to regulate risky activities sufficiently could result in severe harm to human health or the environment, and that "overregulation" causes little or no harm. Brandishing the precautionary principle, environmental groups have prevailed upon governments in recent decades to assail the chemical industry and, more recently, the food industry.

Potential risks should, of course, be taken into consideration before proceeding with any new activity or product, whether it is the siting of a power plant or the introduction of a new drug into the pharmacy. But the precautionary principle focuses solely on the *possibility* that technologies could pose unique, extreme, or unmanageable risks, even after considerable testing has already been conducted. What is missing from precautionary calculus is an acknowledgment that even when technologies introduce new risks, most confer net benefits—that is, their use reduces many other, often far more serious, hazards. Examples include blood transfussions, MRI scans, and automobile air bags, all of which offer immense benefits and only minimal risk.

Several subjective factors can cloud thinking about risks and influence how nonexperts view them. Studies of risk perception have shown that people

From Henry I. Miller and Gregory Conko, "The Perils of Precaution," *Policy Review* (June & July 2001), pp. 25–39. Copyright (c) 2001 by The Hoover Institution, Stanford University. Reprinted by permission.

tend to overestimate risks that are unfamiliar, hard to understand, invisible, involuntary, and/or potentially catastrophic—and vice versa. Thus, they *overes*timate invisible "threats" such as electromagnetic radiation and trace amounts of pesticides in foods, which inspire uncertainty and fear sometimes verging on superstition. Conversely, they tend to *under*estimate risks the nature of which they consider to be clear and comprehensible, such as using a chain saw or riding a motorcycle.

These distorted perceptions complicate the regulation of risk, for if democracy must eventually take public opinion into account, good government must also discount heuristic errors or prejudices. Edmund Burke emphasized government's pivotal role in making such judgments: "Your Representative owes you, not only his industry, but his judgment; and he betrays, instead of serving you, if he sacrifices it to your opinion." Government leaders should *lead*; or putting it another way, government officials should make decisions that are rational and in the public interest even if they are unpopular at the time. This is especially true if, as is the case for most federal and state regulators, they are granted what amounts to lifetime job tenure in order to shield them from political manipulation or retaliation. Yet in too many cases, the precautionary principle has led regulators to abandon the careful balancing of risks *and* benefits—that is, to make decisions, in the name of precaution, that cost real lives due to forgone benefits.

The Danger of Precaution

The danger in the precautionary principle is that it distracts consumers and policymakers from known, significant threats to human health and diverts limited public health resources from those genuine and far greater risks. Consider, for example, the environmental movement's campaign to rid society of chlorinated compounds.

By the late 1980s, environmental activists were attempting to convince water authorities around the world of the possibility that carcinogenic byproducts from chlorination of drinking water posed a potential cancer risk. Peruvian officials, caught in a budget crisis, used this supposed threat to public health as a justification to stop chlorinating much of the country's drinking water. That decision contributed to the acceleration and spread of Latin America's 1991–96 cholera epidemic, which afflicted more than 1.3 million people and killed at least 11,000.

Activists have since extended their antichlorine campaign to so-called "endocrine disrupters," or modulators, asserting that certain primarily man-made chemicals mimic or interfere with human hormones (especially estrogens) in the body and thereby cause a range of abnormalities and diseases related to the endocrine system.

The American Council on Science and Health has explored the endocrine disrupter hypothesis and found that while *high* doses of certain environmental contaminants produce toxic effects in laboratory test animals—in some cases involving the endocrine system—humans' actual exposure to these suspected endocrine modulators is many orders of magnitude lower. It is well documented that while a chemical administered at high doses may cause cancer in certain laboratory animals, it does not necessarily cause cancer in humans—both because of different susceptibilities and because humans are subjected to far lower exposures to synthetic environmental chemicals.

No consistent, convincing association has been demonstrated between real-world exposures to synthetic chemicals in the environment and increased cancer in hormonally sensitive human tissues. Moreover, humans are routinely exposed through their diet to many estrogenic substances (substances having an effect similar to that of the human hormone estrogen) found in many plants. Dietary exposures to these plant estrogens, or phytoestrogens, are far greater than exposures to supposed synthetic endocrine modulators, and no adverse health effects have been associated with the overwhelming majority of these dietary exposures.

Furthermore, there is currently a trend toward *lower* concentrations of many contaminants in air, water, and soil—including several that are suspected of being endocrine disrupters. Some of the key research findings that stimulated the endocrine disrupter hypothesis originally have been retracted or are not reproducible. The available human epidemiological data do not show any consistent, convincing evidence of negative health effects related to industrial chemicals that are suspected of disrupting the endocrine system. In spite of that, activists and many government regulators continue to invoke the need for precautionary (over-) regulation of various products, and even outright bans.

Antichlorine campaigners more recently have turned their attacks to phthalates, liquid organic compounds added to certain plastics to make them softer. These soft plastics are used for important medical devices, particularly fluid containers, blood bags, tubing, and gloves; children's toys such as teething rings and rattles; and household and industrial items such as wire coating and flooring. Waving the banner of the precautionary principle, activists claim that phthalates *might* have numerous adverse health effects—even in the face of significant scientific evidence to the contrary. Governments have taken these unsupported claims seriously, and several formal and informal bans have been implemented around the world. As a result, consumers have been denied product choices, and doctors and their patients deprived of life-saving tools.

In addition to the loss of beneficial products, there are more indirect and subtle perils of government overregulation established in the name of the precautionary principle. Money spent on implementing and complying with regulation (justified or not) exerts an "income effect" that reflects the correlation between wealth and health, an issue popularized by the late political scientists Aaron Wildavsky. It is no coincidence, he argued, that richer societies have lower mortality rates than poorer ones. To deprive communities of wealth, therefore, is to enhance their risks.

Wildavsky's argument is correct: Wealthier individuals are able to purchase better health care, enjoy more nutritious diets, and lead generally less stressful lives. Conversely, the deprivation of income itself has adverse health effects—for example an increased incidence of stress-related problems including ulcers, hypertension, heart attacks, depression, and suicides.

It is difficult to quantify precisely the relationship between mortality and the deprivation of income, but academic studies suggest, as a conservative estimate, that every $7.25 million of regulatory costs will induce one additional fatality through this "income effect." The excess costs in the tens of billions of dollars required annually by precautionary regulation for various classes of consumer products would, therefore, be expected to cause thousands of deaths per year. These are the real costs of "erring on the side of safety." The expression "regulatory overkill" is not merely a figure of speech.

Rationalizing Precaution

During the past few years, skeptics have begun more actively to question the theory and practice of the precautionary principle. In response to those challenges, the European Commission (EC), a prominent advocate of the precautionary principle, last year published a formal communication to clarify and to promote the legitimacy of the concept. The EC resolved that, under its auspices, precautionary restrictions would be "proportional to the chosen level of protection," "non-discriminatory in their application," and "consistent with other similar measures." The commission also avowed that EC decision makers would carefully weigh "potential benefits and costs." EC Health Commissioner David Byrne, repeating these points [recently] in an article on food and agriculture regulation in *European Affairs*, asked rhetorically, "How could a Commissioner for Health and Consumer Protection reject or ignore well-founded, independent scientific advice in relation to food safety?"

Byrne should answer his own question: The ongoing dispute between his European Commission and the United States and Canada over restrictions on hormone-treated beef cattle is exactly such a case of rejecting or ignoring well-founded research. The EC argued that the precautionary principle permits restriction of imports of U.S. and Canadian beef from cattle treated with certain growth hormones.

In their rulings, a WTO [World Trade Organization] dispute resolution panel and its appellate board both acknowledged that the general "look before you leap" sense of the precautionary principle could be found within WTO agreements, but that its presence did not relieve the European Commission of its obligation to base policy on the outcome of a scientific risk assessment. And the risk assessment clearly favored the U.S.-Canadian position. A scientific committee assembled by the WTO dispute resolution panel found that even the scientific studies cited by the EC in its own defense did not indicate a safety risk when the hormones in question were used in accordance with accepted animal husbandry practices. Thus, the WTO ruled in favor of the United States and Canada because the European Commission had failed to demonstrate a real or imminent harm. Nevertheless, the EC continues to enforce restrictions on hormone-treated beef, a blatantly unscientific and protectionist policy that belies the commission's insistence that the precautionary principle will not be abused.

Precaution Meets Biotech

Perhaps the most egregious application by the European Commission of the precautionary principle is in its regulation of the products of the new biotechnology, or gene-splicing. By the early 1990s, many of the countries in Western Europe, as well as the EC itself, had erected strict rules regarding the testing and commercialization of gene-spliced crop plants. In 1999, the European Commission explicitly invoked the precautionary principle in establishing a moratorium on the approval of all new gene-spliced crop varieties, pending approval of an even more strict EU-wide regulation.

Notwithstanding the EC's promises that the precautionary principle would not be abused, all of the stipulations enumerated by the commission have been flagrantly ignored or tortured in its regulatory approach to gene-spliced (or in their argot, "genetically modified" or "GM") foods. Rules for gene-spliced plants and microorganisms are inconsistent, discriminatory, and bear no proportionality to risk. In fact, there is arguably *inverse* proportionality to risk, in that the more crudely crafted organisms of the old days of mutagenesis and gene transfers are subject to less stringent regulation than those organisms more precisely crafted by biotech. This amounts to a violation of a cardinal principle of regulation: that the degree of regulatory scrutiny should be commensurate with risk.

Dozens of scientific bodies—including the U.S. National Academy of Sciences (NAS), the American Medical Association, the UK's Royal Society, and the World Health Organization—have analyzed the oversight that is appropriate for gene-spliced organisms and arrived at remarkably congruent conclusions: The newer molecular techniques for genetic improvement are an extension, or refinement, of earlier, far less precise ones; adding genes to plants or microorganisms does not make them less safe either to the environment or to eat; the risks associated with gene-spliced organisms are the same in kind as those associated with conventionally modified organisms and unmodified ones; and regulation should be based upon the risk-related characteristics of individual products, regardless of the techniques used in their development.

An authoritative 1989 analysis of the modern gene-splicing techniques published by the NAS's research arm, the National Research Council, concluded that "the same physical and biological laws govern the response of organisms modified by modern molecular and cellular methods and those produced by classical methods," but it went on to observe that gene-splicing is more precise, circumscribed, and predictable than other techniques.

> [Gene-splicing] methodology makes it possible to introduce pieces of DNA, consisting of either single or multiple genes, that can be defined in function and even in nucleotide sequence. With classical techniques of gene transfer, a variable number of genes can be transferred, the number depending on the mechanism of transfer; but predicting the precise number or the traits that have been transferred is difficult, and we cannot always predict the [characteristics] that will result. With organisms modified by molecular methods, we are in a better, if not perfect, position to predict the [characteristics].

In other words, gene-splicing technology is a refinement of older, less precise techniques, and its use generates less uncertainty. But for gene-spliced plants, both the fact and degree of regulation are determined by the production methods—that is, if gene-splicing techniques have been used, the plant is immediately subject to extraordinary pre-market testing requirements for human health and environmental safety, regardless of the level of risk posed. Throughout most of the world, gene-spliced crop plants such as insect-resistant corn and cotton are subject to a lengthy and hugely expensive process of mandatory testing before they can be brought to market, while plants with similar properties but developed with older, less precise genetic techniques are exempt from such requirements....

Another striking example of the disproportionate regulatory burden borne only by gene-spliced plants involves a process called induced-mutation breeding, which has been in common use since the 1950s. This technique involves exposing crop plants to ionizing radiation or toxic chemicals to induce random genetic mutations. These treatments most often kill the plants (or seeds) or cause detrimental genetic changes, but on rare occasions, the result is a desirable mutation—for example, one producing a new trait in the plant that is agronomically useful, such as altered height, more seeds, or larger fruit. In these cases, breeders have no real knowledge of the exact nature of the genetic mutation(s) that produced the useful trait, or of what other mutations might have occurred in the plant. Yet the approximately 1,400 mutation-bred plant varieties from a range of different species that have been marketed over the past half century have been subject to no formal regulation before reaching the market—even though several, including two varieties of squash and one of potato, have contained dangerous levels of endogenous toxins and had to be banned afterward.

What does this regulatory inconsistency mean in practice? If a student doing a school biology project takes a packet of "conventional" tomato or pea seed to be irradiated at the local hospital x-ray suite and plants them in his backyard in order to investigate interesting mutants, he need not seek approval from any local, national, or international authority. However, if the seeds have been modified by the addition of one or a few genes via gene-splicing techniques—and even if the genetic change is merely to remove a gene—this would-be Mendel faces a mountain of bureaucratic paperwork and expense (to say nothing of the very real possibility of vandalism, since the site of the experiment must be publicized and some opponents of biotech are believers in "direct action"). The same would apply, of course, to professional agricultural scientists in industry and academia. In the United States, Department of Agriculture requirements for paperwork and field trial design make field trials with gene-spliced organisms 10 to 20 times more expensive than the same experiments with virtually identical organisms that have been modified with conventional genetic techniques.

Why are new genetic constructions crafted with these older techniques exempt from regulation, from the dirt to the dinner plate? Why don't regulatory regimes require that new genetic variants made with older techniques be evaluated for increased weediness or invasiveness, or for new allergens that could show up in food? The answer is based on millennia of experience with genetically improved crop plants from the era before gene-splicing: Even the use of relatively crude and unpredictable genetic techniques for the improvement of crops and microorganisms poses minimal—but, as noted above, not zero—risk to human health and the environment.

If the proponents of the precautionary principle were applying it rationally and fairly, surely greater precautions would be appropriate not to gene-splicing but to the cruder, less precise, less predictable "conventional" forms of genetic modifications. Furthermore, in spite of the assurance of the European Commission and other advocates of the precautionary principle, regulators of gene-spliced products seldom take into consideration the potential risk-*reducing* benefits of the technology. For example, some of the most successful of the gene-spliced crops, especially cotton and corn, have been constructed by splicing in a bacterial gene that produces a protein toxic to predatory insects, but not to people or other mammals. Not only do these gene-spliced corn varieties repel pests, but grain obtained from them is less likely to contain *Furarium*, a toxic fungus often carried into the plants by the insects. That, in turn, significantly reduces the levels of the fungal toxin fumonisin, which is known to cause fatal diseases in horses and swine that eat infected corn, and esophageal cancer in humans. When harvested, these gene-spliced varieties of grain also end up with lower concentrations of insect parts than conventional varieties. Thus, genes-spliced corn is not only cheaper to produce but yields a higher quality product and is a potential boon to public health. Moreover, by reducing the need for spraying chemical pesticides on crops, it is environmentally friendly.

Other products, such as gene-spliced herbicide-resistant crops, have permitted farmers to reduce their herbicide use and to adopt more environment-friendly no-till farming practices. Crops now in development with improved yields would allow more food to be grown on less acreage, saving more land area for wildlife or other uses. And recently developed plant varieties with enhanced levels of vitamins, minerals, and dietary proteins could dramatically improve the health of hundreds of millions of malnourished people in developing countries. These are the kinds of tangible environmental and health benefits that invariably are given little or no weight in precautionary risk calculations.

In spite of incontrovertible benefits and greater predictability and safety of gene-spliced plants and foods, regulatory agencies have regulated them in a discriminatory, unnecessarily burdensome way. They have imposed requirements that could not possibly be met for conventionally bred crop plants. And, as the European Commission's moratorium on new product approvals demonstrates, even when that extraordinary burden of proof is met via monumental amounts of testing and evaluation, regulators frequently declare themselves unsatisfied.

Biased Decision Making

While the European Union is a prominent practitioner of the precautionary principle on issues ranging from toxic substances and the new biotechnology to climate change and gun control, U.S. regulatory agencies also commonly practice excessively precautionary regulation. The precise term of art "precautionary principle" is not used in U.S. public policy, but the regulation of such products as pharmaceuticals, food additives, gene-spliced plants and microorganisms, synthetic pesticides, and other chemicals is without question "precautionary" in nature. U.S. regulators actually appear to be more precautionary than the Europeans towards several kinds of risks, including the licensing of new medicines, lead in gasoline, nuclear power, and others. They have also been highly precautionary towards gene-splicing, although not to the extremes of their European counterparts. The main difference between precautionary regulation in the United States and the use of the precautionary principle in Europe is largely a matter of degree—with reference to products, technologies, and activities—and of semantics.

In both the United States and Europe, public health and environmental regulations usually require a risk assessment to determine the extent of potential hazards and of exposure to them, followed by judgments about how to regulate. The precautionary principle can distort this process by introducing a systematic bias into decision making. Regulators face an asymmetrical incentive structure in which they are compelled to address the potential harms from new products, but are free to discount the hidden risk-reducing properties of unused or underused ones. The result is a lopsided process that is inherently biased against change and therefore against innovation.

To see why, one must understand that there are two basic kinds of mistaken decisions that a regulator can make: First, a harmful product can be approved for marketing—called a Type I error in the parlance of risk analysis. Second, a useful product can be rejected or delayed, can fail to achieve approval at all, or can be inappropriately withdrawn from the market—a Type II error. In other words, a regulator commits a Type I error by permitting something harmful to happen and a Type II error by preventing something beneficial from becoming available. Both situations have negative consequences for the public, but the outcomes for the regulator are very different.

Examples of this Type I-Type II error dichotomy in both the U.S. and Europe abound, but it is perhaps illustrated most clearly in the FDA's [Food and Drug Administration] approval process for new drugs. A classic example is the FDA's approval in 1976 of the swine flu vaccine—generally perceived as a Type I error because while the vaccine was effective at preventing influenza, it had a major side effect that was unknown at the time of approval: A small number of patients suffered temporary paralysis from Guillain-Barré Syndrome. This kind of mistake is highly visible and has immediate consequences: The media pounce and the public and Congress are roused, and Congress takes up the matter. Both the developers of the product and the regulators who allowed it to be marketed are excoriated and punished in such modern-day pillories as congressional hearings, television newsmagazines, and newspaper editorials. Because a regulatory official's career might be damaged irreparably by his good-faith but mistaken approval of a high-profile product, decisions are often made defensively—in other words, above all to avoid Type I errors.

Former FDA Commissioner Alexander Schmidt aptly summarized the regulator's dilemma:

> In all our FDA history, we are unable to find a single instance where a Congressional committee investigated the failure of FDA to approve a new drug. But, the times when hearings have been held to criticize our approval of a new drug have been so frequent that we have not been able to count them. The message to FDA staff could not be clearer. Whenever a controversy over a new drug is resolved by approval of the drug, the agency and the individuals involved likely will be investigated. Whenever such a drug is disapproved, no inquiry will be made. The Congressional pressure for *negative* action is, therefore, intense. And it seems to be ever increasing....

Although they can dramatically compromise public health, Type II errors caused by a regulator's bad judgment, timidity, or anxiety seldom gain public attention. It may be only the employees of the company that makes the product and a few stock market analysts and investors who are knowledgeable about unnecessary delays. And if the regulator's mistake precipitates a corporate decision to abandon the product, cause and effect are seldom connected in the public mind. Naturally, the companies themselves are loath to complain publicly about a mistaken FDA judgment, because the agency has so much discretionary control over their ability to test and market products. As a consequence, there may be no direct evidence of, or publicity about, the lost societal benefits, to say nothing of the culpability of regulatory officials.

Exceptions exist, of course. A few activists, such as the AIDs advocacy groups that closely monitor the FDA, scrutinize agency review of certain products and aggressively publicize Type II errors. In addition, congressional oversight *should* provide a check on regulators' performance, but as noted above by former FDA Commissioner Schmidt, only rarely does oversight focus on their Type II errors. Type I errors make for more dramatic hearings, after all, including injured patients and their family members. And even when such mistakes are exposed, regulators frequently defend Type II errors as erring on the side of caution—in effect, invoking the precautionary principle....Too often this euphemism is accepted uncritically by legislators, the media, and the public, and our system of pharmaceutical oversight becomes progressively less responsive to the public interest.

The FDA is not unique in this regard, of course. All regulatory agencies are subject to the same sorts of social and political pressures that cause them to be castigated when dangerous products accidentally make it to market (even if, as is often the case, those products produce net benefits) but to escape blame when they keep beneficial products out of the hands of consumers. Adding the precautionary principle's bias against new products into the public policy mix further encourages regulators to commit Type II errors in their frenzy to avoid Type I errors. This is hardly conducive to enhancing overall public safety.

Extreme Precaution

For some antitechnology activists who push the precautionary principle, the deeper issue is not really safety at all. Many are more antibusiness and antitechnology than they are pro-safety. And in their mission to oppose business interests and disparage technologies they don't like or that they have decided we just don't need, they are willing to seize any opportunity that presents itself.

These activists consistently (and intentionally) confuse *plausibility* with *provability*. Consider, for example, *Our Stolen Future*, the bible of the proponents of the endocrine disrupter hypothesis discussed above. The book's premise— that estrogen-like synthetic chemicals damage health in a number of ways—is not supported by scientific data. Much of the research offered as evidence for its arguments has been discredited. The authors equivocate wildly: "Those exposed prenatally to endocrine-disrupting chemicals *may* have abnormal hormone levels as adults, and they *could* also pass on persistent chemicals they themselves have inherited—both factors that *could* influence the development of their own children [emphasis added]." The authors also assume, in the absence of any actual evidence, that exposures to small amounts of many chemicals create a synergistic effect—that is, that total exposure constitutes a kind of witches' brew that is far more toxic than the sum of the parts. For these anti-innovation ideologues, the mere fact that such questions have been asked requires that inventors or producers expend time and resources answering them. Meanwhile, the critics move on to yet another frightening plausibility and still more questions. No matter how outlandish the claim, the burden of proof is put on the innovator.

Whether the issue is environmental chemicals, nuclear power, or gene-spliced plants, many activists are motivated by their own parochial vision of what constitutes a "good society" and how to achieve it. One prominent biotechnology critic at the Union of Concerned Scientists rationalizes her organization's opposition to gene-splicing as follows: "Industrialized countries have few genuine needs for innovative food stuffs, regardless of the method by which they are produced"; therefore, society should not squander resources on developing them. She concludes that although "the malnourished homeless" are, indeed, a problem, the solution lies "in resolving income disparities, and educating ourselves to make better choices from among the abundant foods that are available."

Greenpeace, one of the principal advocates of the precautionary principle, offered in its 1999 IRS filings the organization's view of the role in society of safer, more nutritious, higher-yielding, environment-friendly, gene-spliced plants: There isn't any. By its own admission, Greenpeace's goal is not the prudent, safe use of gene-spliced foods or even their mandatory labeling, but rather these products' "complete elimination [from] the food supply and the environment." Many of the groups, such as Greenpeace, do not stop at demanding illogical and stultifying regulation or outright bans on product testing and commercialization; they advocate and carry out vandalism of the very field trials intended to answer questions about environmental safety.

Such tortured logic and arrogance illustrate that the metastasis of the precautionary principle generally, as well as the pseudocontroversies over the testing and use of gene-spliced organisms in particular, stem from a social vision that is not just strongly antitechnology, but one that poses serious challenges to academic, commercial, and individual freedom.

The precautionary principle shifts decision making power away from individuals and into the hands of government bureaucrats and environmental activists. Indeed, that is one of its attractions for many NGOs [nongovernmental organizations]. Carolyn Raffensperger, executive director of the Science and Environmental Health Network, a consortium of radical groups, asserts that discretion to apply the precautionary principle "is in the hands of the people." According to her, this devolution of power is illustrated by violent demonstrations against economic globalization such as those in Seattle at the 1999 meeting of the World Trade Organization. "This is [about] how they want to live their lives," Raffensperger said.

To be more precise, it is about how small numbers of vocal activists want the rest of us to live *our* lives. In other words, the issue here is freedom and its infringement by ideologues who disapprove, on principle, of a certain technology, or product, or economic system....

Precaution *v.* Freedom

History offers compelling reasons to be cautious about societal risks, to be sure. These include the risk of incorrectly assuming the absence of danger (false negatives), overlooking low probability but high impact events in risk assessments, the danger of long latency periods before problems become apparent, and the lack of remediation methods in the event of an adverse event. Conversely, there are compelling reasons to be wary of excessive precaution, including the risk of too eagerly detecting a nonexistent danger (false positives), the financial cost of testing for or remediating low-risk problems, the opportunity costs of forgoing net-beneficial activities, and the availability of a contingency regime in case of an adverse event. The challenge for regulators is to balance these competing risk scenarios in a way that reduces overall harm to public health. This kind of risk balancing is often conspicuously absent from precautionary regulation.

It is also important that regulators take into consideration the degree of restraint generally imposed by society on individuals' and companies' freedom to perform legitimate activities (e.g., scientific research). In Western democratic societies, we enjoy long traditions of relatively unfettered scientific research and development, except in the very few cases where bona fide safety issues are raised. Traditionally, we shrink from permitting small, authoritarian minorities to dictate our social agenda, including what kinds of research are permissible and which technologies and products should be available in the marketplace.

Application of the precautionary principle has already elicited unscientific, discriminatory policies that inflate the costs of research, inhibit the development of new products, divert and waste resources, and restrict consumer choice. The excessive and wrong-headed regulation of the new biotechnology is one particularly egregious example. Further encroachment of precautionary regulation into other areas of domestic and international health and safety standards will create a kind of "open sesame" that government officials could invoke whenever they wish arbitrarily to introduce new barriers to trade, or simply to yield disingenuously to the demands of antitechnology activists. Those of us who both value the freedom to perform legitimate research and believe in the wisdom of market processes must not permit extremists acting in the name of "precaution" to dictate the terms of the debate.

POSTSCRIPT

Is the Precautionary Principle a Sound Basis for International Policy?

In their definition of the precautionary principle, Miller and Conko emphasize supposition: Precautions must be taken whenever there might be a problem. Martuzzi and Bertollini emphasize uncertainty: Lack of full scientific certainty shall not be used as a reason for postponing cost-effective precautions. The same tension is visible in Kenneth R. Foster, Paolo Vecchia, and Michael H. Repacholi, "Science and the Precautionary Principle," *Science* (May 12, 2000).

Other writers who oppose the precautionary principle have approached it in much the same way as Miller and Conko. For instance, Ronald Bailey, in "Precautionary Tale," *Reason* (April 1999), defines the precautionary principle as "precaution in the face of any actions that may affect people or the environment, no matter what science is able—or unable—to say about that action." "No matter what science says" is not quite the same thing as "lack of full scientific certainty." Indeed, Bailey turns the precautionary principle into a straw man and thereby endangers whatever points he makes that are worth considering. One of those points is that widespread use of the precautionary principle would hamstring the development of the Third World. Bonner R. Cohen, in "The Safety Nazis," *American Spectator* (July/August 2001), echoes this point, calling the precautionary principle a massive threat to human health in the less developed countries. Jonathan Adler, in "The Precautionary Principle's Challenge to Progress," in Ronald Bailey, ed., *Global Warming and Other Eco-Myths* (Prima, 2002), argues that because the precautionary principle does not adequately balance risks and benefits, "the world would be safer without it."

Yet the 1992 Rio Declaration emphasized that the precautionary principle should be "applied by States according to their capabilities" and that it should be applied in a cost-effective way. These provisions would seem to preclude the draconian interpretations that most alarm the critics. Yet, say David Kriebel et al., in "The Precautionary Principle in Environmental Science," *Environmental Health Perspectives* (September 2001), "environmental scientists should be aware of the policy uses of their work and of their social responsibility to do science that protects human health and the environment." Businesses are also conflicted, writes Arnold Brown in "Suitable Precautions," *Across the Board* (January/February 2002), because the precautionary principle tends to slow decision making, but he maintains that "we will all have to learn and practice anticipation." Roger Scruton, in "The Cult of Precaution," *National Interest* (Summer 2004), calls the Precautionary Principle "a meaningless nostrum" that is used to avoid risk and says it "clearly presents an obstacle to innovation and experiment," which are essential, Bernard D. Goldstein and Russellyn S. Carruth remind us in "Implications of the Precautionary Principle: Is It a Threat to Science?" *International Journal of Occupational Medicine and Environmental Health* (January 2004), for proper assessment of risk.

ISSUE 2

Is Sustainable Development Compatible With Human Welfare?

YES: Dinah M. Payne and Cecily A. Raiborn, from "Sustainable Development: The Ethics Support the Economics," *Journal of Business Ethics* (July 2001)

NO: Ronald Bailey, from "Wilting Greens," *Reason* (December 2002)

ISSUE SUMMARY

YES: Professor of management Dinah M. Payne and professor of accounting Cecily A. Raiborn argue that environmental responsibility and sustainable development are essential parts of modern business ethics and that only through them can both business and humans thrive.

NO: Environmental journalist Ronald Bailey states that sustainable development results in economic stagnation and threatens both the environment and the world's poor.

Over the last 30 years, many people have expressed concerns that humanity cannot continue indefinitely to increase population, industrial development, and consumption. The trends and their impacts on the environment are amply described in numerous books, including historian J. R. McNeill's *Something New Under the Sun: An Environmental History of the Twentieth-Century World* (W. W. Norton, 2000).

"Can we keep it up?" is the basic question behind the issue of sustainability. In the 1960s and 1970s, this was expressed as the "Spaceship Earth" metaphor, which said that we have limited supplies of energy, resources, and room and that we must limit population growth and industrial activity, conserve, and recycle in order to avoid crucial shortages. "Sustainability" entered the global debate in the early 1980s, when the United Nations secretary general asked Gro Harlem Brundtland, a former prime minister and minister of environment in Norway, to organize and chair the World Commission on Environment and Development and produce a "global agenda for change." The resulting report, *Our Common Future* (Oxford University Press, 1987), defined *sustainable development* as "development that meets the needs of the present without compromising the ability of future generations to meet their own needs." It recognized that limits on population size and resource use cannot be known precisely; that problems may arise not suddenly but rather gradually, marked by rising costs; and that limits may be redefined by changes in technology. The report also recognized that limits exist and must be taken into account when governments, corporations, and individuals plan for the future.

The Brundtland report led to the UN Conference on Environment and Development held in Rio de Janeiro in 1992. The Rio conference set sustainability firmly on the global agenda and made it an essential part of efforts to deal with global environmental issues and promote equitable economic development. In brief, sustainability means such things as cutting forests no faster than they can grow back, using groundwater no faster than it is recharged by precipitation, stressing renewable energy sources rather than exhaustible fossil fuels, and farming in such a way that soil fertility does not decline. In addition, economics must be revamped to take into account environmental costs as well as capital, labor, raw materials, and energy costs. Many add that the distribution of the Earth's wealth must be made more equitable as well.

Given continuing growth in population and demand for resources, sustainable development is clearly a difficult proposition. Some think it can be done, but others think that for sustainability to work, either population or resource demand must be reduced. Not surprisingly, many people see sustainable development as in conflict with business and industrial activities, private property rights, and such human freedoms as the freedoms to have many children, to accumulate wealth, and to use the environment as one wishes. Economics professor Jacqueline R. Kasun, in "Doomsday Every Day: Sustainable Economics, Sustainable Tyranny," *The Independent Review* (Summer 1999), goes so far as to argue that sustainable development will require sacrificing human freedom, dignity, and material welfare on a road to tyranny.

In the following selections, Dinah M. Payne and Cecily A. Raiborn argue that environmental responsibility and sustainable development are essential parts of modern business ethics. Because the consequence of the activities of all members of society, including business, "is an environment that is either habitable or one that is not," all people have a responsibility toward the environment and each other; the basic issue is "life versus death." Ronald Bailey argues that preserving the environment, eradicating poverty, and limiting economic growth are incompatible goals. Indeed, vigorous economic growth provides wealth for all and leads to environmental protection.

Dinah M. Payne and
Cecily A. Raiborn

Sustainable Development: The Ethics Support the Economics

Introduction

The field of business ethics is rampant with diverse issues and dilemmas. One critical ethical issue has, for many years, received significantly less attention than it merited: the responsibility of business organizations to their environments. Organizations world-wide have created and have faced resource depletion and pollution. However, there now seems to be a distinct and overt embracing of environmental social responsibility by many companies. This new-found interest may have been generated, in part, by gatherings such as the Rio de Janeiro (Earth) Summit and Kyoto Protocol. But, more importantly, these gatherings have spawned a plethora of groups focused on the issue of environmental social responsibility and, specifically, the issue of sustainable development. What is this concept and why should it concern businesses and their managers? Why should sustainable development be viewed as an ethical responsibility of businesses? To what extent should businesses attempt to engage in sustainable development activities? And what actions, beyond legal requirements, can be and are being taken by businesses to promote this concept with its resultant benefit to all business stakeholders?

Issue Definition and Identification

The term *sustainable development* was introduced in the 1970s, but actually became part of mainstream vocabulary during and after the 1987 World Commission on Environment and Development (also known as the Brundtland Commission). The Commission defined sustainable development as "development that meets the needs of the present without compromising the ability of future generations to meet their own needs." On the surface, this definition seems to be fairly simplistic, but the issue's breadth and depth create complexities.

To more fully and meaningfully refine the concept, the Earth Council indicated that such development should be economically viable, socially just, and environmentally appropriate. An additional expansion suggested that sustainable development should mean that the basic needs of all are met and that all should have the opportunity to fulfill their aspirations for a better life. The definition postulated by the World Business Council for Sustainable Development is that sustainable development is "the integration of economic development with environmental protection and social equity." Several complicated and sensitive issues are inherent in these definitions.

First, how can the "needs" of the present be differentiated from the "wants" of the present as well as how can the needs of the future be ascertained currently? Into this debate fall questions such as how can nonexistent future generations be protected and to what extent should today's civilization be sacrificed to protect future generations? Although the answers to these questions are arguably unanswerable, it is apparent that business have some responsibility to provide goods and services to the world. The free market helps "push" businesses to produce the goods and services currently desired for purchase (whether these goods are needed or simply wanted). Additionally, businesses partially establish future needs and wants of consumers through product development in response to current external pressures (desires communicated from consumers) as well as current internal abilities (research and scientific discoveries). In responding to these current pressures or abilities, many businesses utilize life-cycle analysis to assess potential future environmental impacts of product design, manufacturability, and recyclability.

Second, relative to what context or benchmark should "economically viable" be determined? This term could mean radically different things between businesses in developing and in developed nations, between start-up and long-standing businesses, or between business having significant environmental impacts and those having minimal environmental impacts. In each of these three scenarios, the cost of sustainable development would generally be more expensive (in relative cost to revenue proportions) to the former companies than to the latter. Thus, what might be deemed economically viable for a large retailer in England might mean financial ruin for a small mining company in Haiti.

There is a clear trend in the developing world towards better environmental policies that include the pursuit of economic development alternatives that minimize negative environmental impacts. Evidence also exists to indicate that "through technological change, substitution between resources, and higher prices for goods that pollute, environmental objectives and economic growth can be made more compatible." In regard to technological change, it is generally true that as technology advances, it becomes more efficient. Thus, because the industrialization process in developing countries often begins with the use of outdated technology, production may be environmentally expensive (the lower efficiency contributes to increased resource depletion and less emphasis on pollution control). As technology becomes more sophisticated, efficiency increases causing an increase in productive activity with fewer defects and spoilage, and thus a decline in the rate at which resource depletion occurs. Additionally, as the country advances, less environmental pollution may be tolerated. In the last stage of industrialization, organizations use advanced (more efficient and cleaner) technology, causing a net decline in resource depletion and pollution. Per capita income and social and governmental consciousness about the environment also rise; more "green" laws are written and enforced. Thus, an inverted U-shaped curve can be used to represent the changes in a society that starts at a point without environmental quality, rapidly advances, and then slows and turns around when that society has the time and/or money to spend to protect the environment.

Third, how and by what party should "socially just" development or "social equity" be determined? These factors would depend on who was obtaining the benefit from the development, what form that benefit took, what level of economic development existed in the area, whether resources consumed were replenishable, and what political and social issues were being faced or remedied.

Last, how and by what party is "environmentally appropriate" development to be judged? This judgment must reflect the answer to whether the environment should be protected for its own sake and/or for the sake of human inhabitants. Ecological ethicists argue that non-human inhabitants are intrinsically valuable and, thus, deserve respect and that humans have duties of preservation towards them. Alternatively, even if the intrinsic value of the environment and its non-human inhabitants is refuted, a livable environment is owed to all humans so that they may be permitted to fulfill their capacities as rational and free beings. Healy asserts that future sustainability will require a reorientation away from the human-centered (or anthropocentric) anthropological view towards more nature-centered (or ecocentric) view. Thus, determination of "environmentally appropriate" would commonly be more an issue of perspective than one of specific activity. Some individuals and businesses will take a broad perspective and assess the impact of an activity on the overall current and future physical environment (not just that part inhabited or used by humans). Other individuals and businesses will take a narrow perspective and assess the impact of the activity on the surrounding environment in the here-and-now.

Regardless of the definition or the diverse possible answers to definitional issues, it is clear that all publics (businesses, consumers, regulatory agencies, scientists, communities, and governments) are touched by the concept of sustainable development. All of these publics interact, directly or indirectly, and face the same outcome, which will not be locale-by-locale, industry-by-industry, or political party-by-political-party based. The long-term consequence of the activities of all publics is an environment that is either habitable or one that is not. That being the case, each public separately and all publics collectively have a responsibility towards the environment and each other to better understand sustainable development and to strive to achieve meaningful progress towards its attainment. Thus, the ethical issue in sustainable development is the basic issue of life versus death; if business and all other publics do not begin practicing the tenets of sustainable development, life as it currently exists will be extinct.

Businesses and their managers should be concerned about sustainable development for many reasons. Economic pragmatists would base their arguments on the simple fact that, without sustainable development, neither businesses nor the societies in which they exist will have a long-run future. Others believe that engaging in sustainable development will be a megatrend that will enhance organizational reputations. Others believe that sustainable development can be used by businesses as a unique core competency to obtain a strategic competitive advantage. All three rationales are valid and serve to stress the need for responsible business to pursue sustainable development in the current competitive reality.

Sustainable Development as an Ethical Issue

A 1996 survey of American and Canadian corporate executives included the question, "Why does, or will, your company practice sustainable development?" On a 10-point scale of level of importance, the responses of (1) promoting good relations and (2) creating shareholder value scored, respectively, 8.1 and 7.3. However, more importantly, the two most highly ranked responses were (1) to comply with legal regulations and (2) a moral commitment to environmental stewardship (8.8 and 8.5, respectively). Thus, there is evidence that business executives recognize that sustainable development can and should be viewed as part of the interwoven framework of business ethics. Ethicists would applaud such a view and could use the theories of utilitarianism, rights/duties, and the categorical imperative to provide the underlying support.

In making a utilitarian analysis of businesses' implementation of sustainable development concepts, the "greatest good or least harm for the greatest number" principle can be easily envisioned. The stakeholders involved are all the earth's inhabitants, both human and non-human. Sustainable development would create the greatest good or least harm by allowing those inhabitants (and potential offspring) to exist in a world where the air is breathable, the water is drinkable, the soil is fertile, and renewable resources thrive. It is difficult to use traditional monetary cost-benefit analysis to determine whether sustainable development is worthwhile. First, although many current and future costs could be estimated and discounted back to present values, it is probably impossible to even comprehend what types and amounts of costs might be necessary in the future. Second, the benefits of sustainable development are significantly more qualitative than monetarily quantitative; for example, how can the value of a living species be estimated? But, even without finances attached, the result would be undeniably conclusive: no matter how high the costs of sustainable development are, the benefits of current and continued existence by the earth's species *must* exceed that cost. Ethically, the benefits of life outweigh the costs to obtain it.

Analyzing sustainable development activities by business entities using the theory of right/duties addresses the issue of whether an inhabitable environment is a moral right.... Blackstone postulated that access to livable environment is a human right because such an environment is essential for humans to fulfill their capacities. Thus, everyone has the correlative moral obligation to respect that right. Rawls and Kant would support this concept because of the rationality of people being entitled to rights that do not infringe upon others' rights. A human's inhabitable environment includes other living creatures, flora, fauna, and resources (e.g., air, water, and minerals). These non-human elements of the planet are not responsible for, nor can they correct, the ecologically damaging discharges of pollution or disproportionate use of resources created by humans. Thus, businesses, as collections of human beings, have the duty to engage in sustainable development activities so as to mitigate their environmental impacts and help in providing, protecting, and preserving a livable environment.

In its determination of morality as objectively and universally binding, Kant's categorical imperative would support businesses' sustainable development actions. Proponents of Kantianism, however, would be quick to point out that sustainable development activities should be performed from duty, not simply from inclination or self-interest. In other words, businesses should not engage in sustainable development because such activities will reduce costs, increase revenues, or provide an advantageous reputation. Businesses should engage in sustainable development because, in the minds of all rational people, reclaiming and preserving the earth's environment as well as limiting pollution and resource depletion is the "right" thing to do. In the final analysis, sustainable development represents an action that would be right and valid "even if everyone were to violate it in actual conduct."

Sustainable development is, then, an important and ethical value to be upheld by businesses. But some aspects of sustainable development are more clearly pursued, or pursued to different degrees, by some publics than by others.

Level of Sustainable Development Efforts for Businesses

From the standpoint of businesses, it is important to ascertain which sustainable development issues can and cannot be addressed. Businesses cannot pass laws or treaties to protect the environment, enact land reforms, or control populations. Businesses cannot force consumers to recycle, reuse, or slow consumption. Businesses, in general, cannot produce the scientific knowledge that will end global warming, save the rain forests, or eliminate pollution. Businesses cannot stop societal development. And businesses cannot decide to pursue totally altruistic environmental goals without any concern for profitability or longevity. (To do so would be to guarantee organizational failure: owners would remove financial backing because they could not achieve a reasonable return on investment; employees would look elsewhere for jobs because they could not rely on continued employment; and suppliers would limit or revoke credit because they could not be assured of payment.)

Although businesses cannot do any of the things mentioned above unilaterally, there are many things that they can do. Businesses can influence passage of laws through lobbying and other efforts. They can influence consumer behavior (through product development and packaging, encouraging consumer recycling and reuse, and community awareness activities). Businesses can (through research agendas and new product discovery and development) help reduce or eliminate pollution causes. Businesses can also influence how societal development will occur and what the impact of that development will be through their location and technological investment choices. And businesses can undertake a strategy of pursuing sustainable development in conjunction with profitability and longevity to the benefit of all organizational stakeholders. Such a strategy would focus on both current and future eco-efficiencies.

Given the myriad of opportunities for engaging in environmentally "correct" or, at a higher level, sustainable development activities, how should a business determine its participation? One possible technique would be the use of the hierarchy of ethical behavior suggested by Raiborn and Payne. The hierarchy consists of four degrees of achievement:

- basic (reflects minimally acceptable behavior that complies with the letter, but not the spirit, of the law);
- currently attainable (reflects behavior deemed moral, but not laudable, by society);
- practical (reflects extreme diligence toward moral behavior; achievable but difficult); and
- theoretical (reflects the highest potential for good or the spirit of morality).

Basic Level of Behavior

A business operating at the basic level of behavior would merely comply with the laws of the jurisdictions in which it operates. Such an organization would make no sustainable development efforts because the concept is not embedded into the law in any country in the world. This organization would remain within legally acceptable pollution levels, although it would possibly view those levels as hindrances to productive activities. Such organizations ... would more than likely espouse (although quietly) the following beliefs: *We recognize that the environment is not a "free and unlimited" good. However, environmental laws cost money that could be going to support the economic goal of increased shareholder value. We will operate within the law, but will not seek environmental improvements beyond the law.* Thus, these companies' behaviors would be deemed legal, but not necessarily ethical.

Currently Attainable Level of Behavior

A business operating at the currently attainable level of behavior would acknowledge that some benefits do arise from engaging in environmentally-friendly activities that are not legally mandated. These organizations, however, probably engage in such activities for the "wrong" reasons (according to the categorical imperative): cost reduction, revenue enhancement, or reputation improvement. In other words, the activities are likely to provide short-term monetary benefits greater than their costs.... These companies would more than likely espouse the following belief: *We recognize that the environment is not a "free and unlimited" good. Environmental laws are necessary because business should be held responsible to remove the damaging effects they have had and to reduce or limit the future impacts they will have on the earth's ecosystems in their role as society's major tangible goods producers. We will operate within the law and will seek to find environmental improvements that reduce costs or improve productive activities so that short-term profits are enhanced and shareholder value is increased.* These organizations may be viewed by society as environmentally-conscious companies that are operating for the greater good ... but, in reality, the greater good is primarily that of the organization.

Practical Level of Behavior

A business operating at the practical level of behavior would also acknowledge that benefits arise from engaging in environmentally-friendly activities. These organizations, however, would strive to do the "right" thing relative to the environment because it is "right" rather than because of short-term profits or reputation. These businesses and their managers recognize the need for, and worth of, environmentally sound production and marketing practices. These organizations would attempt, in their varying activities, to engage in environmental innovations that might be expensive but that would provide the most beneficial future outcomes. In doing so, the businesses would hope that consumers would recognize the benefits of such innovative practices are worth purchasing at a higher cost than those of less environmentally sensitive competitors. There should be no question that these businesses are profit-motivated: management has a fiduciary duty towards a number of groups (among which are shareholders, creditors, employees, and consumers) to maximize profits and, therefore, efficiency. "For both infrastructure and services, it has to be recognized that private sector participation will be achieved only on the basis of an acceptable expected revenue scheme."

... These companies would more than likely espouse the following belief: *We recognize that the environment must be protected, not only through laws but also through our own proactive involvement. We will find and implement environmental improvements and innovations for our products and processes, knowing that consumers will recognize the long-run benefits of our actions and be willing to support those actions with their purchasing decisions. Through this strategy, we believe that we will provide high quality products that have the least detrimental environmental impact on our local and global community.* Thus, these organizations view themselves as forerunners in the area of environmental protection, for the sake of all stakeholders. But these companies have not crossed the line from overt environmental concern to cutting edge, world-class leadership in sustainability.

Theoretical Level of Behavior

A business operating at the theoretical level of behavior would have incorporated the idea of sustainable development into its organizational strategy. There would be no "piecemeal projects aimed at controlling or preventing pollution. Focusing on sustainability requires putting business strategies to a new test. Taking the entire planet as the context in which they do business, companies must ask whether they are part of the solution to social and environmental problems or part of the problem." These organizations ... would more than likely espouse the following belief: *The new paradigm must view the environment as fundamental to the business', society's, and the earth's continued existence. It is to be protected and replenished through all human and machine investments that are necessary to secure our place and the place of others (both human and nonhuman) on this planet. In doing so, our organization will be cost efficient from waste reduction and resource productivity maximization. Our business will be respected by our stakeholders; our products and services will be desired and recognized as value-added; and our eco-efficiency will enhance organizational profitability and promote organizational longevity.* These organizations take the concept of "walking the talk" completely literally.

What Actions Can and Are Being Taken by Businesses?

One statistic starkly exhibits the crisis that looms: "By the year 2030, world population will double from 5.5 billion to 11 billion.... To provide basic amenities to all people, it is estimated that production of goods and energy will need to increase 5 to 35 times today's levels." Such changes will cause further environmental strain and perhaps irreparable damage. Can the earth assimilate the massive pollution and resource depletion inherent in such growth? Should economic growth be pitted against environmental and human health? Will implementation of sustainable development activities require a change in consumption habits and, if so, what habits of whom should be altered? Will technological innovation arise as the hoped-for panacea, such that consumption habits may remain unchanged? Can stakeholders accept, encourage, and reward through product/service purchases and organizational investment business actions toward sustainable development? Answers to these questions would obviously ameliorate the chance for efficacious solutions. Unfortunately, only simple answers can be provided for these complex questions at this time. Significant research needs to be performed to ascertain the answers that are the most ethical and the most eco-efficient. But one thing is clear: if businesses, as the manufacturers and providers of the world's products and services, do not begin individually and collectively to immediately work toward a solution, after some point there will be no solution to achieve.

Businesses should not be considered as irresponsible entities that must be forced into doing the ethical thing with regard to environmental protection or sustainable development. Businesses recognize the symbiotic relationship between the environment, consumers' demands, and the provision of goods and services to the world's communities. Businesses also recognize the synergistic relationship between them and the environment/society in which they operate. It would be irrational to suggest that business could exist without society and equally irrational to suggest that society could exist as well as, better, or at all in the absence of business. In other words, business and society need each other for practical reasons: businesses want to provide goods and services that society needs and/or wants....

Shrivastava has suggested that, as a beginning, businesses strive to attain various goals that are commensurate with the goals of sustainable development. He suggests that energy conservation techniques could be employed that would have a positive impact on pollution and resource depletion. Businesses could also engage in resource regeneration aimed specifically at the reduction of resource depletion. Additionally, he promotes environmental preservation, which strengthens and is strengthened by arguments that the environment itself is worthy of care and protection, aside from its human-associated values. To implement these three goals, businesses can improve processes, educate employees, provide consumer advice, perform research, be prepared for emergencies, and listen openly to concerns.

A final, but very important method by which businesses can strive toward sustainable development is to join with others to form organizations focused on this goal. Some of these organizations include the World Trade Organization's Committee on Trade and Environment, the World Business Council for Sustainable Development, the International Chamber of Commerce's Commission on Environment, and the United Nations Environment Program. As aptly stated in the International Chamber of Commerce Commitment to Sustainable Development:

> all sectors of society, including government, business, public interest groups and consumers, have a role to play in contributing to sustainable development, and they must work in partnership, bringing their values and experience to bear on the challenge. Sustainable development will only be achieved if each one plays its part. Each sector should focus on what it can do best, but, through partnerships, local, national or even global, we can build on the strengths of each group.... Business is best suited to contributing to sustainable development in the economic sphere— through the creation of wealth in an environmentally sound manner.

Conclusions

Businesses need to assert their commitment to sustainable development over and above environmental legalities. As indicated by Porter and van der Linde, "Regulators tend to set regulations in ways that deter innovation. Companies, in turn, oppose and delay regulations instead of innovating to address them. The whole process has spawned an industry of litigators and consultants that drains resources away from real solutions."

Who, in business, should lead the way in the pursuit of sustainable development goals? The easiest answer is that global, multinationals based in highly developed countries should be the leaders; some of these entities have already begun the journey. Another answer is that those entities creating the biggest environmental problems should lead the way. The most appropriate answer, however, is that organizations whose stakeholders recognize the necessity of sustainable development as part and parcel of the company's need to act ethically should be the role models.

Businesses, acting alone, cannot create sustainability. If the internal and external stakeholders are not willing to adopt the concept of sustainability as a long term necessity, then should businesses view the idea as not worthy and expunge it from the organizational strategy? Absolutely not! As indicated within the paper, there is significant interaction between and among all value chain constituents. And, similar to the spread of high product and service quality as a priority among value chain members, as one member of the value chain demands a view of sustainable development, so will others. In some cases, there will be a trickle-down effect; in others, there will be a waterfall.

It is time that businesses realized that environmental responsibility and sustainable development are part and parcel of business ethics. Rules can be written and laws can be passed about pollution control or environmental degradation, but the framework to which these are bound is the minimum or basic level of acceptable behavior. Like a corporate code of ethics, an environmental policy will reflect the corporate culture from which it stems. The companies that move in a continuous path up the hierarchy of ethical behavior from merely complying with legalities to integrating sustainable development concepts into strategic initiatives and mission statements are companies whose managers understand, espouse, advocate, and uphold the fundamentals of business ethics. These are also the companies and managers that are well aware that ethical business is good business. These are the long term survivors.

NO

Ronald Bailey

Wilting Greens

It's clear that we've suffered a number of major defeats," declared Andrew Hewett, executive director of Oxfam Community Aid, at the conclusion of the World Summit on Sustainable Development, held in Johannesburg, South Africa, in September. Greenpeace climate director Steve Sawyer complained, "What we've come up with is absolute zero, absolutely nothing." The head of an alliance of European green groups proclaimed, "We barely kept our heads above water."

It wasn't supposed to be this way. Environmental activists hoped the summit would set the international agenda for sweeping environmental reform over the next 15 years. Indeed, they hoped to do nothing less than revolutionize how the world's economy operates. Such fundamental change was necessary, said the summiteers, because a profligate humanity consumes too much, breeds too much, and pollutes too much, setting the stage for a global ecological catastrophe.

But the greens' disappointment was inevitable because their major goals—preserving the environment, eradicating poverty, and limiting economic growth—are incompatible. Economic growth is a prerequisite for lessening poverty, and it's also the best way to improve the environment. Poor people cannot afford to worry much about improving outdoor air quality, let alone afford to pay for it. Rather than face that reality, environmentalists increasingly invoke "sustainable development." The most common definition of the phrase comes from the 1987 United Nations report *Our Common Future*: development that "meets the needs of the present without compromising the ability of future generations to meet their own needs."

For radical greens, sustainable development means economic stagnation. The Earth Island Institute's Gar Smith told Cybercast News, "I have seen villages in Africa ... that were disrupted and destroyed by the introduction of electricity." Apparently, the natives no longer sang community songs or sewed together in the evenings. "I don't think a lot of electricity is a good thing," Smith added. "It is the fuel that powers a lot of multinational imagery." He doesn't want poor Africans and Asians "corrupted" by ads for Toyota and McDonald's, or by Jackie Chan movies.

Indian environmentalist Sunita Narain decried the "pernicious introduction of the flush toilet" during a recent PBS/BBC television debate hosted by Bill Moyers. Luckily, most other summiteers disagreed with Narain's curious disdain for sanitation. One of the few firm goals set at the confab was that adequate sanitation should be supplied by 2015 to half of the 2.2 billion people now lacking it.

Sustainable development boils down to the old-fashioned "limits to growth" model popularized in the 1970s. Hence Daniel Mittler of Friends of the Earth International moaned that "the summit failed to set the necessary economic and ecological limits to globalization." The *Jo'burg Memo*, issued by the radical green Heinrich Böll Foundation before the summit, summed it up this way: "Poverty alleviation cannot be separated from wealth alleviation."

The greens are right about one thing: The extent of global poverty is stark. Some 1.1 billion people lack safe drinking water, 2.2 billion are without adequate sanitation, 2.5 billion have no access to modern energy services, 11 million children under the age of 5 die each year in developing countries from preventable diseases, and 800 million people are still malnourished, despite a global abundance of food. Poverty eradication is clearly crucial to preventing environmental degradation, too, since there is nothing more environmentally destructive than a hungry human.

Most summit participants from the developing world understood this. They may be egalitarian, but unlike their Western counterparts they do not aim to make everyone equally poor. Instead, they want the good things that people living in industrialized societies enjoy.

That explains why the largest demonstration during the summit, consisting of more than 10,000 poor and landless people, featured virtually no banners or chants about conventional environmentalist issues such as climate change, population control, renewable resources, or biodiversity. Instead, the issues were land reform, job creation, and privatization.

The anti-globalization stance of rich activists widens this rift. Environmentalists claim trade harms the environment and further impoverishes people in the developing world. They were outraged by the dominance of trade issues at the summit.

"The leaders of the world have proved that they work as employees for the transnational corporations," asserted Friends of the Earth Chairman Ricardo Navarro. Indian eco-feminist Vandana Shiva added, "This summit has become a trade summit, it has become a trade show." Yet the U.N.'s own data underscore how trade helps the developing world. As fact sheets issued by the U.N. put it, "During the 1990s the economies of developing countries that were integrated into the world economy grew more than twice as fast as the rich countries. The 'non-globalizers' grew only half as fast and continue to lag further behind."

By invoking a zero sum version of sustainable development, environmentalists not only put themselves at odds with the developing world; they ignore the way in which economic growth helps protect the environment. The real commons from which we all draw is the growing pool of scientific, technological, and institutional concepts, and the capital they create. Past generations have left us far more than they took, and the result has been an explosion in human well-being, longer life spans, less disease, more and cheaper food, and expanding political freedom.

Such progress is accompanied by environmental improvement. Wealthier is healthier for both people and the environment. As societies become richer and more technologically adept, their air and water become cleaner, they set aside more land for nature, their forests expand, they use less land for agriculture, and more people cherish wild species. All indications suggest that the 21st century will be the century of ecological restoration, as humanity uses physical resources ever more efficiently, disturbing the natural world less and less.

In their quest to impose a reactionary vision of sustainable development, the disappointed global greens will turn next to the World Trade Organization, the body that oversees international trade rules. During the summit, the WTO emerged as the greens' bête noire. As Friends of the Earth International's Daniel Mittler carped, "Instead of using the [summit] to respond to global concerns over deregulation and liberalization, governments are pushing the World Trade Organization's agenda." "See you in Cancun!" promised Greenpeace's Steve Sawyer, referring to the location of the next WTO ministerial meeting in September 2003. That confab will build on the WTO's Doha Trade Round, launched last year, which is aimed at reducing the barriers to trade for the world's least developed countries.

The WTO may achieve worthy goals that eluded the Johannesburg summit, such as eliminating economically and ecologically ruinous farm and energy subsidies and opening developed country markets to the products of developing nations. Free marketeers and greens might even form an alliance on those issues.

But environmentalists want to use the WTO to implement their sustainable development agenda: global renewable energy targets, regulation based on the precautionary principle, a "sustainable consumption and production project," a worldwide eco-labeling scheme. According to Greenpeace's Sawyer, nearly everyone at the Johannesburg summit agreed "there is something wrong with unbridled neoliberal capitalism."

Let's hope the greens fail at the WTO just as they did at the U.N. summit. Their sustainable development agenda, supposedly aimed at improving environmental health, instead will harm the natural world, along with the economic prospects of the world's poorest people. The conflicting goals on display at the summit show that at least some of the world's poor are wise to that fact.

POSTSCRIPT

Is Sustainable Development Compatible With Human Welfare?

The first of the Rio Declaration's 22 principles states, "Human beings are at the centre of concerns for sustainable development. They are entitled to a healthy and productive life in harmony with nature." Any solution to the sustainability problem therefore should not infringe human welfare. This makes any solution that involves limiting or reducing human population or blocking improvements in standard of living very difficult to sell. Yet solutions may be possible. David Malin Roodman suggests in *The Natural Wealth of Nations: Harnessing the Market for the Environment* (W. W. Norton, 1998) that taxing polluting activities instead of profit or income would stimulate corporations and individuals to reduce such activities or to discover nonpolluting alternatives. In "Building a Sustainable Society," *State of the World 1999* (W. W. Norton, 1999), he adds recommendations for citizen participation in decision making, education efforts, and global cooperation, without which we are heading for "a world order [that] almost no one wants." (He is referring to a future of environmental crises, not the "new world order" feared by many conservatives, in which national policies are dictated by international [UN] regulators.)

Julie Davidson, in "Sustainable Development: Business as Usual or a New Way of Living?" *Environmental Ethics* (Spring 2000), notes that efforts to achieve sustainability cannot by themselves save the world. But such efforts may give us time to achieve new and more suitable values. It is thus heartening to see that the UN World Summit on Sustainable Development was held in Johannesburg, South Africa, in August 2002. Its aim was to strengthen partnerships between governments, business, nongovernmental organizations, and other stakeholders and to seek to eradicate poverty and make more equal the distribution of the benefits of globalization. See Gary Gardner, "The Challenge for Johannesburg: Creating a More Secure World," *State of the World 2002* (W. W. Norton, 2002), and the United Nations Environmental Programme's Global Environmental Outlook 3 (Earthscan, 2002), prepared as a "global state of the environment report" in preparation for the Johannesburg Summit.

The World Council of Churches brought to the Johannesburg Summit an emphasis on social justice. Martin Robra, in "Justice—The Heart of Sustainability," *Ecumenical Review* (July 2002), writes that the dominant stress on economic growth "has served, first and foremost, the interests of the powerful economic players. It has further marginalized the poor sectors of society, simultaneously undermining their basic security in terms of access to land, water, food, employment, and other basic services and a healthy environment."

Is social justice or equity worth this emphasis? Or is sustainability more a matter of population control, of shielding the natural environment from human impacts, or of economics? A. J. McMichael, C. D. Butler, and Carl Folke, in "New Visions for Addressing Sustainability," *Science* (December 12, 2003), argue that it is wrong to separate—as did the Johannesburg Summit— achieving sustainability from other goals such as reducing fertility and poverty and improving social equity, living conditions, and health. They observe that human population and lifestyle affect ecosystems, ecosystem health affects human health, human health affects population and lifestyle. "A more integrated ... approach to sustainability is urgently needed," they say, calling for more collaboration among researchers and other fields.

An attempt to achieve a somewhat different kind of integration is visible in Thomas Prugh and Erik Assadourian, "What Is Sustainability, Anyway?" *World Watch* (September–October 2003). The authors define sustainability as having four components—human survival, biodiversity, equity, and life quality. Survival, they say, must come first because without it the rest do not matter. Yet, they conclude, "human environmental blunders and excesses are not likely to threaten us as a species." The other three components thus become the more important, even though all are very closely related.

ISSUE 3

Should a Price Be Put on the Goods and Services Provided by the World's Ecosystems?

YES: Janet N. Abramovitz, from "Putting a Value on Nature's 'Free' Services," *World Watch* (January/February 1998)

NO: Marino Gatto and Giulio A. De Leo, from "Pricing Biodiversity and Ecosystem Services: The Never-Ending Story," *BioScience* (April 2000)

ISSUE SUMMARY

YES: Janet N. Abramovitz, a senior researcher at the Worldwatch Institute, argues that if we fail to attach economic value to supposedly free services provided by nature, we are more likely to misuse and destroy the ecosystems that provide those services.

NO: Professors of applied ecology Marino Gatto and Giulio A. De Leo contend that the pricing approach to valuing nature's services is misleading because it falsely implies that only economic values matter.

Human activities frequently involve trading a swamp or forest or mountainside for a parking lot or housing development or farm. People generally agree that the parking lot, housing development, or farm is a worthwhile project, for it has obvious benefits—expressible in economic terms—to human beings. But are there costs as well? Construction costs, labor costs, and material costs can easily be calculated, but what about the swamp? The forest? The species living there?

How much is a species worth? One approach to answering this question is to ask people how much they would be willing to pay to keep a species alive. If the question is asked when there are a million species in existence, few people will likely be willing to pay much. But if the species is the last one remaining, they might be willing to pay a great deal. Most people would agree that both answers fail to get at the true value of a species, for nature is not expressible solely in terms of cash values. Yet some way must be found to weigh the effects of human activities on nature against the benefits gained from those activities. If it is not, we will continue to degrade the world's ecosystems and threaten our own continued well-being.

Traditional economics views nature as a "free good." That is, forests generate oxygen and wood, clouds bring rain, and the sun provides warmth, all without charge to the humans who benefit. At the same time, nature has provided ways for people to dispose of wastes—such as dumping raw sewage into rivers or emitting smoke into the air—without paying for the privilege. This "free" waste disposal has turned out to have hidden costs in the form of the health effects of pollution (among other things), but it has been up to individuals and governments to bear the costs associated with those effects. The costs are real, but in general, they have not been borne by the businesses and other organizations that produced them. They have thus come to be known as "external" costs.

Environmental economists have recognized the problem of external costs, and government regulators have devised a number of ways to make those who are responsible accept the bill, such as instituting requirements for pollution control and fining those who exceed permitted emissions. Yet some would say that this approach does not help enough.

The *ecosystem services* approach recognizes that undisturbed ecosystems do many things that benefit us. A forest, for instance, slows the movement of rain and snowmelt into streams and rivers; if the forest is removed, floods may follow (a connection that recently forced China to deemphasize forest exploitation). Swamps filter the water that seeps through them. Food chains cycle nutrients necessary for the production of wood and fish and other harvests. Bees pollinate crops and make food production possible. These services are valuable—even essential—to us, and anything that interferes with them must be seen as imposing costs just as significant as the illnesses associated with pollution.

How can those costs be assessed? In 1997 Robert Costanza and his colleagues published an influential paper entitled "The Value of the World's Ecosystem Services and Natural Capital" in the May 15 issue of the journal *Nature*. In it, the authors listed a variety of ecosystem services and attempted to estimate what it would cost to replace those services if they were somehow lost (such as by building a sewage treatment plant to replace a swamp). The total bill for the entire biosphere came to $33 trillion (the middle of a $16–54 trillion range), compared to a global gross national product of $25 trillion. Costanza et al. stated that this was surely an underestimate.

What good is such an estimate? Perhaps it could motivate governments to greater efforts to protect the environment in order to avoid colossal financial (or other) future consequences. This point is made in the following selection by Janet N. Abramovitz, who argues that nature's services are responsible for the vast bulk of the value in the world's economy and that attaching economic value to those services may encourage their protection. In the second selection, Marino Gatto and Giulio A. De Leo argue that the pricing approach to valuing nature's services is misleading because it ignores equally important "nonmarket" values.

Janet N. Abramovitz

Putting a Value on Nature's "Free" Services

During the last half of 1997, massive fires swept through the forests of Sumatra, Borneo, and Irian Jaya, which together form a stretch of the Indonesian archipelago as wide as all of Europe. By November, almost 2 million hectares had burned, leaving the region shrouded in haze and more than 20 million of its people breathing hazardous air. Tens of thousands of people had been treated for respiratory ailments. Hundreds had died from illness, accidents and starvation. The fires, though by then out of control, had been set deliberately and systematically—not by small farmers, and not by El Niño, but by commercial outfits operating with implicit government approval. Strange as this immolation of some of the world's most valuable natural assets may seem, it was not unique. The same year, a large part of the Amazon Basin in Brazil was blanketed by smoke for similar reasons. The fires in the Amazon have been set annually, but in 1997 they destroyed over 50 percent more forest than the year before, which in turn had recorded five times as many fires (some 19,115 fires during a single six-week period) as in 1995.

For the timber and plantation barons of Indonesia, as for the cattle ranchers and frontier farmers of Amazonia, setting fires to clear forests has become standard practice. To them, the natural rainforests are an obstruction that must be sold or burned to make way for their profitable pulp and palm oil plantations. Yet, these are the same forests that for many others serve as both homes and livelihoods. For the hundreds of millions who live in Indonesia and in the neighboring nations of Malaysia, Singapore, Brunei, southern Thailand and the Philippines, it is becoming painfully apparent that without healthy forests, it is difficult to remain healthy people.

As this issue of WORLD WATCH went to press [December 1997], the fires in Southeast Asia were still generating enough smoke to be visible from space. Some relief was expected with the arrival of the seasonal rains, but those rains were past due—in part because of an unusually strong El Niño effect. Along with the trees, the region's large underground peat deposits have caught on fire, and such fires are perniciously difficult to put out; they can continue smoldering for years.

When the smoke does finally clear, Southeast Asia—and the world—will attempt to tally the costs. There are the costs of impaired health and sometimes death, from both lung diseases and accidents caused by poor visibility. There is the productivity that was lost as factories, schools, roads, docks, and airports were shut down (over 1,000 flights in and out of Malaysia were cancelled in September alone); there are the crop yields that fell as haze kept the region in day-long twilight, and the harvests of forest products that were wiped out. Timber (some of the most valuable species in the world) and wildlife (some of the most endangered in the world) are still being consumed by flames. Over three-fourths of the world's remaining wild orangutans live on the fire-ravaged provinces of Sumatra and Kalimantan. Some of them, caught fleeing the flames, have become part of the illegal trade. Because of their location, the Indonesian fires, like those in the Amazon, have dealt a heavy blow to the biodiversity of the earth as a whole.

As the smoke billowed dramatically from Southeast Asia, a much less visible—but similarly costly—ecological loss was taking place in a very different kind of location. While the Indonesian haze was being photographed from satellites, this other loss might not be noticed by a person standing within an arm's length of the evidence—yet, in its implications for the human future, it is a close cousin of the Asian catastrophe. In the United States, more than 50 percent of all honeybee colonies have disappeared in the last 50 years, with half of that loss occurring in just the last 5 years. Similar losses have been observed in Europe. Thirteen of the 19 native bumblebee species in the United Kingdom are now extinct. These bees are just two of the many kinds of pollinators and their decline is costing farmers, fruit growers, and beekeepers hundreds of millions of dollars in losses each year.

What the ravaged Indonesian forests and disappearing bees have in common is that they are both examples of "free services" that are provided by nature and consumed by the human economy—services that have immense economic value, but that go largely unrecognized and uncounted until they have been lost. Many of those services are indispensable to the people who exploit them, yet are not counted as real benefits, or as a part of GNP.

Though widely taken for granted, the "free" services provided by the natural world form the invisible foundation that supports all societies and economies. We rely on the oceans to provide abundant fish, on forests for wood and new medicines, on insects and other creatures to pollinate our crops, on birds and frogs to keep pests in check, and on forests and rivers to supply clean water. We take it for granted that when we need timber we can cut trees, or that when we need water we can find a spring or drill a well. We assume that clean air will blow the smog out of our cities, that the climate will be stable and predictable, and that the mounting quantity of waste we generate will continue to disappear, if we can just get it out of sight. Nature's services have always been there, free for the taking, and our expectations—and economies—are based on the premise that they always will be. A timber magnate or farmer may have to pay a price for the land, but assumes that what happens naturally on the land—the growing of trees, or pollinating of crops by wild bees, or filtering of fresh water—usually happens for free. We are like young children who think that food comes from the refrigerator, and who do not yet understand that what now seems free is not.

Ironically, by undervaluing natural services, economies unwittingly provide incentives to misuse and destroy the very systems that produce those services; rather than protecting their assets, they squander them. Nature, in turn becomes increasingly less able to supply the prolific range of services that the earth's expanding population and economy demand.... It is no exaggeration to suggest that the continued erosion of natural systems threatens not only the continuing viability of today's human enterprise, but ultimately the prospects for our continued existence.

Underpinning the steady stream of services nature provides to us, there is a more fundamental service these systems provide—a kind of self-regulating process by which ecosystems and the biosphere are kept relatively stable and resilient. The ability to withstand disturbances like fires, floods, diseases, and droughts, and to rebound from the shocks these events inflict, is essential to keeping the life-support system operating. As systems are simplified by monoculture or cut up by roads, and the webs that link systems become disconnected, they become more brittle and vulnerable to catastrophic, irreversible decline. We are being confronted by ample evidence, now—from the breakdown of the ozone layer to the increasingly severity of fires, floods and droughts, to the diminished productivity of fruit and seed sets in wild and agricultural plants—that the biosphere is becoming less resilient.

Unfortunately, much of the human economy is based on practices that convert natural systems into something simpler, either for ease of management (it's easier to harvest straight rows of trees that are all the same age than to harvest carefully from complex forests) or to maximize the production of a desired commodity (like corn). But simplified systems lack the resilience that allows them to survive short-term shocks such as outbreaks of diseases or pests, or forest fires, or even longer-term stresses such as that of global warming. One reason is that the conditions within these simplified systems are not hospitable to all of the numerous organisms and processes needed to keep such systems running. A tree plantation or fish farm may provide some of the products we need, but it cannot supply the array of services that natural diverse systems do—and must do—in order to survive over a range of conditions. To keep our own economies sustainable, then, we need to use natural systems in ways that capitalize on, rather than destroy, their regenerative capacity. For humans to be healthy and resilient, nature must be too.

Resiliency is destroyed by fragmentation, as well as by simplification. Fires in healthy rainforests are very rare. By nature, they are too wet to burn. But as they are opened up and fragmented by roads and logging and pasture, they become drier and more prone to fire. When fire strikes forests that are not adapted to fire (as is the case in the rainforests of both Brazil and Indonesia), it is exceptionally destructive and tends to kill a majority of the trees. The fires in Southeast Asia's peat swamp rainforests bring further disruption, by releasing long-sequestered carbon into the atmosphere.

The fires in Indonesia are not being started by poor slash-and-burn peasants, but by "slash-and-burn industrialists"—owners of rubber, palm oil, rice, and timber plantations who have been taking advantage of a dry year to clear as much natural forest as they can. Though it issued a recent law forbidding the burning, the government in Indonesia is in fact pushing for higher production levels from these export sectors. In both the rainforests and the peat swamps, it has given the plantation owners large concessions to encourage continued "conversion" to one-crop commodities. And the government continues to push costly agricultural settlements into peat forests ill-suited to rice. After the fires became a serious regional problem (and international embarrassment), the government revoked the permits of 29 companies, but such actions were too little, too late.

The current fires are not the first to ravage parts of Southeast Asia; extensive logging in Indonesia and Malaysia led to a major conflagration in 1983 that burned over 3 million hectares and wiped out $5 billion worth of standing timber in Indonesia alone. After 1983, fires that had once been rare became a common occurrence. The 1997 fire will likely turn out to have been the most costly yet. Unless policies change, the fires will be reignited this year.

What Forests Do

Around the world, the degradation, fragmentation, and simplification—or "conversion"—of ecosystems is progressing rapidly. Today, only 1 to 5 percent of the original forest cover of the United States and Europe remains. One-third of Asia's forest has been lost since 1960, and half of what remains is threatened by the same industrial forest activities responsible for the Indonesian fires. In the Amazon, 13 percent of the natural cover has already been cleared, mostly for cattle pasture. In many countries, including some of the largest, more than half of the land has been converted from natural habitat to other uses that are less resilient. In countries that stayed relatively undisturbed until the 1980s, significant portions of remaining ecosystems have been lost in the last decade. These trends have been accelerating everywhere. As the natural ecosystems disappear, so do many of the goods and services they provide.

That may seem to contradict the premise that people want those goods and services and would not deliberately destroy them. But there's a logical explanation: governments and business owners typically perceive that the way they can make the most profit from an ecosystem is to maximize its production of a single commodity, such as timber from a forest. For the community (or society) as a whole, however, that is often the least profitable or sustainable use. The economic values of other uses, and the number of people who benefit, added up, can be enormous. A forest, if not cut down to make space for a one-commodity plantation, can produce a rich variety of nontimber forest products (NTFPs) on one hand, while providing essential watershed protection and climate regulation, on the other. These uses not only have more immediate economic value but can also be sustained over a longer term and benefit more people.

In 1992, alternative management strategies were reviewed for the mangrove forests of Bintuni Bay in Indonesia. When nontimber uses such as fish, locally used products, and erosion control were included in the calculations, the researchers found that the most economically profitable strategy was to keep the forest standing with only a modest amount of timber cutting—yielding $4,800 per hectare. If the forest was managed only for timber-cutting, it would yield only $3,600 per hectare. Over the longer term, it was calculated that keeping the forest intact would ensure continued local uses of the area worth $10 million a year (providing 70 percent of local income) and protect fisheries worth $25 million a year—values that would be lost if the forest were cut.

The variety and value of goods produced and collected from forests, and their importance to local livelihoods and national economies, is an economic reality worldwide. For instance, rattan—a vine that grows naturally in tropical forests—is widely used to make furniture. Global trade in rattan is worth $2.7 billion in exports each year, and in Asia it employs a half-million people. In Thailand, the value of rattan exports is equal to 80 percent of the legal timber exports. In India, such "minor" products account for three fourths of the net export earnings from forest produce, and provide more than half of the formal employment in the forestry sector. And in Indonesia, hundreds of thousands of people make their livelihoods collecting and processing NTFPs for export, a trade worth at least $25 million a year. Many of these forests were destroyed in the fire.

Even so, non-timber commodities are only part of what is lost when a forest is converted to a one-commodity industry. There is a nexus between the two catastrophes of the Indonesian fires and the North American and European bee declines, for example, since forests provide habitat for bees and other pollinators. They also provide habitat for birds that control disease-carrying and agricultural pests. Their canopies break the force of the winds and reduce rainfall's impact on the ground, which lessens soil erosion. Their roots hold soil in place, further stemming erosion. In purely monetary terms, a forest's capacity to protect a watershed alone can exceed the value of its timber. Forests also act as effective water-pumping and recycling machinery, helping to stabilize local climate. And, through photosynthesis, they generate enough of the planet's oxygen, while absorbing and storing so much of its carbon (in living trees and plants), that they are essential to the stability of climate worldwide.

Beyond these general functions, there are services that are specific to particular kinds of forests. Mangrove forests and coastal wetlands, notably, play critical roles in linking land and sea. They buffer coasts from storms and erosion, cycle nutrients, serve as nurseries for coastal and marine fisheries, and supply critical resources to local communities. For flood control alone, the value of mangroves has been calculated at $300,000 per kilometer of coastline in Malaysia—the cost of the rock walls that would be needed to replace them. Protecting coasts from storms will be especially important as climate change makes storms more violent and unpredictable. One force driving the accelerated loss of these mangroves in the last two decades has been the explosive growth of intensive commercial aquaculture, especially for shrimp export. Another has been the excess diversion of inland rivers and streams, which reduces downstream flow and allows the coastal waters to become too salty to support the coastal forests.

The planet's water moves in a continuous cycle, falling as precipitation and moving slowly across the landscape to streams and rivers and ultimately to the sea, being absorbed and recycled by plants along the way. Yet, human actions have changed even that most fundamental force of nature by removing natural plant cover, draining swamps and wetlands, separating rivers from their floodplains, and paving over land. The slow natural movement of water across the landscape is also vital for refilling nature's underground reservoirs, or aquifers, from which we draw much of our water. In many places, water now races across the landscape much too quickly, causing flooding and droughts, while failing to adequately recharge aquifers.

The value of a forested watershed comes from its capacity to absorb and cleanse water, recycle excess nutrients, hold soil in place, and prevent flooding. When plant cover is removed or disturbed, water and wind not only race across the land, but carry valuable topsoil with them. According to David Pimentel, an agricultural ecologist at Cornell University, exposed soil is eroded at several thousand times the natural rate. Under normal conditions, each hectare of land loses somewhere between 0.004 and 0.05 tons of soil to erosion each year—far less than what is replaced by natural soil building processes. On lands that have been logged or converted to crops and grazing, however, erosion typically takes away 17 tons in a year in the United States or Europe, and 30 to 40 tons in Asia, Africa, or South America. On severely degraded land, the hemorrhage can rise to 100 tons in a year. The eroded soil carries nutrients, sediments, and chemicals valuable to the system it leaves, but often harmful to the ultimate destination.

One way to estimate the economic value of an ostensibly free service like that of a forested watershed is to estimate what it would cost society if that service had to be replaced. New York City, for example, has always relied on the natural filtering capacity of its rural watersheds to cleanse the water that serves 10 million people each day. In 1996, experts estimated that it would cost $7 billion to build water treatment facilities adequate to meet the city's future needs. Instead, the city chose a strategy that will cost it only one-tenth that amount: simply helping upstream counties to protect the watersheds around its drinking water reservoirs.

Even an estimate like that tends to greatly understate the real value, however, because it covers the replacement cost of only one of the many services the ecosystem provides. A watershed, for example, also contributes to the regulation of the local climate. After forest cover is removed, an area can become hotter and drier, because water is no longer cycled and recycled by plants (it has been estimated that a single rainforest tree pumps 2.5 million gallons of water into the atmosphere during its lifetime.) Ancient Greece and turn-of-the-century Ethiopia, for example, were moister, wooded regions before extensive deforestation, cultivation, and the soil erosion that followed transformed them into the hot, rocky countries they are today. The global spread of desertification offers brutal evidence of the toll of lost ecosystem services.

The cumulative effects of local land use changes have global implications. One of the planet's first ecosystem services was the production of oxygen over billions of years of photosynthetic activity, which allowed oxygen-breathing organisms—such as ourselves—to evolve. Humans have begun to unbalance the global climate regulation system, however, by generating too much carbon dioxide and reducing the capacity of ecosystems to absorb it. Burning forests and peat deposits only makes the problem worse. The fires in Asia sent about as much carbon into the atmosphere last year as did all of the factories, power plants, and vehicles in the United Kingdom. For carbon sequestration alone, economists have been able to estimate the value of intact forests at anywhere from several hundred to several thousand dollars per hectare. As the climate changes the value of being able to regulate local and global climates will only increase.

What Bees Do

If we are often blind to the value of the free products we take from nature, it is even easier to overlook the value of those products we don't harvest directly—but without which our economies could not function. Among these less conspicuous assets are the innumerable creatures that keep potentially harmful organisms in check, build and maintain soils, and decompose dead matter so it can be used to build new life, as well as those that pollinate crops. These various birds, insects, worms, and microorganisms demonstrate that small things can have hugely disproportionate value. Unfortunately, their services are in increasingly short supply because pesticides, pollutants, disease, hunting, and habitat fragmentation or destruction have drastically reduced their numbers and ability to function. As Stephen Buchmann and Gary Paul Nabhan put it in a recent book on pollinators, "nature's most productive workers [are] slowly being put out of business."

Pollinators, for example, are of enormous value to agriculture and the functioning of natural ecosystems. Without them, plants cannot produce the seeds that ensure their survival—and ours. Unlike animals, plants cannot roam around looking for mates. To accomplish sexual reproduction and ensure genetic mixing, plants have evolved strategies for moving genetic material from one plant to the next, sometimes over great distances. Some rely on wind or water to carry pollen to a receptive female, and some can self-pollinate. The most highly evolved are those that use flowers, scents, oils, pollens, and nectars to attract and reward animals to do the job. In fact, more than 90 percent of the world's quarter-million flowering plant species are animal-pollinated. When animals pick up the flower's reward, they also pick up its pollen on various body parts—faces, legs, torsos. Laden with sticky yellow cargo, they can appear comical as they veer through the air—but their evolutionary adaptations are uncannily potent.

Developing a mutually beneficial relationship with a pollinator is a highly effective way for a plant to ensure reproductive success, especially when individuals are isolated from each other. Spending energy producing nectars and extra pollen is a small price to pay to guarantee reproduction. Performing this matchmaking service are between 120,000 and 200,000 animal species, including bees, beetles, butterflies, moths, ants, and flies, along with more than 1,000 species of vertebrates such as birds, bats, possums, lemurs, and even geckos. New evidence shows that many more of these pollinator species than previously believed are threatened with extinction.

Eighty percent of the world's 1,330 cultivated crop species (including fruits, vegetables, beans and legumes, coffee and tea, cocoa, and spices) are pollinated by wild and semi-wild pollinators. One-third of U.S. agricultural output is from insect-pollinated plants (the remainder is from wind-pollinated grain plants such as wheat, rice, and corn). In dollars, honeybee pollination services are 60 to 100 times more valuable than the honey they produce. The value of wild blueberry bees is so great, with each bee pollinating 15 to 19 liters (about 40 pints) of blueberries in its life, that they are viewed by farmers as "flying $50 bills."

Without pollinator services, crops would yield less, and wild plants would produce few seeds—with large economic and ecological consequences. In Europe, the contribution of honey bee pollination to agriculture was estimated to be worth $100 billion in 1989. In the Piedmont region of Italy, poor pollination of apple and apricot orchards cost growers $124 million in 1996. The most pervasive threats to pollinators include habitat fragmentation and disturbance, loss of nesting and over-wintering sites, intense exposure of pollinators to pesticides and of nectar plants to herbicides, breakdown of "nectar corridors" that provide food sources to pollinators during migration, new diseases, competition from exotic species, and excessive hunting. The rapid spread of two parasitic mites in the United States and Europe has wiped out substantial numbers of honeybee colonies. A "forgotten pollinators" campaign was recently launched by the Arizona Sonoran Desert Museum and others, to raise awareness of the importance and plight of these service providers.

Ironically, many modern agricultural practices actually limit the productivity of crops by reducing pollination. According to one estimate, for example, the high levels of pesticides used on cotton reduce annual yields by 20 percent (worth $400 million) in the United States alone by killing bees and other insect pollinators. One-fifth of all honeybee losses involve pesticide exposure, and honeybee poisonings may cost agriculture hundreds of millions of dollars each year. Wild pollinators are particularly vulnerable to chemical poisoning because their colonies cannot be picked up and moved in advance of spraying the way domesticated hives can. Herbicides can kill the plants that pollinators need to sustain themselves during the "off-season" when they are not at work pollinating crops. Plowing to the edges of fields to maximize planting area can reduce yields by disturbing pollinator nesting sites. Just one hectare of unplowed land, for example, provides nesting habitat for enough wild alkali bees to pollinate 100 hectares of alfalfa.

Domesticated honeybees cannot be expected to fill the gap left when wild pollinators are lost. Of the world's major crops, only 15 percent are pollinated by domesticated and feral honeybees, while at least 80 percent are serviced by wild pollinators. Honeybees do not "fit" every type of flower that needs pollination. And because honeybees visit so many different plant species, they are not very "efficient"—that is, there is no guarantee that the pollen will be carried to a potential mate of the same species and not deposited on a different species.

Many plants have developed interdependencies with particular species of pollinators. In peninsular Malaysia, the bat *Eonycteris spelea* is thought to be the exclusive pollinator of the durian, a large spiny fruit that is highly valued in Southeast Asia. The bats' primary food supply is a coastal mangrove that flowers continuously throughout the year. The bats routinely fly tens of kilometers from their roost sites to the mangrove stands, pollinating durian trees along the way. However, mangrove stands in Malaysia and elsewhere are under siege, as are the inland forests. Without both, the bats are unlikely to survive.

Pollinators that migrate long distances, such as bats, monarch butterflies, and hummingbirds, need to follow routes that offer a reliable supply of nectar-providing plants for the full journey. Today, however, such nectar corridors are being stretched increasingly thin and are breaking. When the travelers cannot rest and "refuel" every day, they may not survive the journey.

The migratory route followed by long-nosed bats from their summer breeding colonies in the desert regions of the U.S. Southwest to winter roosts in central Mexico illustrates the problems faced by many service providers. To fuel trips of up to 150 kilometers a night, these bats rely on the sequential flowering of at least 16 plant species—particularly century plants and columnar cacti. Along much of the migratory route, the nectar corridor is being fragmented. On both U.S. and Mexican rangelands, ranchers are converting native vegetation into exotic pasture grasses for grazing cattle. In the Mexican state of Sonora, an estimated 376,000 hectares have been stripped of nectar source plants. In parts of the Sierra Madre, the bat-pollinators are threatened by competition from human bootleggers, who have been over-harvesting century plants to make the alcoholic beverage mescal. And the latest threat comes from dynamiting and burning of bat roosts by Mexican ranchers attempting to eliminate vampire bats that feed on cattle and spread livestock diseases. The World Conservation Union estimates that worldwide, 26 percent of bat species are threatened with extinction.

Many of the disturbances that have harmed pollinators are also hurting creatures that provide other beneficial services, such as biological control of pests and disease. Much of the wild and semi-wild habitat inhabited by beneficial predators such as birds has been wiped out. The "pest control services" that nature provides are incalculable, and do not have the fundamental flaws of chemical pesticides (which kill beneficial insects along with the pests and harm people). Individual bat colonies in Texas can eat 250 tons of insects each night. Without birds, leaf-eating insects are more abundant and can slow the growth of trees or damage crops. Biologists Paul and Anne Ehrlich speculate that without birds, insects would have become so dominant that humans might never have been able to achieve the agricultural revolution that set the stage for the rise of civilization.

It is not too late to provide essential protections to the providers of such essential services—by using no-till farming to reduce soil erosion and allow nature's underground economy to flourish, by cutting back on the use of toxic agricultural chemicals, and by protecting migratory routes and nectar corridors to ensure the survival of wild pollinators and pest control agents.

Buffer areas of native vegetation and trees can have numerous beneficial effects. They can serve as havens for resident and migratory insects and animals that pollinate crops and control pests. They can also help to reduce wind erosion, and to absorb nutrient pollution that leaks from agricultural fields. Such zones have been eliminated from many agricultural areas that are modernized to accommodate new equipment or larger field sizes. The "sacred groves" in South Asian and African villages—natural areas intentionally left undeveloped—still provide such havens. Where such buffers have been removed, they can be reestablished; they can be added not only around farmers' fields, but along highways and river banks, links between parks, and in people's back yards.

People can also encourage pollinators by providing nesting sites, such as hollow logs, or by ensuring that pollinators have the native plants they need during the "off-season" when they are not working on the agricultural crops. Changing some prevalent cultural or industrial practices, too, can help. There is the practice, for example, of growing tidy rows of cocoa trees. These may make for a handsome plantation. But midges, the only known pollinator of cultivated cacao (the source of chocolate), prefer an abundance of leaf litter and trees in a more natural array. Plantations that encourage midges can have ten times the yield of those that don't.

Scientists have begun to ratchet up their study of wild pollinators and to domesticate more of them. The bumblebee, for example, was domesticated ten years ago and is now a pollinator of valuable greenhouse grown crops.

The Other Service Economy

Natural services have been so undervalued because, for so long, we have viewed the natural world as an inexhaustible resource and sink. Human impact has been seen as insignificant or beneficial. The tools used to gauge the economic health and progress of a nation have tended to reinforce and encourage these attitudes. The gross domestic product (GDP), for example, supposedly measures the value of the goods and services produced in a nation. But the most valuable goods and services—the ones provided by nature, on which all else rests—are measured poorly or not at all.... The unhealthy dynamic is compounded by the fact that activities that pollute or deplete natural capital are counted as contributions to economic wellbeing. As ecologist Norman Myers puts it, "Our tools of economic analysis are far from able to apprehend, let alone comprehend, the entire range of values implicit in forests."

When economies and societies use misleading signals about what is valuable, people are encouraged to make decisions that run counter to their own long-range interests—and those of society and future generations. Economic calculations grossly underestimate the current and future value of nature. While a fraction of nature's goods are counted when they enter the marketplace, many of them are not. And nature's services—the life-support systems—are not counted at all. When the goods are considered free and therefore valued at zero, the market sends signals that they are only economically valuable when converted into something else. For example, the profit from deforesting land is counted as a plus on a nation's ledger sheet, because the trees have been converted to saleable lumber or pulp, but the depletions of the timber stock, watershed, and fisheries are not subtracted.

Last year, an international team of researchers led by Robert Costanza of the University of Maryland's Institute for Ecological Economics, published a landmark study on the importance of nature's services in supporting human economies. The study provides, for the first time, a quantification of the current economic value of the world's ecosystem services and natural capital. The researchers synthesized the findings of over 100 studies to compute the average per hectare value for each of the 17 services that world's ecosystems provide. They concluded that the current economic value of the world's ecosystem services is in the neighborhood of $33 trillion per year, exceeding the global GNP of $25 trillion.

Placing a monetary value on nature in this way has been criticized by those who believe that it commoditizes and cheapens nature's infinite value. But in practice, we all regularly assign value to nature through the choices we make. The problem is that in normal practice, many of us don't assign such value to nature until it is converted to something man-made—forests to timber, or swimming fish to a restaurant meal. With a zero value, it's easy to see why nature has almost always been the loser in standard economic equations. As the authors of the Costanza study note, " ... the decisions we make about ecosystems imply valuations (although not necessarily expressed in monetary terms). We can choose to make these valuations explicit or not ... but as long as we are forced to make choices, we are going through the process of valuation." The study is also raising a powerful new challenge to those traditional economists who are accustomed to keeping environmental costs and benefits "external" to their calculations.

While some skeptics will doubtless argue that the global valuation reported by Costanza and his colleagues overestimates the current value of nature's services, if anything it is actually a very conservative estimate. As the authors point out, values for some biomes (such as mountains, arctic tundra, deserts, urban parks) were not included. Further, they note that as ecosystem services become scarcer, their economic value will only increase.

Clearly, failure to value nature's services is not the only reason why these services are misused. Too often, illogical and inequitable resource use continues—even in the face of evidence that it is ecologically, economically, and socially unsustainable—because powerful interests are able to shape policies by legal or illegal means. Frequently, some individuals or entities get the financial benefits from a resource while the losses are distributed across society. Economists call this "socializing costs." Stated simply, the people who get the benefits are not the ones who pay the costs. Thus, there is little economic incentive for those exploiting a resource to use it judiciously or in a manner that maximizes public good. Where laws are lax or are ignored, and where people do not have an opportunity for meaningful participation in decision-making, such abuses will continue.

The liquidation of 90 percent of the Philippines' forest during the 1970s and 1980s under the Ferdinand Marcos dictatorship, for example, made a few hundred families over $42 billion richer. But 18 million forest dwellers became much poorer. The nation as a whole went from being the world's second largest log exporter to a net importer. Likewise, in Indonesia today, the "benefits" from burning the forest will enrich a relatively few well-connected individuals and companies but tens of millions of others are bearing the costs. Even in wealthy nations, such as Canada, the forest industry wields heavy influence over how the forests are managed, and for whose benefit.

We have already seen that the loss of ecosystem services can have severe economic, social, and ecological costs even though we can only measure a fraction of them. The loss of timber and lives in the Indonesian fires, and the lower production of fruits and vegetables from inadequate pollination, are but the tip of the iceberg. The other consequences for nature are often unforeseen and unpredictable. The loss of individual species and habitat, and the degradation and simplification of ecosystems, impair nature's ability to provide the services we need. Many of these changes are irreversible, and much of what is lost is simply irreplaceable.

By reducing the number of species and the size and integrity of ecosystems, we are also reducing nature's capacity to evolve and create new life. Almost half of the forests that once covered the Earth are now gone, and much of what remains is in fragmented patches. In just a few centuries we have gone from living off nature's interest to spending down the capital that has accumulated over millions of years of evolution. At the same time we are diminishing the capacity of nature to create new capital. Humans are only one part of the evolutionary product. Yet we have taken on a major role in shaping its future production course and potential. We are pulling out the threads of nature's safety net even as we depend on it to support the world's expanding human population and economy.

In that expanding economy, consumers now need to recognize that it is possible to reduce and reverse the destructive impact of our activities by consuming less and by placing fewer demands on those services we have so mistakenly regarded as free. We can, for example, reduce the high levels of waste and overconsumption of timber and paper. We can also increase the efficiency of water and energy use. In agricultural fields we can leave hedgerows and unplowed areas that serve as nesting and feeding sites for pollinators. We can sharply reduce reliance on agricultural chemicals, and improve the timing of their application to avoid killing pollinators.

Maintaining nature's services requires looking beyond the needs of the present generation, with the goal of ensuring sustainability for many generations to come. We have no honest choice but to act under the assumption that future generations will need at least the same level of nature's services as we have today. We can neither practically nor ethically decide what future generations will need and what they can survive without.

**Marino Gatto and
Giulio A. De Leo**

Pricing Biodiversity and Ecosystem Services: The Never-Ending Story

In 1844, the French engineer Jules Juvénal Dupuit introduced cost–benefit analysis to evaluate investment projects.... The application of cost–benefit analysis to ecological issues fell out of favor three decades ago, and it was gradually replaced by multicriteria analysis in the decision-making process for projects that have an impact on the environment. Although multicriteria analysis is currently used for environmental impact assessments [EIA] in many nations, [recently] the concept of cost–benefit analysis has again become fashionable, along with the various pricing techniques associated with it, such as contingent valuation methods, hedonic prices, and costs of replacement of ecological services.... Economists have generated a wealth of virtuosic variations on the theme of assessing the societal value of biodiversity, but most of these techniques are invariably based on price—that is, on a single scale of values, that of goods currently traded on world markets.

Perhaps the most famous recent study on the issue of pricing biodiversity and ecological services is that by Costanza et al., who argued that if the importance of nature's free benefits could be adequately quantified in economic terms, then policy decisions would better reflect the value of ecosystem services and natural capital. Drawing on earlier studies aimed at estimating the value of a wide variety of ecosystem goods and services, Costanza et al. estimated the current economic value of the entire biosphere at $16–54 trillion per year, with an average value of approximately $33 trillion per year. By contrast, the gross national product of the United States totals approximately $18 trillion per year. The paper, as its authors intended, stimulated much discussion, media attention, and debate. A special issue of *Ecological Economics* (April 1998) was devoted to commentaries on the paper, which, with few exceptions, were laudatory. Some economists have questioned the actual numbers, but many scientists have praised the attempt to value biodiversity and ecosystem functions.

Although Costanza et al. acknowledged that their estimates were crude and imperfect, they also pointed the way to improved assessments. In particular, they noted the need to develop comprehensive ecological economic models that could adequately incorporate the complex interdependencies between ecosystems and economic systems, as well as the complex individual dynamics of both types of systems. Despite the authors' caveats and the fact that many economists have been circumspect in applying their own tools to decisions regarding natural systems, the monetary approach is perceived by scientists, policymakers, and the general public as extremely appealing; a number of biologists are also of the opinion that attaching economic values to ecological services is of paramount importance for preserving the biosphere and for effective decision-making in all cases where the environment is concerned.

In this article, we espouse a contrary view, stressing that, for most of the values that humans attach to biodiversity and ecosystem services, the pricing approach is inadequate—if not misleading and obsolete—because it implies erroneously that complex decisions with important environmental impacts can be based on a single scale of values. We contend that the use of cost–benefit analysis as the exclusive tool for decision-making about environmental policy represents a setback relative to the existing legislation of the United States, Canada, the European Union, and Australia on environmental impact assessment, which explicitly incorporates multiple criteria (technical, economic, environmental, and social) in the process of evaluating different alternatives. We show that there are sound methodologies, mainly developed in business and administration schools by regional economists and by urban planners, that can assist decision-makers in evaluating projects and drafting policies while accounting for the nonmarket values of environmental services.

The Limitations of Cost–Benefit Analysis and Contingent Valuation Methods

Historically, the first important implementation of cost–benefit analysis at the political level came in 1936, with passage of the US Flood Control Act. This legislation stated that a public project can be given a green light if the benefits, to whomsoever they accrue, are in excess of estimated costs. This concept implies that all benefits and costs are to be considered, not just actual cash flows from and to government coffers. However, public agencies (e.g., the US Army Corps of Engineers) quickly ran into a problem: They were not able to give a monetary value to many environmental effects, even those that were predictable in quantitative terms. For instance, engineers could calculate the reduction of downstream water flow resulting from construction of a dam, and biologists could predict the river species most likely to become extinct as a consequence of this flow reduction. However, public agencies were not able to calculate the cost of each lost species. Therefore, many ingenious techniques for the monetary valuation of environmental goods and services have been devised since the 1940s. These techniques fall into four basic categories.

- **Conventional market approaches.** These approaches, such as the replacement cost technique, use market prices for the environmental service that is affected. For example, degradation of vegetation in developing countries leads to a decrease in available fuelwood. Consequently, animal dung has to be used as a fuel instead of a fertilizer, and farmers must therefore replace dung with chemical fertilizers. By computing the cost of these chemical fertilizers, a monetary value for the degradation of vegetation can then be calculated.

· **Household production functions.** These approaches, such as the travel cost method, use expenditures on commodities that are substitutes or complements for the environmental service that is affected. The travel cost method was first proposed in 1947 by the economist Harold Hotelling, who, in a letter to the director of the US National Park Service, suggested that the actual traveling costs incurred by visitors could be used to develop a measure of the recreation value of the sites visited.

· **Hedonic pricing.** This form of pricing occurs when a price is imputed for an environmental good by examining the effect that its presence has on a relevant market-priced good. For instance, the cost of air and noise pollution is reflected in the price of plots of land that are characterized by different levels of pollution, because people are willing to pay more to build their houses in places with good air quality and little noise....

· **Experimental methods.** These methods include contingent valuation methods, which were devised by the resource economist Siegfried V. Ciriacy-Wantrup. Contingent valuation methods require that individuals express their preferences for some environmental resources by answering questions about hypothetical choices. In particular, respondents to a contingent valuation methods questionnaire will be asked how much they would be willing to pay to ensure a welfare gain from a change in the provision of a nonmarket environmental commodity, or how much they would be willing to accept in compensation to endure a welfare loss from a reduced provision of the commodity.

Among these pricing techniques, the contingent valuation methods approach is the only one that is capable of providing an estimate of existence values, in which biologists have a special interest. Existence value was first defined by Krutilla as the value that individuals may attach to the mere knowledge that rare and diverse species, unique natural environments, or other "goods" exist, even if these individuals do not contemplate ever making active use of or benefiting in a more direct way from them. The name "contingent valuation" comes from the fact that the procedure is contingent on a constructed or simulated market, in which people are asked to manifest, through questionnaires and interviews, their demand function for a certain environmental good (i.e., the price they would pay for one extra unit of the good versus the availability of the good)....

The limits of cost–benefit analysis were discussed in the 1960s, after more than two decades of experimentation. In particular, many authors pointed out that cost–benefit analysis encouraged policymakers to focus on things that can be measured and quantified, especially in cash terms, and to disregard problems that are too large to be assessed easily. Therefore, the associated price might not reflect the "true" value of social equity, environmental services, natural capital, or human health. In particular, economists themselves recognize that the increasingly popular contingent valuation methods are undermined by several conceptual problems, such as free-riding, overbidding, and preference reversal.

When it comes to monetary valuation of the goods and services provided by natural ecosystems and landscapes specifically, a number of additional problems undermine the effectiveness of pricing techniques and cost–benefit analysis. These problems include the very definition of "existence" value, the dependence of pricing techniques on the composition of the reference group, and the significance of the simulated market used in contingent valuation.

The definition of "existence" value   A classic example of contingent valuation methods is to ask for the amount of money individuals are willing to pay to ensure the continued existence of a species such as the blue whale. However, the existence value of whales does not take into account potential indirect services and benefits provided by these mammals. It is just the value of the existence of whales for humans, that is, the satisfaction that the existence of blue whales provides to people who want them to continue to exist. Therefore, there is a real risk that species with very low or no aesthetic appeal or whose biological role has not been properly advertised will be given a low value, even if they play a fundamental ecological function. Without adequate information, most people do not understand the extent, importance, and gravity of most environmental problems. As a consequence, people may react emotionally and either underestimate or overestimate risks and effects.

Therefore, it is not surprising that five of the seven guidelines issued by the National Oceanic and Atmospheric Administration [NOAA] about how to conduct contingent valuation discuss how to properly inform and question respondents to produce reliable estimates (e.g., in-person interviews are preferred to telephone surveys to elicit values). Of course, acquisition of reliable and complete information is always possible in theory, but in practice strict adherence to NOAA guidelines makes contingent valuation methods expensive and time consuming.

Difficulties with the reference group for pricing   Pricing techniques such as contingent valuation methods provide information about individual willingness to pay or willingness to accept, which must be summed up in the final balance of cost–benefit analysis. Therefore, the outcome of cost–benefit analysis depends strongly on the group of people that is taken as a reference for valuation—particularly on their income. Van der Straaten noted that the Exxon *Valdez* oil spill in 1989 provides a good example of this dependence. The population of the United States was used as a reference group to calculate the damage to the existence value of the affected species and ecosystems using contingent valuation methods. Exxon was ultimately ordered to pay $5 billion to compensate the people of Alaska for their losses. This huge figure was a consequence of the high income of the US population. If the same accident had occurred in Siberia, where salaries are lower, the outcome would certainly have been different.

This example shows that contingent valuation methods simply provide information about the preferences of a particular group of people but do not necessarily reflect the ecological importance of ecosystem goods and services. Moreover, the outcome of cost–benefit analysis depends on which individual willingness to pay or willingness to accept are included in the cost–benefit analysis. If the quality of the Mississippi River is at issue, should the analysis be restricted to US citizens living close to the river, or should the willingness to pay of Californians and New Yorkers be included too? According to Krutilla's definition of existence value, for many environmental goods and ecological services that may ultimately affect ecosystem integrity at the global level, the preferences of the entire human population should potentially be considered in the analysis. Because practical reasons obviously preclude doing so, contingent valuation methods will inevitably only provide information about the preferences of specific groups of people. For many of the ecological services that may be considered the heritage of humanity, contingent valuation methods analyses performed locally in a particular economic situation should be extrapolated only with great caution to other areas. The process of placing a monetary value on biodiversity and ecosystem functioning through nonuser willingness to pay is performed in the same way as for user willingness to pay, but the identification of people who do not use an environmental good directly and still have a legitimate interest in its preservation is problematic.

Significance of the simulated market   Contingent valuation methods are contingent on a market that is constructed or simulated, not real. It is difficult to believe in the efficiency of what Adam Smith called the "invisible hand" of the market for a process that is the artificial production of economic advisors and does not possess the dynamic feedback that characterizes real competitive markets. Is it even possible to simulate a market where units of biodiversity are bought and sold? As Friend stated, "these contingency evaluation methods (CVM) tend to create an illusion of choice based on psychology (willingness) and ideology (the need to pay) which is supposed, somewhat mysteriously, to reflect an equilibrium between the consumer demand for and producer supply of environmental goods and services."

Many additional criticisms of pricing ecological services are more familiar to biologists. For many ecological services, there is simply no possibility of technological substitution. Moreover, the precise contribution of many species is not known, and it may not be known until the species is close to extinction.... In addition, specific ecosystem services, as evaluated by Costanza et al., should not be separated from one another and valued individually because the importance of any piece of biodiversity cannot be determined without considering the value of biodiversity in the aggregate. And finally, the use of marginal value theory may be invalidated by the erratic and catastrophic behavior of many ecological systems, resulting in potentially detrimental effects on the health of humans, the productivity of renewable resources, and the vitality and stability of societies themselves.

Despite the efforts of many economists, we believe that some goods and services, especially those related to ecosystems, cannot reasonably be given a monetary value, although they are of great value to humans. Economists coined the term "intangibles" to define these goods. Cost–benefit analysis cannot easily deal with intangibles. As Nijkamp wrote, more than 20 years ago, "the only reasonable way to take account of intangibles in the traditional cost–benefit analysis seems to be the use of a balance with a debit and a credit side in which all intangible project effects (both positive and negative) are represented in their own (qualitative or quantitative) dimensions" as secondary information. In other words, the result of cost–benefit analysis is primarily a single number, the net monetary benefit that comprises all the effects that can be sensibly converted into monetary returns and costs.

Commensurability of Different Objectives and Multicriteria Analysis

Cost–benefit analysis includes intangibles in the decision-making process only as ancillary information, with the main focus being on those effects that can be converted to monetary value. This approach is not a balanced solution to the problem of making political decisions that are acceptable to a wide number of social groups with a range of legitimate interests....

However, even if the attempt to put a price on everything is abandoned, it is not necessary to give up the attempt to reconcile economic issues with social and environmental ones. Social scientists long ago developed multicriteria techniques to reach a decision in the face of multiple different and structurally incommensurable goals. The most important concept in multicriteria analysis was actually conceived by an Italian economist, Vilfredo Pareto, at the end of the nineteenth century. It is best explained by a simple example. Suppose that a natural area hosting several rare species is a target for the development of a mining activity. Alternative mining projects can have different effects in terms of profits from mining (measured in dollars) and in terms of sustained biodiversity (measured in suitable units, for instance, through the Shannon index). Profit from mining can be corrected using welfare economics to include those environmental and social effects that can be priced (e.g., the benefit of providing jobs to otherwise unemployed people, the cost of treating lung disease of miners, and the cost of the loss of the tourists who used to visit the natural area)....

The methods of multicriteria analysis are intended to assist the decision-maker in choosing among ... alternatives ... (a task that is particularly difficult when there are several incommensurable objectives, not just two). Nevertheless, the initial step of determining [these] alternatives is of enormous importance, for three reasons. First, [doing so] makes perfect sense even if there is no way of pricing a certain environmental good because each objective can be expressed in its own proper units without reduction to a common scale. Second, the determination of all the feasible alternatives ... requires the joint effort of a multidisciplinary team that includes, for example, economists, engineers, and biologists and that must predict the effects of alternative decisions on all of the different environmental and social components to which humans are sensitive and which, therefore, deserve consideration. Third, the determination of [feasible alternatives] allows the objective elimination of inadequate alternatives because [they are] independent of the subjective perception of welfare ... [and] in essence describe the tradeoff between the various incommensurable objectives when every effort is made to achieve the best results in all respects; the attention of the authority that must make the final decision is thus directed toward genuine potential solutions because nonoptimal decisions have already been discarded.

It should be noted that a cost–benefit analysis does not elicit tradeoffs between incommensurable goods because it also gives a green light to projects ..., provided that the benefits that can be converted into a monetary scale exceed the costs.... Cost–benefit analysis, however, is not useful for eliciting the tradeoffs between two incommensurable goods, neither of which is monetary. For instance, there might be a conflict between the goals of preserving wildlife within a populated area and minimizing the risk that wild animals are vectors of dangerous diseases. A multicriteria analysis can describe this tradeoff, whereas a cost–benefit analysis cannot.

Another philosophical point concerning the issue of commensurability is the question of implicit pricing. Economists often argue that to make a decision is to put an implicit price on such intangibles as human life or aesthetics and, therefore, to reduce their value to a common scale (as pointed out also by Costanza et al.)....

Environmental Impact Assessment and Multiattribute Decision-Making

Because of the flaws of cost–benefit analysis, many countries have taken a different approach to decision-making through the use of environmental impact assessment legislation (e.g., the United States in 1970, with the signing of the National Environmental Policy Act, NEPA; France in 1976, with the act 76/629; the European Union in 1985, with the directive 85/337). Environmental impact assessment procedures, if properly carried out, represent a wiser approach than setting an a priori value of biodiversity and ecosystem services because these procedures explicitly recognize that each situation, and every regulatory decision, responds to different ethical, economic, political, historical, and other conditions and that the final decision must be reached by giving appropriate consideration to several different objectives. As Canter noted, all projects, plans, and policies that are expected to have a significant environmental impact would ideally be subject to environmental impact assessment.

The breadth of goals embraced by environmental impact assessment is much wider than that of cost–benefit analysis. Environmental impact assessment provides a conceptual framework and formal procedures for comparing different alternatives to a proposed project (including the possibilities of not development a site, employing different management rules, or using mitigation measures); for fostering interdisciplinary team formation to investigate all possible environmental, social, and economic consequences of a proposed activity; for enhancing administrative review procedures and coordination among the agencies involved in the process; for producing the necessary documentation to enhance transparency in the decision-making process and the possibility of reviewing all the objective and subjective steps that resulted in a given conclusion; for encouraging broad public participation and the input of different interest groups; and for including monitoring and feedback procedures. Classical multiattribute analysis can be used to rank different alternatives.... Ranking usually requires the use of value functions to transform environmental and other indicators (e.g., biological oxygen demand or animal density) to levels of satisfaction on a normalized scale, and the weighting of factors to combine value functions and to rank the alternatives. These weights explicitly reflect the relative importance of the different environmental, social, and economic compartments and indicators.

A wide range of software packages for decision support can assist experts in organizing the collected information; in documenting the various phases of EIA; in guiding the assignment of importance weights; in scaling, rating, and ranking alternatives; and in conducting sensitivity analysis for the overall decision-making process. This last step, of testing the robustness and consistency of multiattribute analysis results, is especially important because it shows how sensitive the final ranking is to small or large changes in the set of weights and value functions, which often reflect different and subjective perspectives. It is important to stress that, although the majority of environmental impact assessments have been conducted on specific projects, such as road construction or the location of chemical plants, there is no conceptual barrier to extending the procedure to evaluation of plans, programs, policies, and regulations. In fact, according to NEPA, the procedure is mandatory for any federal action with an important impact on the environment. The extension of environmental impact assessment to a level higher than a single project is termed "strategic environmental assessment" and has received considerable attention.

Conclusions

An impressive literature is available on environmental impact assessment and multiattribute analysis that documents the experience gained through 30 years of study and application. Nevertheless, these studies seem to be confined to the area of urban planning and are almost completely ignored by present-day economists as well as by many ecologists. Somewhere between the assignment of a zero value to biodiversity (the old-fashioned but still used practice, in which environmental impacts are viewed as externalities to be discarded from the balance sheet) and the assignment of an infinite value (as advocated by some radical environmentalists), lie more sensible methods to assign value to biodiversity than the price tag techniques suggested by the new wave of environmental economists. Rather than collapsing every measure of social and environmental value onto a monetary axis, environmental impact assessment and multiattribute analysis allow for explicit consideration of intangible nonmonetary values along with classical economic assessment, which, of course, remains important. It is, in fact, possible to assess ecosystem values and the ecological impact of human activity without using prices. Concepts such as Odum's eMergy [the available energy of one kind previously required to be used up directly and indirectly to make the product or service] and Rees' ecological footprint [the area of land and water required to support a defined economy or population at a specified standard of living], although perceived by some as naive, may aid both ecologists and economists in addressing this important need.

To summarize our viewpoint, economists should recognize that cost–benefit analysis is only part of the decision-making process and that it lies at the same level as other considerations. Ecologists should accept that monetary valuation of biodiversity and ecosystem services is possible (and even helpful) for part of its value, typically its use value. We contend that the realistic substitute for markets, when they fail, is a transparent decision-making process, not old-style cost–benefit analysis. The idea that, if one could get the price right, the best and most effective decisions at both the individual and public levels would automatically follow is, for many scientists, a sort of Panglossian obsession. In reality, there is no simple solution to complex problems. We fear that putting an a priori monetary value on biodiversity and ecosystem services will prevent humans from valuing the environment other than as a commodity to be exploited, thus reinvigoraing the old economic paradigm that assumes a perfect substitution between natural and human-made capital. As Rees wrote, "for all its theoretical attractiveness, ascribing money values to nature's services is only a partial solution to the present dilemma and, if relied on exclusively, may actually be counterproductive."

POSTSCRIPT

Should a Price Be Put on the Goods and Services Provided by the World's Ecosystems?

In "Can We Put a Price on Nature's Services?" *Report From the Institute for Philosophy and Public Policy* (Summer 1997), Mark Sagoff objects that trying to attach a price to ecosystem services is futile because it legitimizes the accepted cost-benefit approach and thereby undermines efforts to protect the environment from exploitation. The March 1998 issue of *Environment* contains environmental economics professor David Pearce's detailed critique of the 1997 Costanza et al. study. Pearce objects chiefly to the methodology, not the overall goal of attaching economic value to ecosystem services. Costanza et al. reply to Pearce's objections in the same issue. Pearce and Edward B. Barbier have published *Blueprint for a Sustainable Economy* (Earthscan, 2000), in which they discuss how governments worldwide are now applying economics to environmental policy.

Despite the controversy over the worth of assigning economic values to various aspects of nature, researchers continue the effort. Gretchen C. Daily et al., in "The Value of Nature and the Nature of Value," *Science* (July 21, 2000), discuss valuation as an essential step in all decision making and argue that efforts "to capture the value of ecosystem assets ... can lead to profoundly favorable effects." Daily and Katherine Ellison continue the theme in *The New Economy of Nature: The Quest to Make Conservation Profitable* (Island Press, 2002). In "What Price Biodiversity?" *Ecos* (January 2000), Steve Davidson describes an ambitious program funded by the Commonwealth Scientific and Industrial Research Organization (CSIRO) and the Myer Foundation that is aimed at developing principles and methods for objectively valuing "ecosystem services—the conditions and processes by which natural ecosystems sustain and fulfil human life—and which we too often take for granted. These include such services as flood and erosion control, purification of air and water, pest control, nutrient cycling, climate regulation, pollination, and waste disposal."

R. Kerry Turner et al., in "Valuing Nature: Lessons Learned and Future Research Directions," *Ecological Economics* (October 2003), attempt to assess how useful the ecosystem valuation approach has been so far. They conclude that "studies which seek to capture the 'before and after' states as environmental changes take place are rare [but] are most important as aids to more rational decision making" when conservation interests confront development interests.

On the Internet . . .

ECOLEX: A Gateway to Environmental Law

This site, sponsored by the United Nations and the World Conservation Union is a comprehensive resource for environmental treaties, national legislation, and court decisions.

http://www.ecolex.org/ecolex/

EarthTrends

The World Resources Institute offers data on biodiversity, fisheries, agriculture, population, and a great deal more.

http://earthtrends.wri.org/

Environmental Defense

Environmental Defense (once The Environmental Defense Fund) is dedicated to "protecting the environmental rights of all people, including future generations." Guided by science, Environmental Defense "evaluates environmental problems and works to create and advocate solutions that win lasting political, economic and social support because they are nonpartisan, cost-efficient and fair."

http://www.environmentaldefense.org/home.cfm

Office of Environmental Justice

The U.S. Environmental Protection Agency (EPA) pursues environmental justice under the Office of Enforcement and Compliance Assurance as part of its "firm commitment to the issue of environmental justice and its integration into all programs, policies, and activities, consistent with existing environmental laws and their implementing regulations."

http://www.epa.gov/compliance/environmentaljustice/index.html

The Heritage Foundation

The Heritage Foundation is a think tank whose mission is to formulate and promote conservative public policies on many issues, including environmental ones. It bases its work on the principles of free enterprise, limited government, individual freedom, and traditional American values.

http://www.heritage.org/

PART 2

Principles Versus Politics

In many environmental issues, it is easy to tell what basic principles apply and therefore determine what is the right thing to do. Ecology is clear on the value of species to ecosystem health. Sociology and politics have agreed that racism is an evil to be avoided. Medicine makes no bones about the ill effects of pollution. But are the environmental problems so bad that we must act immediately? How much of the "right thing" should we do? How should we do it? Such questions arise in connection with every environmental issue, not just the four issues raised in this section, but they will serve to introduce the theme of principles versus politics.

- Is Biodiversity Overprotected?

- Should Environmental Policy Attempt to Cure Environmental Racism?

- Can Pollution Rights Trading Effectively Control Environmental Problems?

- Do Environmentalists Overstate Their Case?

ISSUE 4

Is Biodiversity Overprotected?

YES: David N. Laband, from "Regulating Biodiversity: Tragedy in the Political Commons," *Ideas on Liberty* (September 2001)

NO: Howard Youth, from "Silenced Springs: Disappearing Birds," *Futurist* (July/August 2003)

ISSUE SUMMARY

YES: Professor of economics David N. Laband argues that the public demands excessive amounts of biodiversity largely because decision makers and voters do not have to bear the costs of producing it.

NO: Wildlife conservation researcher and writer Howard Youth argues that the actions needed to protect biodiversity not only have economic benefits but also are the same actions needed to ensure a sustainable future for humanity.

Extinction is normal. Indeed, 99.9 percent of all the species that have ever lived are extinct, according to some estimates. But the process is normally spread out over time, with the formation of new species by mutation and selection balancing the loss of old ones to disease, new predators, climate change, habitat loss, and other factors. Today, human activities are an important cause of species loss mostly because humans destroy or alter habitat but also because of hunting, the introduction of competitors, and the introduction of diseases. According to Martin Jenkins, "Prospects for Biodiversity," *Science* (November 14, 2003), some 350 (3.5 percent) of the world's bird species may vanish by 2050. Other categories of living things may suffer greater losses, leading to a "biologically impoverished" world. He states that the consequences for human life are "unforeseeable but probably catastrophic."

Awareness of the problem has been growing. In 1973 the United States adopted the Endangered Species Act to protect species that were so reduced in numbers or restricted in habitat that a single untoward event could wipe them out. The act barred construction projects that would further threaten endangered species. In one famous case, construction on the Tellico Dam on the Little Tennessee River in Loudon County, Tennessee, was halted because it threatened the snail darter, a small fish. Another case involved the spotted owl, which was threatened by logging in the Northwest. Those in favor of the dam or the timber industry felt that the endangered species was trivial compared to the human benefits at stake. Those in favor of the act argued that the loss of a single species might not matter to the world, but where one species went, others would follow. Protecting one species also protects others. However, the number of threatened and endangered species has not diminished. In fact, that number has increased more than sevenfold, from 174 in 1976 to 1,244 as of November 2000.

Internationally, species protection is covered by the Convention on International Trade in Endangered Species of Wild Fauna and Flora (CITES). This agreement has banned trade in such natural products as elephant ivory to prevent the continued slaughter of elephants. Less successfully, it has also tried to protect rhinoceroses (killed for their horns) and about 5,000 other species of animals and 25,000 species of plants, including some whole groups, such as primates, cetaceans (whales, dolphins, and porpoises), sea turtles, parrots, corals, cacti, and orchids.

Is it enough to stop construction projects and ban trade? Some argue that efforts should be made to undo some of the damage that has already been done. For example, there is a movement to tear down dams that block the path of migratory fish, such as salmon and shad, so that they may once more breed and multiply (see http://www.amrivers.org). For another example, urbanization and agricultural development have greatly altered the Everglades in Florida: rivers have been straightened, water has been diverted, and land has been drained for farms. This activity has resulted in low water tables, increased fire danger, smaller bird populations, and the decline of the Florida panther, among other negative impacts. Currently, the Army Corps of Engineers is planning to undo some of the changes to the Everglades' water flow in order to restore the natural habitat as much as possible. The Comprehensive Everglades Restoration Plan (CERP) was approved by Congress in 2000, will cost almost $8 billion, and will take more than 30 years. See Phyllis McIntosh, "Reviving the Everglades," *National Parks* (January 2002).

Is too much being done to protect the species with which we share the earth? In the following selection, David N. Laband argues that it is, largely because the people who set environmental policy do not need to pay for protection efforts themselves. Instead, the costs of protecting biodiversity are unfairly laid upon landowners. In the second selection, Howard Youth reviews the declining state of the world's birds, describes what is causing the decline and what can be done, and argues that actions that protect birds can have economic benefits. These actions are also needed to ensure a sustainable future for humanity.

David N. Laband  **YES**

Regulating Biodiversity:
Tragedy in the Political Commons

Last summer, lightning struck and killed an enormous pine tree on one side of my backyard. At about the same time, voracious pine bark beetles girdled and killed an equally impressive pine tree on the other side. Now bereft of needles, these two arboreal giants pose a potential threat to my house: if they were to fall at just the right angle, the damage could be substantial. In the interest of safety, my wife wants to have the trees removed; for the sake of promoting biodiversity on my two-acre lot, I do not.

Our personal dilemma mirrors a much larger struggle that quietly threatens to destroy the rights of private timberland owners across the United States—the desire of urban dwellers to have their cake and eat it too. They demand houses made of wood, wood furniture, paper and paper products, and so on, while also demanding environmental amenities such as aesthetically pleasing landscape views, biodiversity, and animal habitat. At a personal level this can't be done. If the trees are removed, my wife has peace of mind, but the many animals that depend on dead pine trees for their existence, either directly or indirectly, will vanish. If the trees stay, we will be promoting the ecological diversity of our property, but my wife will worry about our house with every gust of wind. We can't have it both ways. Similarly, at a macro level, there is a tradeoff between production/consumption of timber and production/consumption of related environmental amenities.

The Role of Intensively Managed Forests

The problem of how to grow and harvest increasing amounts of timber while simultaneously producing a steadily increasing array and level of environmental amenities associated with forested land has resulted in an industry-wide discussion of how to simultaneously achieve both objectives. There is a growing appreciation within the forestry community for the prospect that intensively managed forests may yield increasing amounts of wood while minimizing the total acreage from which wood is harvested. This maximizes the amount of acreage available to meet other demands—such as agricultural production, animal habitat, and other environmental amenities associated with natural forests.

However, intensively managed forests have come under heavy fire from self-proclaimed environmentalists. In these so-called plantation forests, man, not nature, regenerates the trees, which accordingly grow in even-aged stands. Their well-being is affected by the application of herbicides and pesticides, as well as by occasional thinning and fire management. In contrast to naturally (re)generated timberland, plantation timberland has been described as an "ecological desert," with the stated or implied conclusion that the nature and extent of biological diversity associated with natural forests is both greater and therefore more desirable than that associated with plantation forests.[1]

The Threat to Private Landowners and Social Welfare

Such pejorative rhetoric is both misleading and counterproductive. The unfortunate but nonetheless compelling truth is that we can't have our cake and eat it too. We must make responsible choices about what to produce and how to produce it. A serious threat to private landowners develops when citizens living in urban areas demand that private owners of timberland (definitionally located in rural areas) produce environmental amenities such as aesthetically pleasing views, biodiversity, animal habitat, and the like, *provided the urbanites don't have to pay for it.*

Further, they seek to enforce their demands by using the political process to pass regulations that require landowners disproportionately to bear the cost of producing these environmental amenities. For example, Oregon law requires private timberland owners to replant within two years areas from which they cut trees. Other regulations forbid clearcutting of timberland. Federal regulations pertaining to endangered species are incredibly restrictive and intrusive with respect to an individual's property rights. The pursuit of environmental amenities that we are told are vital to some vaguely defined public interest through policies that impose virtually all the costs on relatively small numbers of private landowners generates what might be termed a "tragedy of the political commons."

Garrett Hardin introduced us to the tragedy of the commons.[2] Hardin developed a stylized example of a communal pasture open to all comers. There are no private property rights to the pasture, or rules, customs, or norms for shared use. In this setting, each shepherd, seeking to maximize the value of his holdings, keeps adding sheep to his flock as long as doing so adds an increment of gain. Further, the shepherds graze their sheep on the commons as long as the pasture provides any sustenance. Ignorant of the effects of their individual actions on the others, the shepherds collectively (and innocently) destroy the pasture. As Hardin concludes: "Therein is the tragedy. Each man is locked into a system that compels him to increase his herd without limit—in a world that is limited. Ruin is the destination toward which all men rush, each pursuing his own best interest in a society that believes in freedom of the commons."

Man's exploitation of the political commons is analogous to his exploitation of natural-resource commons. Our majority-rule voting process, which permits a majority of citizens to impose differential costs on the minority, encourages overprotection of endangered species, and overproduction of biodiversity, animal habitat, and landscape views. This occurs because each individual who bears a negligible portion of the costs of providing environmental amenities has a private incentive to keep demanding additional environmental protections as long as there is *any* perceived marginal benefit. As with the overgrazed pasture, the result of overprotecting Bambi is, as has become apparent all over the eastern United States, disastrous. Moreover, and not surprisingly, we are starting to hear real concern voiced about the recent proliferation of other animal species such as black bears, mountain lions, and coyotes. We are creating social tragedies that result from the political commons.

The tragedy is compounded by the incentives generated for private landowners by the heavy hand of command-and-control policies. When government abrogates property rights without compensation, landowners have strong incentives to mitigate their expected losses. They can do so by changing their land use from timber production to housing or commercial development. There is no incentive to promote habitat for endangered species; doing so means only that use of one's land will be seriously compromised by the highly restrictive provisions of the Endangered Species Act. Instead, a landowner who finds a member of an endangered species on his property has a well-understood incentive to "shoot, shovel, and shut up." Such behaviors are not likely to further environmental objectives.

Other People's Costs

It is relatively easy to demonstrate that because private timberland owners bear the cost of producing biodiversity, nonland-owners demand excessive amounts of it. The first point to be made in this regard is that urbanites do not in fact place a high value on biodiversity. One need look no further than the readily observable behavior of urbanites for proof of this claim. Urbanites have the ability and prerogative to produce biodiversity on their own residential property. That is, they could let their residential lots grow wild with natural flora and fauna. This would, without question, promote ecological diversity. In practice, virtually no residential property owners, living anywhere in the United States, do this. Instead, they invest (implicitly through their time and explicitly by purchase) hundreds, if not thousands, of dollars annually in the care and maintenance of their lawns and grounds in a decidedly unnatural state. Like owners of intensively managed timberland, owners of residential property chemically treat and harvest the growth on their property. In so doing, they create a landscape with relatively little floral or faunal diversity. What this behavior reveals, of course, is that urban dwellers place a higher value on having their own aesthetically pleasing ecological deserts than on personally promoting local biodiversity, even when the latter would save them hundreds, perhaps thousands, of dollars each year. The clear implication is that urbanites simply do not attach much importance to biodiversity.

This leads directly to a second point: notwithstanding that biodiversity is of little importance to them personally, urbanites may favor local, state, and federal statutes that ostensibly enhance biodiversity, provided such statutes impose the cost burden on rural landowners. The feel-good benefit of such regulation may be small, but with no personal costs to worry about, urbanites can be convinced to vote for them. However, if there were even a moderate cost to urban dwellers, we can be reasonably certain that restrictive regulations would not be passed. This explains why, for example, Oregon's replanting regulations are not imposed on owners of residential properties who cut down trees.

Earth's limited resources cannot provide all things to all people simultaneously. For that matter, the earth cannot provide all things just to self-proclaimed environmentalists. Consequently, responsible choices about the use of resources must be made. It is irresponsible to enact environmental policies that impose costs disproportionately on private timberland owners. Such policies lead to overproduction of environmental protection because urban voters who place little value on environmental amenities support regulations that impose little or no cost on themselves. Further, these policies create incentives for private timberland owners to minimize, not maximize, their production of environmental amenities. This problem of incompatible incentives makes it less likely that public policy will actually attain its stated objectives.

Notes

1. National Audubon Society, www.audubon.org/campaign/fh/chipmills.htm, no date.
2. Garrett Hardin, "The Tragedy of the Commons," *Science* (162), 1968, pp. 1243–48; see www.dieoff.org/page95.htm.

NO 

Howard Youth

Silenced Spring: Disappearing Birds

Almost 1,200 bird species—about 12% of those remaining in the world—may face extinction within the next century. Most struggle against a deadly mixture of threats, including habitat loss, human disasters, and disease. Although some bird extinctions now seem imminent, many can still be avoided with deep commitment to conservation as an integral part of a sustainable development strategy. Such a commitment would be in humanity's best interest.

What ornithologists have already tallied is alarming. Human-related factors threaten 99% of the species in greatest danger. Bird extinctions are increasing, already topping 50 times the natural rate of loss. At least 128 species vanished over the last 500 years—103 of which became extinct since 1800.

If we focus solely on the prospects of extinction, we partly miss the point. From an ecological perspective, extinction is but the last stage in a spiraling degeneration that sends a thriving species slipping toward oblivion. Species stop functioning as critical components of their ecosystems well before they completely disappear.

Although birds are probably the best-studied animal class, a great deal remains to be learned about them—from their life histories to their vulnerability to environmental change. In the tropics, where both avian diversity and habitat loss are greatest, experts just do not know the scope of bird declines because many areas remain poorly surveyed, if at all. Species may vanish even before scientists can classify them or study their behavior, let alone determine their ecological importance.

Several new bird species are described every year. One of this century's earliest was an owl discovered in Sri Lanka in 2001, the first new bird species found there in 132 years. These scarce and newly described birds sit at a crossroads, as does humanity. One path leads toward continued biodiversity and sustainability. The other heads toward extinction.

Habitat Loss: The Greatest Threat

Many problems faced by birds and other wildlife stem from how we handle our real estate. The human population explosion from 1.6 billion to 6 billion during the last century fueled widespread habitat loss that chiseled once-extensive wilderness into habitat islands. Today, loss or damage to species' living spaces poses by far the greatest threat to birds and biodiversity in general.

Forests. Timber operations, farms, pastures, and settlements have already claimed almost half the world's forests. Between the 1960s and 1990s, about 4.5 million square kilometers (1,100 acres), or 20% of the world's tropical forest cover, were cut or burned. Habitat loss jeopardizes 1,008 (85%) of the world's most-threatened bird species.

Foresters herald regrowth of temperate forests as an environmental success story, and in recent decades, substantial reforestation did take place in the eastern United States, China, and Europe. Forest management profoundly affects diversity and natural balances, however, and satellite images of tree cover do not tell us how much regrown habitat is indeed quality habitat. In the southeastern United States, foresters replace the clearcut area with rows of same-age, same-species pine saplings. For many native animals and plants, simplified plantation monocultures are no substitute for more-complex natural forests.

Grasslands. Once cloaking more than a third of the earth's surface, grasslands sustain bird populations found nowhere else, but they also host almost one-sixth of the human population. Few large, undisturbed grassland areas remain. In North America, where less than 4% remains, bird populations continue to shrivel. According to the U.S. Geological Survey, between 1966 and 1998, 15 of 28 grassland bird species steadily declined. The last strongholds for many grassland species in Europe face severe pressure from increased irrigation and modernization programs subsidized by the European Union's common agricultural policy.

Wetlands. Draining, filling, and conversion to farmlands or cities destroyed an estimated half of the world's wetlands during the twentieth century. Estimates within individual countries are often much higher. Spain, for instance, has lost an estimated 60%–70% of its wetland area since the 1940s. Even wilderness areas such as the Everglades and Spain's Donana National Park have not been spared.

Outside protected areas, changes have been far more dramatic. Over the last 70 years, Armenia's Lake Sevan suffered dramatic lowering due to water diversion, and Lake Gill was drained entirely. With their vital wetlands destroyed, at least 31 locally breeding bird species abandoned the lakes.

Mountains. Mountains often hold their habitats longer against human endeavors. In many countries, including Jamaica and Mexico, much of the remaining habitat is found only in prohibitively steep terrain. Once targeted, though, mountain habitats and wildlife are extremely vulnerable. Altitude and moisture levels dictate vegetation and wildlife occurrence there, creating narrow ribbons of habitat. Humans and migrant birds alike particularly favor temperate and rainsoaked middle elevations. In the Andes, Himalayas, and Central American highlands, among other areas, middle-elevation forests are highly degraded, creating severe erosion problems, fouling watersheds vital to human populations, and providing less area for wintering and resident birds.

Exotic Animals and Plants

Even in otherwise undisturbed wildlife habitats, a new order is taking hold as exotic (nonnative) species are introduced. Today, exotics threaten birds and their ecosystems in myriad ways, constituting the second most intense threat to birds worldwide, after habitat loss and degradation. Once introduced, some exotic predators became all the more lethal on islands, where endemic species evolved with few or no defenses against such hunters. To date, 93% of bird extinctions have occurred on islands, where extremely vulnerable endemic species succumbed to habitat loss, hunting, and, in most cases, exotic species that unsettle unique island ecological balances.

The yellow crazy ant, for example, a frenetic, fast-multiplying insect, is marching across Australia's Christmas Island following its introduction there during the twentieth century. As they spread across the island, crazy ants will likely kill young native birds, including those of two critically endangered species, the Christmas Island hawk owl and Abbott's booby, a seabird that nests nowhere else but in the island's forest canopy. Both species are expected to decline 80% due to ant invasion.

Exotic birds compete with native birds both genetically and directly. People around the world have dumped familiar domesticated mallard ducks into ponds and other wetlands, where they vigorously interbreed with closely related species. Such hybridization affects South Africa's yellow-billed ducks, endangered Hawaiian ducks, American black ducks, and mottled ducks.

Introduced plants change birds' habitats until they are eventually uninhabitable. Exotic plant species have gone wild in many parts of the world at the expense of birds and other wildlife. Cheatgrass brought over from Eurasia has spread far and wide since its introduction to North America in the late 1800s. As it overtakes sagebrush and bunchgrass habitats, cheatgrass fuels the decline of birds that nest among sagebrush shrubs and depend on them for food. Cheatgrass now covers an area of North America larger than the area of Germany. Perhaps 5% of 283 million hectares (700 million acres) of public land is "seriously infested" in the United States, where at least 400 exotic plant species have gone out of control.

Dealing with exotic introductions often requires active management, including hunting, poisoning, herbicide spraying, and in some cases introducing natural predators of the out-of-control exotic—activities that can also potentially disturb or harm native birds and other wildlife. In the United States alone, estimates of the annual cost of damage caused by exotics and the measures to control them reach an estimated $137 billion.

Bullets, Cages, Chemicals, and Climate

Other threats come from human activities, such as unregulated hunting and trapping. On the island of Malta, hunters take aim at island-hopping birds during migration. Officially protected birds, such as swallows, bee-eaters, harriers, and herons, fall to Maltese shooters in staggering numbers. Most of this hunting is just target practice and hurts already declining European nesting bird populations.

The wild bird pet trade has also been devastating. Parrots have long been loved by people the world over for their colorful plumages, potential affection toward their owners, and, in many species, adept "talking" abilities. Almost a third of the world's 330 parrot species are threatened with extinction due to habitat loss and collecting pressures, part of a burgeoning illegal wildlife trade valued at billions of dollars a year.

Other human activities have had indirect but nonetheless harmful effects on birds and their habitats, such as oil spills, factory effluents, and lead poisoning, as well as tall buildings, towers, and power lines interfering with birds' migratory paths or daily movements.

Various solutions to the problems posed by human attention to resident and migrant birds have met with some success. Over the last decade, for example, protection measures helped reduce the international trade in wild parrots. But protection laws in many parrot-rich countries often go unheeded, and parrot poaching and smuggling remain widespread.

Since Canadian and U.S. efforts to stem industrial contaminants such as PCBs [polychlorinated biphenyls] began in the late 1970s, herring gull and double-crested cormorant populations have grown, and the bald eagle returned to the Great Lakes region. After U.S. law banned the pesticide DDT in 1972, the country's peregrine falcon, bald eagle, osprey, and brown pelican populations rebounded. Similar rebounds occurred in Britain in such raptors as sparrowhawks after a ban was initiated there. In 2001, 120 countries signed a pesticide treaty that included phasing out DDT except for limited use in controlling malaria.

Global warming is another threat to birds. Global warming is hastened by many of the same activities that destroy habitat: forest clearing, rampant forest fires, road building, and urban expansion. Scientists estimate that the earth's climate warmed 0.3–0.6 [degrees]C over the past century, and that temperature change will continue and possibly intensify Already, ecological changes seem to be under way in ecosystems around the world. Temperate fauna and flora seem to be changing their schedules. Over the past few decades, scientists have documented earlier flower blooming, butterfly emergence, and frog calling—and earlier bird migration and egg-laying dates in Europe and North America.

Global climate change will also likely increase the frequency and severity of weather anomalies that pound bird populations. El Niño events, for example, could finish off rare, localized, and declining species such as the Galapagos penguin, which has evolved and thrived on an equatorial archipelago flushed by cool, fish-rich currents. In addition, intensified and more-frequent droughts and fires could accompany El Niño and other cycles, both in the tropics and as far-north as Canada's boreal forests.

Biological Hot Spots

Decades of fieldwork, computer modeling, and satellite imagery analysis have pinpointed "hot spots"—areas that harbor disproportionately high diversity and high numbers of imperiled bird species. BirdLife International, working with organizations, agencies, and biologists around the world, identified 7,000 important bird areas in 140 countries—critical bird breeding and migration spots—and 218 endemic bird areas, which are places with the highest numbers of restricted-range and endemic species. While not conferring formal protection, these designations offer a framework from which to set international, national, and local protection priorities.

Linking these bird hot spots and other key habitats and striking a balance between developed and undeveloped areas will be key in saving birds in our ever more crowded world. Over the past 20 years, the emergence of the multi-disciplinary field of conservation biology has changed the focus of biodiversity protection efforts from the park to the landscape level, incorporating not just protected areas but also adjacent lands and water resources and the people who inhabit and use them. This landscape focus increasingly brings conservation goals alongside business plans.

The approach is not only progressive but also pragmatic, since most of the world's remaining wild areas remain in private hands or are managed by no one at all. All told, between 6.4% and 8.8% of the earth's land area falls under some category of formal habitat protection. These areas are sprinkled across the globe, many are quite small, and management varies from one place to the next. In general, the largest and most biologically diverse parks are the least well staffed and protected, as they are in some of the world's poorest regions. An example is Peru's Manu National Park, where up to 1,000 species—about 10% of the world's bird species—have been recorded. Local support is critical for these areas—and for the buffer zones and green corridors needed to protect them adequately.

Most of the world remains open to alteration, and people who are hungry and lack alternatives cannot embrace or focus on efforts to protect natural resources unless they clearly benefit in the bargain. Boosting economic prospects and educational opportunities will allow local people to focus on saving birds and other natural resources for the future.

Conservation Programs That Work

The growing awareness that biodiversity protections can be combined with moneymaking ventures seems to be bringing enterprise and environmentalism together. Shade-grown coffee is increasingly popular, for instance. This crop is grown the traditional way, beneath a tropical forest canopy that also shelters resident and migratory birds. In addition, cultivating various fruits, cork, cacao, and other crops supports many bird species. Farm operations that minimize use of harmful pesticides provide more diverse food sources and safer habitats for birds.

Some successful incentive programs pay farmers to set aside land for wildlife, water, and soil conservation purposes. From 2002 to 2007, for example, about 15.9 million hectares will be enrolled in the U.S. Department of Agriculture's Conservation Reserve Program. Hundreds of thousands of farmers enroll land for 10 to 15 years—taking it out of production, planting grasses and trees, restoring wetlands, or grazing or harvesting hay in a way compatible with wildlife and erosion control. Although some of the grasses used in this program are invasive exotics, since its inception in 1985, the program has helped many declining grassland birds regain ground.

In the Netherlands, a program set up by Dutch biologists offers dairy farmers payments to protect and encourage nesting birds as a farm product. An experiment conducted between 1993 and 1996 found that it was cheaper to pay farmers to monitor and manage breeding wild birds as if they were a crop rather than compensate them for restricting farming practices for the sake of bird protection. The project resulted in increased breeding success of meadow-nesting lapwings, godwits, ruffs, and redshanks, while not interrupting the dairy business. By 2002, about 36,000 hectares of Dutch farmland were enrolled in this program.

Ecotourism, which first arose in Costa Rica and Kenya in the early 1980s, is loosely defined as nature-oriented travel that does not harm the environment and that benefits both the traveler and the local host community. Most nations now court ecotourists. Although nature-oriented tourism is not always light on the environment, this industry shows signs of improving and is often an economically viable alternative to resource extraction.

Success in Florida

The increasingly crowded state of Florida provides a compelling example of how local, state, federal, and private concerns prioritized conservation while struggling with relentless development and population growth. Florida is one of the most biologically diverse and environmentally challenged U.S. states. Fortunately, since the 1980s, careful study and planning have been hallmarks of growing conservation efforts there.

One 2000 study by three University of Florida biologists plotted out an interconnected web of wildlife habitat called the Florida Ecological Network, which embraces the state's most-diverse remaining habitats and wildlife. More than half of the network is already under protected status. With the most-critical areas mapped out and many of them targeted, planners should be better able to steer and concentrate development into the many areas outside the park and corridor network and incorporate protected lands into landscapes that combine compatible forms of agriculture.

Another state study plotted private lands needed to ensure a secure future for the most threatened wildlife, including Florida's 117 rare and endangered listed animals. The researchers deduced that a specifically targeted 33% of the state's land area would need protection to lower significantly the chances of rare species extinctions. They included the 20% of the state that already falls under protection. Florida has identified at least 6% more land for future acquisition or protection through easements.

As prime wild real estate becomes more expensive and hard to find, conservationists have stepped up efforts to secure targeted Florida lands. In 2001, The Nature Conservancy announced that it had helped protect its millionth Florida acre. This organization secures funding to buy acreage that is later turned over to government protection or kept as private preserves.

Meanwhile, Florida's government runs a land-buying program called Florida Forever, an aggressive 10-year effort that targets properties most in need of conservation. Under this program, the state spends about $105 million each year to acquire critical conservation lands, protect watersheds, restore polluted or degraded areas, and provide public recreation. Some properties are held in conservation easements, under which property owners receive state payments or tax incentives in return for managing property as wildlife habitat.

A good part of Florida's economy derives from tourism, and more than 40 million people flood into the state each year on vacation. Meanwhile, almost 20% of the state's population is over age 65, and many are retired and are frequent visitors to state tourist attractions. Combining its huge tourism infrastructure and highway system with a newly honed focus on wild places, the state identified nature watching as vital tourism with The Great Florida Birding Trail. Slated for completion in 2005, this sign-marked driving route of some 3,000 kilometers (1,864 miles) winds its way past most of the state's bird hot spots, including county parks, ranches, state forests, private preserves, an alligator farm or two, and federal lands.

Bird Watching

Birding trails follow decades of growing interest in birding, a hobby that turns most of its participants into supporters of conservation efforts that protect birds and other wildlife. One survey noted that more than 66 million Americans aged 16 or older observed, fed, or photographed wildlife (particularly birds) during the year, spending an estimated $40 billion on equipment and travel expenses. Another report estimated that at least 70.4 million U.S. residents 16 or older go outdoors to watch birds sometime during the year, and that these numbers more than doubled between 1983 and 2001. Surveys conducted in Britain yielded similar results.

Economic impact aside, the burgeoning ranks of birders also provide a powerful infusion of eyes and ears that assist scientists in monitoring bird and other wildlife populations around the world. For example, more than 50,000 volunteers a year now participate in the 103-year-old National Audubon Society Christmas Bird Count, the largest and probably longest-running bird census. These knowledgeable birders identify and tally birds wintering at more than 1,800 local census sites throughout North America and in an increasing number of other regions as well. More than a century's worth of wintering bird data gives ornithologists a telling picture of bird abundance and distribution. As bird surveyors note, many bird species are in decline, and prospects remain bleak for many of the world's most-threatened bird species. Governmental and private efforts to save some, however, are bearing fruit, setting good examples for future endeavors elsewhere:

- The Seychelles magpie-robin is rebounding after being reintroduced to predator-free islands and after reductions in pesticide use in its habitat.
- The whooping crane has been a hallmark of conservation efforts between Canada and the United States—up to about 200 birds after a

low of 14 adults in 1938. A nonmigratory population was reintroduced to Florida, providing an extra hedge against extinction.
- In 1999, the peregrine falcon was lifted from the U.S. endangered species list following the ban on DDT in the 1970s and decades of protection, captive breeding, and reintroduction programs.
- Protection combined with apparent adaptability to changed landscapes enabled red kites to return to former haunts in western Europe.
- Four threatened parrot species on three Caribbean islands are inching back from the brink thanks to government and other protections, public-education campaigns, and some captive-breeding efforts.
- On the island of Mauritius, habitat protection and exotic plant and animal eradication efforts benefit now-growing populations of the endemic Mauritius cuckoo-shrike and Mauritius kestrel, a species that also benefited from captive breeding and release programs until the early 1990s.
- The bright blue Lear's macaw, a rare parrot of northeastern Brazil, appears to be steadily rising in number, from about 170 in the late 1990s to about 250. A local landowner, Brazilian conservation organizations, the World Parrot Trust, and funding from the Disney Conservation Initiative help conservationists plant essential foods for the birds, monitor the population, and protect nest sites.

The actions needed to ensure a secure future for birds are the same ones needed to achieve a sustainable human future: preserving and restoring ecosystems, cleaning up polluted areas, reducing use of harmful pesticides, reversing global climate change, restoring ecological balances, and controlling exotic species. Wildlife conservation must be compatible with rural, suburban, and urban planning efforts that improve prospects for the world's poor while making our cities and industries safer for all living beings.

POSTSCRIPT

Is Biodiversity Overprotected?

There is debate over whether the best way to protect biodiversity is to protect individual endangered species or to protect habitat, which must also shield all the species that share that habitat. In 1998 a bill that would have substituted habitat conservation plans for the protection of individual species was proposed in the U.S. Senate but never came to a vote. It was opposed by environmentalists and House representatives, who argued that habitat conservation does not adequately protect endangered species, and by conservatives who wanted the bill to compensate property owners who lost property value or income opportunities because of habitat protection restrictions.

Mark L. Shaffer, J. Michael Scott, and Frank Casey, in "Noah's Options: Initial Cost Estimates of a National System of Habitat Conservation Areas in the United States," *Bioscience* (May 2002), state, "Solving the habitat portion of the endangered species and biodiversity conservation problems is neither trivial nor overwhelming. A national system of habitat conservation areas in the United States could be secured for an initial annual investment between $5 billion and $8 billion, sustained over 30 years, or roughly one-fourth to one-third the cost of maintaining our national highway system over the same period."

The sixth International Conference on Biological Diversity was held in The Hague in April 2002. Sponsored by the United Nations Environment Programme, it emphasized efforts to protect genetic resources and addressed the hazards posed to biodiversity by exotic (nonnative) species, a topic that is itself controversial. Peter Warshall, in "Green Nazis?" *Whole Earth* (March 2001), says that conservatives are attacking attempts to control or eradicate nonnative species as motivated by a kind of environmental racism.

Although the Endangered Species Act was due for reauthorization in 1993, the necessary legislation has not yet been enacted. Useful articles in favor of reauthorization include John Volkman's "Making Room in the Ark," *Environment* (May 1992) and T. H. Watkins's "What's Wrong With the Endangered Species Act?" *Audubon* (January/February 1996). Bonnie B. Burgess discusses the history and future of the Endangered Species Act in *Fate of the Wild: The Endangered Species Act and the Future of Biodiversity* (University of Georgia Press, 2001). Burgess feels that the act is itself endangered because of attacks from conservatives and obstacles erected by the government. According to Margaret Kritz, "Species Act in Their Sights," *National Journal* (December 20, 2003), the Bush administration, conservative Republicans, and business groups want to change the Endangered Species Act to give greater weight to the economic impact of species protections and less to protecting species. Finally, Omar N. White addresses the question of the act's constitutionality in "The Endangered Species Act's Precarious Perch: A Constitutional Analysis Under the Commerce Clause and the Treaty Power," *Ecology Law Quarterly* (February 2000).

ISSUE 5

Should Environmental Policy Attempt to Cure Environmental Racism?

YES: Robert D. Bullard, from "Environmental Justice for All," *Crisis (The New)* (January/February 2003)

NO: David Friedman, from "The Environmental Racism' Hoax," *The American Enterprise* (November/December 1998)

ISSUE SUMMARY

YES: Professor of sociology Robert D. Bullard argues that even though environmental racism has been recognized for a quarter of a century, it remains a problem, and Bush administration policies threaten to undo what progress has been achieved.

NO: Writer and social analyst David Friedman denies the existence of environmental racism. He argues that the environmental justice movement is a government-sanctioned political ploy that will hurt urban minorities by driving away industrial jobs.

Archeologists delight in our forebears' habit of dumping their trash behind the house or barn. Today, however, most people try to arrange for their junk to be disposed of as far away from home as possible. Landfills, junkyards, recycling centers, and other operations with large negative environmental impacts tend to be sited in low-income and minority areas. Boston's Great Molasses Flood (see http://www.mv.com/ipusers/arcade/molasses.htm and Stephen Puleo, *Dark Tide: The Great Boston Molasses Flood of 1919* [Beacon Press, 2003]) happened when a two-million-gallon molasses storage tank burst; the tank had been built in a neighborhood crowded with immigrant laborers who lacked the political influence to say "Not in My Back Yard." Was such a location mere coincidence? Or was it deliberate?

Does the paucity of poor people and minorities in the environmental movement indicate that these people do not really care? (See Robert Emmett Jones, "Blacks Just Don't Care: Unmasking Popular Stereotypes About Concern for the Environment Among African-Americans," *International Journal of Public Administration* [vol. 25, nos. 2 & 3, 2002]).

The environmental movement has, in fact, been charged with having been created to serve the interests of white middle- and upper-income people. Native

Americans, blacks, Hispanics, and poor whites were not well represented among early environmental activists. It has been suggested that the reason for this is that these people were more concerned with more basic needs, such as jobs, food, health, and safety. However, the situation has been changing. In 1982, for example, in Warren County, North Carolina, poor black and Native American communities held demonstrations in protest of a poorly planned PCB (polychlorinated biphenyl) disposal site. This incident kicked off the environmental justice movement, which has since grown to include numerous local, regional, national, and international groups. The movement's target is systematic discrimination in the setting of environmental goals and in the siting of polluting industries and waste disposal facilities—also known as environmental racism. The global reach of the problem is discussed by Jan Marie Fritz in "Searching for Environmental Justice: National Stories, Global Possibilities," *Social Justice* (Fall 1999).

In 1990 the Environmental Protection Agency (EPA) published "Environmental Equity: Reducing Risks for All Communities," a report that acknowledged the need to pay attention to many of the concerns raised by environmental justice activists. At the 1992 United Nations Earth Summit in Rio de Janeiro, a set of "Principles of Environmental Justice" was widely discussed. In 1993 the EPA opened an Office of Environmental Equity (now the Office of Environmental Justice) with plans for cleaning up sites in several poor communities. In February 1994 President Bill Clinton made environmental justice a national priority with an executive order. Since then, many complaints of environmental discrimination have been filed with the EPA under Title VI of the federal Civil Rights Act of 1964; and in March 1998 the EPA issued guidelines for investigating those complaints. However, in April 2001 the U.S. Supreme Court ruled that individuals cannot sue states by charging that federally funded policies unintentionally violate the Civil Rights Act of 1964. The decision is expected to limit environmental justice lawsuits (see Franz Neil, "Supreme Court Ruling May Hurt Environmental Justice Claims," *Chemical Week* [May 2, 2001]). At present, "the sole relief available for victims of environmental civil rights violations is through a private action against a state if the community can prove intentional discrimination. To date, no such action has been successful." See the Environmental Justice Coalition's home page at http:// groups.msn.com/environmentaljusticecoalition/mission.msnw.

Critics of the environmental justice movement contend that inequities in the siting of sources of pollution are the natural consequence of market forces that make poor neighborhoods (whether occupied by whites or minorities) the economically logical choice for locating such facilities. Critics also charge that such facilities depress property values and drive more prosperous people away while attracting a poorer population. In the following selections, Robert D. Bullard describes the history of the environmental justice movement; argues that even though environmental racism has been recognized for a quarter of a century, it remains a problem; and calls for government to maintain its commitment to protecting the environment for all Americans. David Friedman, on the other hand, asserts that the environmental justice movement is a politically inspired movement that is unsupported by scientific facts. He calls environmental racism a hoax and argues that attacking it will harm the urban poor by denying them the industrial jobs they need.

Robert D. Bullard

 **YES**

Environmental Justice for All

Almost every day the media discovers an African American community fighting some form of environmental threat from land fills, garbage dumps, incinerators, lead smelters, petrochemical plants, refineries, highways, bus depots, and the list goes on. For years, residents watched helplessly as their communities became dumping grounds.

But citizens didn't remain silent for long. Local activists have been organizing under the mantle of environmental justice since as far back as 1968. In 1979, a landmark environmental discrimination lawsuit filed in Houston, followed by similar litigation efforts in the 1980s, rallied activists to stand up to corporations and demand government intervention. More than a decade ago, environmental activists from across the country came together to pool their efforts and organize.

In 1991, a new breed of environmental activists gathered in Washington D.C., to bring national attention to pollution problems threatening low-income and minority communities. Leaders introduced the concepts of environmental justice, protesting that Black, poor and working-class communities often received less environmental protection than White or more affluent communities. The first National People of Color Environmental Leadership Summit effectively broadened what "the environment" was understood to mean. It expanded the definition to include where we live, work, play, worship and go to school, as well as the physical and natural world. In the process, the environmental justice movement changed the way environmentalism is practiced in the United States and, ultimately, worldwide.

Because many issues identified at the inaugural summit remain unaddressed, the second National People of Color Environmental Leadership Summit was convened in Washington, D.C., this past October [2002]. The second summit was planned for 500 delegates; but more than 1,400 people attended the four-day gathering.

"We are pleased that the Summit II was able to attract a record number of grassroots activists, academicians, students, researchers, planners, policy analysts and government officials. We proved to the world that our movement is alive and well, and growing," says Beverly Wright, director of the Deep South Center for Environmental Justice at Xavier University in New Orleans and chair of the summit.

McGraw-Hill/Dushkin wished to thank The Crisis Publishing Co., Inc., the publisher of the magazine of the National Association for the Advancement of Colored People, for the use of this material first published in the January/February 2003 issue of *Crisis*.

The meeting produced two dozen policy papers that show powerful environmental and health disparities between people of color and Whites.

"This is not rocket science. We live these statistics every day, and the sad thing is many of us have to die to prove the point," says Donele Wilkins, executive director of Detroiters Working for Environmental Justice.

At the 2002 Essence Music Festival in New Orleans, the National Black Environmental Justice Network (NBEJN) launched its Healthy and Safe Communities Campaign, which targets childhood lead poisoning, asthma and cancer in the Black community. "We are dead serious about educating and mobilizing Black people to eliminate these diseases that cause so much pain, suffering and death in the Black community," says Damu Smith, executive director of NBEJN.

Birth of a Movement

More than three decades ago, the concept of environmental justice had not registered on the radar screens of many environmental or civil rights groups. But environmental justice fits squarely under the civil rights umbrella. It should not be forgotten that Dr. Martin Luther King Jr. went to Memphis on an environmental and economic justice mission in 1968, seeking support for striking garbage workers who were underpaid and whose basic duties exposed them to dangerous environmentally hazardous conditions. King was killed before completing his mission, but others continued it.

The first lawsuit to challenge environmental discrimination using civil rights law, *Bean v. Southwestern Waste Management, Inc.*, was filed in Houston in 1979. Black homeowners in a suburban middle-income Houston neighborhood and their attorney, Linda McKeever Bullard (my wife) filed a class action lawsuit challenging a waste facility siting.

From the early 1920's through 1978, more than 80 percent of Houston's garbage landfills and incinerators were located in mostly black neighborhoods—even though Blacks made up only 25 percent of the city's population. The residents were not able to halt the landfill, but they were able to impact the city and state waste facility siting regulations.

The *Bean* lawsuit was filed three years before the environmental justice movement was catapulted into the national limelight after rural and mostly Black Warren County, N.C., was selected as the final resting place for toxic waste. Oil laced with highly toxic PCBs (polychlorinated biphenyls) was illegally dumped along roadways in 14 North Carolina counties in 1978 and cleaned up in 1982. The decision to dispose of the contaminated soil in a landfill in the county sparked protests and more than 500 arrests—marking the first time any Americans had been jailed protesting the placement of waste facility.

The Warren County protesters put "environmental racism" on the map. Environmental racism refers to any environmental policy, practice, or directive that negatively affects (whether intentionally or not) individuals, groups, or communities based on race or color. Although the Warren County protesters were also unsuccessful in blocking the PCB landfill, they galvanized Black church leaders, civil rights organizers, youth and grassroots activists around environmental issues in the Black community.

The events in Warren County prompted District of Columbia Delegate Walter Fauntroy to request a General Accounting Office (GAO) investigation of hazardous waste facilities in the EPA's Region IV—which includes Alabama, Florida, Georgia, Kentucky, Mississippi, South Carolina, North Carolina and Tennessee. The 1983 GAO study found that three of four off-site hazardous waste landfills in the region were located in predominantly Black communities, even though Blacks make up just 20 percent of the region's population.

The events in Warren County led the United Church of Christ (UCC) Commission for Racial Justice to publish its landmark 1987 "Toxic Wastes and Race in the U.S." report. The UCC study documented that three of every five Blacks live in communities with abandoned toxic waste sites.

The publication of my book *Dumping in Dixie: Race, Class and Environmental Quality* in 1990 offered the nation a first-hand glimpse of environmental racism struggles all across the South—a region whose "look-the-other-way" environmental policies allowed it to become the most environmentally befouled part of the country.

Sumter County, Ala., typifies the environmental racism pattern chronicled in *Dumping in Dixie*. The county is 71.8 percent Black, and 35.9 percent of the jurisdiction's residents live below the poverty line. The county is home to the nation's largest hazardous waste landfill, in Emelle, Ala., which is more than 90 percent Black. In 1978, the hazardous waste landfill was lured to Sumter County during a period when no Blacks held public office or served on governing bodies, including the state legislature, county commission, or industrial development board. The landfill, which had been dubbed the "Cadillac of Dumps," accepts hazardous wastes—as much as nearly 800,000 tons per year—from the 48 contiguous states and several foreign countries.

It's About Winning, not Whining

Environmental justice networks and grassroots community groups are making their voices heard loud and clear. In 1996, after five years of organizing, Citizens Against Toxic Exposure convinced the EPA to relocate 358 Pensacola, Fla., families from a dioxin dump, marking the first time a Black community was relocated under the federal government's giant Superfund program.

After eight years in a struggle that began in 1989, Citizens Against Nuclear Trash (CANT) defeated the plans by Louisiana Energy Services (LES) to build the nation's first privately owned uranium enrichment plant in the mostly Black rural communities of Forest Grove and Center Springs, La. On May 1, 1997, a three-judge panel of the Nuclear Regulatory Commission Atomic Safety and Licensing Board ruled that "racial bias played a role in the selection process." The court decision was upheld on appeal April 4, 1998.

In September 1998, after more than 18 months of intense grassroots organizing and legal maneuvering, St. James Citizens for Jobs and the Environment forced the Japanese owned Shintech Inc. to scrap its plan to build a giant polyvinyl chloride (PVC) plant in Convent, La.—a community that is more than 80 percent Black. The Shintech plant would have added 600,000 pounds of air pollutants annually.

In April 2001, a group of 1,500 Sweet Valley/Cobb Town neighborhood plaintiffs in Anniston, Ala., reached a $42.8 million out-of-court settlement with Monsanto. The group filed a class action lawsuit against Monsanto for contaminating the Black community with PCBs. Monsanto manufactured PCBs from 1927 through 1972 for use as insulation in electrical equipment. The Environmental Protection Agency (EPA) banned PCB production in the late 1970s amid questions of health risks.

In June 2002, victory finally came to the Norco, La., community, whose residents are sandwiched between the Shell Oil plant and the Shell/Motiva refinery. Concerned Citizens of Norco and their allies forces Shell to agree to a buyout that allowed residents to relocate. Shell also is considering a $200 million investment in environmental improvements to its facility.

"I am surrounded by 27 petrochemical companies and oil refineries. My house is located only three meters away from the 15-acre Shell chemical plant," says long-time Norco resident Margie Richard. "We are not treated as citizens with equal rights according to U.S. law and international human rights laws."

Getting government to respond to environmental justice problems has not been easy. In 1992, after meeting with environmental justice leaders, the EPA administrator William Reilly (under the first Bush administration) established the Office of Environmental Equity (the name was changed to the Office of Environmental Justice under the Clinton administration) and produced "Environmental Equity: Reducing Risks for All Communities," one of the first comprehensive government reports to examine environmental hazards and social equity.

In response to growing public concern and mounting scientific evidence, President Clinton on Feb. 11, 1994 issued Executive Order 12898, Federal Actions to Address Environmental Justice in Minority Populations and Low-Income Populations. The order attempts to address environmental justice within existing federal laws and regulations.

But the current Bush administration's actions may erode years of modest progress. The new air pollution rules proposed by the administration are alarming. The controversial rule changes, which have been a top priority of the White House, would make it easier for utilities and refinery operators to change operations and expand production without installing new controls to capture the additional pollution. Industry has argued that the old EPA regulations have hindered operations and prevented efficiency improvements.

Three decades after the first Clean Air Act was passed, the nation should be getting tougher on plants originally exempted from those standards, not more lenient. Communities that are already overburdened with dirty air need stronger, not weaker, air quality standards. Reducing emissions is a matter of public health. Rolling back the Clean Air Act and allowing polluters to spew their toxic fumes into the air would spell bad news for asthma sufferers, poor people and people of color who are concentrated in the most polluted urban areas.

If this nation is to achieve environmental justice, the environment in urban ghettos, barrios, rural "poverty pockets" and on reservations must be give the same protection as is provided to affluent suburbs. All communities— Black or White, rich or poor, urban or suburban—deserve to be protected from the ravages of pollution.

NO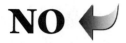

David Friedman

The "Environmental Racism" Hoax

When the U.S. Environmental Protection Agency (EPA) unveiled its heavily criticized environmental justice "guidance" earlier this year, it crowned years of maneuvering to redress an "outrage" that doesn't exist. The agency claims that state and local policies deliberately cluster hazardous economic activities in politically powerless "communities of color." The reality is that the EPA, by exploiting every possible legal ambiguity, skillfully limiting debate, and ignoring even its own science, has enshrined some of the worst excesses of racialist rhetoric and environmental advocacy into federal law.

"Environmental justice" entered the activist playbook after a failed 1982 effort to block a hazardous-waste landfill in a predominantly black North Carolina county. One of the protesters was the District of Columbia's congressional representative, who returned to Washington and prodded the General Accounting Office (GAO) to investigate whether noxious environmental risks were disproportionately sited in minority communities.

A year later, the GAO said that they were. Superfund and similar toxic dumps, it appeared, were disproportionately located in non-white neighborhoods. The well-heeled, overwhelmingly white environmentalist lobby christened this alleged phenomenon "environmental racism," and ethnic advocates like Ben Chavis and Robert Bullard built a grievance over the next decade.

Few of the relevant studies were peer-reviewed; all made critical errors. Properly analyzed, the data revealed that waste sites are just as likely to be located in white neighborhoods, or in areas where minorities moved only after permits were granted. Despite sensational charges of racial "genocide" in industrial districts and ghastly "cancer alleys," health data don't show minorities being poisoned by toxic sites. "Though activists have a hard time accepting it," notes Brookings fellow Christopher H. Foreman, Jr., a self-described black liberal Democrat, "racism simply doesn't appear to be a significant factor in our national environmental decision-making."

This reality, and the fact that the most ethnically diverse urban regions were desperately trying to *attract* employers, not sue them, constrained the environmental racism movement for a while. In 1992, a Democrat-controlled Con-

From David Friedman, "The 'Environmental Racism' Hoax," *The American Enterprise*, vol. 9, no. 6 (November/December 1998). Copyright © 1998 by *The American Enterprise*. Reprinted by permission of *The American Enterprise*, a national magazine of politics, business, and culture.

gress ignored environmental justice legislation introduced by then-Senator Al Gore. Toxic racism made headlines, but not policy.

All of that changed with the Clinton-Gore victory. Vice President Gore got his former staffer Carol Browner appointed head of the EPA and brought Chavis, Bullard, and other activists into the transition government. The administration touted environmental justice as one of the symbols of its new approach.

Even so, it faced enormous political and legal hurdles. Legislative options, never promising in the first place, evaporated with the 1994 Republican takeover in Congress. Supreme Court decisions did not favor the movement.

So the Clinton administration decided to bypass the legislative and judicial branches entirely. In 1994, it issued an executive order—ironically cast as part of Gore's "reinventing government" initiative to streamline bureaucracy—which directed that every federal agency "make achieving environmental justice part of its mission."

At the same time, executive branch lawyers generated a spate of legal memoranda that ingeniously used a poorly defined section of the Civil Rights Act of 1964 as authority for environmental justice programs. Badly split, confusing Supreme Court decisions seemed to construe the 1964 Act's "nondiscrimination" clause (prohibiting federal funds for states that discriminate racially) in such a way as to allow federal intervention wherever a state policy ended up having "disparate effects" on different ethnic groups.

Even better for the activists, the Civil Rights Act was said to authorize private civil rights lawsuits against state and local officials on the basis of disparate impacts. This was a valuable tool for environmental and race activists, who are experienced at using litigation to achieve their ends.

Its legal game plan in place, the EPA then convened an advocate-laden National Environmental Justice Advisory Council (NEJAC), and seeded activist groups (to the tune of $3 million in 1995 alone) to promote its policies. Its efforts paid off. From 1993, the agency backlogged over 50 complaints, and environmental justice rhetoric seeped into state and federal land-use decisions.

◦◉◦

Congress, industry, and state and local officials were largely unaware of these developments because, as subsequent news reports and congressional hearings established, they were deliberately excluded from much of the agency's planning process. Contrary perspectives, including EPA-commissioned studies highly critical of the research cited by the agency to justify its environmental justice initiative in the first place, were ignored or suppressed.

The EPA began to address a wider audience in September 1997. It issued an "interim final guidance" (bureaucratese for regulation-like rules that agencies can claim are not "final" so as to avoid legal challenge) which mandated that environmental justice be incorporated into all projects that file federal environmental impact statements. The guidance directed that applicants pay particular attention to potential "disparate impacts" in areas where minorities live in "meaningfully greater" numbers than surrounding regions.

The new rules provoked surprisingly little comment. Many just "saw the guidance as creating yet another section to add to an impact statement," explains Jennifer Hernandez, a San Francisco environmental attorney. In response, companies wanting to build new plants had to start "negotiating with community advocates and federal agencies, offering new computers, job training, school or library improvements, and the like" to grease their projects through.

In December 1997, the Third Circuit Court of Appeals handed the EPA a breathtaking legal victory. It overturned a lower court decision against a group of activists who sued the state of Pennsylvania for granting industrial permits in a town called Chester, and in doing so the appeals court affirmed the EPA's extension of Civil Rights Act enforcement mechanisms to environmental issues.

(When Pennsylvania later appealed, and the Supreme Court agreed to hear the case, the activists suddenly argued the matter was moot, in order to avoid the Supreme Court's handing down an adverse precedent. This August, the Court agreed, but sent the case back to the Third Circuit with orders to dismiss the ruling. While activists may have dodged a decisive legal bullet, they also wiped from the books the only legal precedent squarely in their favor.)

Two months after the Third Circuit's decision, the EPA issued a second "interim guidance" detailing, for the first time, the formal procedures to be used in environmental justice complaints. To the horror of urban development, business, labor, state, local, and even academic observers, the guidance allows the federal agency to intervene at any time up to six months (subject to extension) after any land-use or environmental permit is issued, modified, or renewed anywhere in the United States. All that's required is a simple allegation that the permit in question was "an act of intentional discrimination or has the effect of discriminating on the basis of race, creed, or national origin."

The EPA will investigate such claims by considering "multiple, cumulative, and synergistic risks." In other words, an individual or company might not itself be in violation, but if, combined with previous (also legal) land-use decisions, the "cumulative impact" on a minority community is "disparate," this could suddenly constitute a federal civil rights offense. The guidance leaves important concepts like "community" and "disparate impact" undefined, leaving them to "case by case" determination. "Mitigations" to appease critics will likewise be negotiated with the EPA case by case.

This "guidance" subjects virtually any state or local land-use decision—made by duly elected or appointed officials scrupulously following validly enacted laws and regulations—to limitless ad hoc federal review, any time there is the barest allegation of racial grievance. Marrying the most capricious elements of wetlands, endangered species, and similar environmental regulations with the interest-group extortion that so profoundly mars urban ethnic politics, the guidance transforms the EPA into the nation's supreme land-use regulator.

⋅⊙⋅

Reaction to the Clinton administration's gambit was swift. A coalition of groups usually receptive to federal interventions, including the U.S. Conference of Mayors, the National Association of Counties, and the National Association of

Black County Officials, demanded that the EPA withdraw the guidance. The House amended an appropriations bill to cut off environmental justice enforcement until the guidance was revised. This August, EPA officials were grilled in congressional hearings led by Democratic stalwarts like Michigan's John Dingell.

Of greatest concern is the likelihood the guidance will dramatically increase already-crippling regulatory uncertainties in urban areas where ethnic populations predominate. Rather than risk endless delay and EPA-brokered activist shakedowns, businesses will tacitly "redline" minority communities and shift operations to white, politically conservative, less-developed locations.

Stunningly, this possibility doesn't bother the EPA and its environmentalist allies. "I've heard senior agency officials just dismiss the possibility that their policies might adversely affect urban development," says lawyer Hernandez. Dingell, a champion of Michigan's industrial revival, was stunned when Ann Goode, the EPA's civil rights director, said her agency never considered the guidance's adverse economic and social effects. "As director of the Office of Civil Rights," she lectured House lawmakers, "local economic development is not something I can help with."

Perhaps it should be. Since 1980, the economies of America's major urban regions, including Cleveland, Chicago, Milwaukee, Detroit, Pittsburgh, New Orleans, San Francisco, Newark, Los Angeles, New York City, Baltimore, and Philadelphia, grew at only one-third the rate of the overall American economy. As the economies of the nation's older cities slumped, 11 million new jobs were created in whiter areas.

Pushing away good industrial jobs hurts the pocketbook of urban minorities, and, ironically, harms their health in the process. In a 1991 *Health Physics* article, University of Pittsburgh physicist Bernard L. Cohen extensively analyzed mortality data and found that while hazardous waste and air pollution exposure takes from three to 40 days off a lifespan, poverty reduces a person's life expectancy by an average of 10 *years*. Separating minorities from industrial plants is thus not only bad economics, but bad health and welfare policy as well.

◈

Such realities matter little to environmental justice advocates, who are really more interested in radical politics than improving lives. "Most Americans would be horrified if they saw NEJAC [the EPA's environmental justice advisory council] in action," says Brookings's Foreman, who recalls a council meeting derailed by two Native Americans seeking freedom for an Indian activist incarcerated for killing two FBI officers. "Because the movement's main thrust is toward ... 'empowerment'..., scientific findings that blunt or conflict with that goal are ignored or ridiculed."

Yet it's far from clear that the Clinton administration's environmental justice genie can be put back in the bottle. Though the Supreme Court's dismissal of the Chester case eliminated much of the EPA's legal argument for the new rules, it's likely that more lawsuits and bureaucratic rulemaking will keep the program alive. The success of the environmental justice movement over the last six years shows just how much a handful of ideological, motivated bureaucrats and their activist allies can achieve in contemporary America unfettered by fact, consequence, or accountability, if they've got a President on their side.

POSTSCRIPT

Should Environmental Policy Attempt to Cure Environmental Racism?

The problems that led to the environmental justice movement have been documented in many reports. For example, in "Who Gets Polluted? The Movement for Environmental Justice," *Dissent* (Spring 1994), Ruth Rosen presents a history of the environmental justice movement, stressing how the movement has woven together strands of the civil rights and environmental struggles. Rosen argues that racial discrimination plays a significant role in the unusually intense exposure to industrial pollutants experienced by disadvantaged minorities, and she expresses the hope that "greening the ghetto will be the first step in greening our entire society." In addition, Bullard's *Dumping in Dixie: Race, Class and Environmental Quality* (Westview Press, 1990, 1994, 2000) has become a standard text in the environmental justice field. Also see his *Unequal Protection: Environmental Justice and Communities of Color* (Sierra Club Books, 1994); Michael Heiman's "Waste Management and Risk Assessment: Environmental Discrimination Through Regulation," *Urban Geography* (vol. 17, no. 5, 1996); and Luke W. Cole and Sheila R. Foster's *From the Ground Up: Environmental Racism and the Rise of the Environmental Justice Movement* (New York University Press, 2000). David W. Allen, in "Social Class, Race, and Toxic Releases in American Counties, 1995," *Social Science Journal* (vol. 38, no. 1, 2001), finds that the data support the existence of environmental racism but that the effect is strongest in the southern portion of the United States (the Sun Belt). Nor is the issue one only of dumps and industrial sites; it also concerns agricultural laborers. See Carlos Marentes, "Farm Workers Fight Against Environmental Racism and Neo-Liberalism," *Synthesis/Regeneration* (vol. 33, Winter 2004) at http://www.greens.org/s-r/33/33-06.html.

Those who criticize the environmental justice movement tend to focus on other studies. In "Green Redlining: How Rules Against 'Environmental Racism' Hurt Poor Minorities Most of All," *Reason* (October 1998), Henry Payne labels the Environmental Protection Agency's efforts to impose environmental equity "redlining" and, like Friedman, argues that the practice reduces job opportunities and economic benefits for minorities.

There is great contrast in the sides to this debate. In such cases, the reader must not ignore the social values and political commitments of the debaters. The reader must also be careful to consider the data relied on by the debaters and to watch for unsupported claims and simplistic explanations for events whose causes are likely to be more complicated.

Where is government policy going? Jim Motavalli, in "Toxic Targets: Polluters That Dump on Communities of Color Are Finally Being Brought to Justice," *E: The Environmental Magazine* (July–August 1998), states that although minorities and the poor have been forced to bear a disproportionate share of the burden of industrial pollution, changes in environmental policy and law are finally offering remedies. And in an August 9, 2001, memorandum regarding the Environmental Protection Agency's stance on environmental justice, EPA administrator Christine Todd Whitman wrote, "The Environmental Protection Agency has a firm commitment to the issue of environmental justice and its integration into all programs, policies, and activities, consistent with existing environmental laws and their implementing regulations.... [E]nvironmental justice is the goal to be achieved for all communities and persons across this Nation. Environmental justice is achieved when everyone, regardless of race, culture, or income, enjoys the same degree of protection from environmental and health hazards and equal access to the decision-making process to have a healthy environment in which to live, learn, and work." However, as Bullard notes, the Bush administration does not seem to be living up to this rhetoric. Indeed, the proposed 2005 federal budget slashes EPA funding by nearly $100 million. See Chris Mooney, "Earth Last," *The American Prospect* (May 2004).

ISSUE 6

Can Pollution Rights Trading Effectively Control Environmental Problems?

YES: Charles W. Schmidt, from "The Market for Pollution," *Environmental Health Perspectives* (August 2001)

NO: Brian Tokar, from "Trading Away the Earth: Pollution Credits and the Perils of 'Free Market Environmentalism,'" *Dollars & Sense* (March/April 1996)

ISSUE SUMMARY

YES: Freelance science writer Charles W. Schmidt argues that economic incentives such as emissions rights trading offer the most useful approaches to reducing pollution.

NO: Author, college teacher, and environmental activist Brian Tokar maintains that pollution credits and other market-oriented environmental protection policies do nothing to reduce pollution while transferring the power to protect the environment from the public to large corporate polluters.

Following World War II the United States and other developed nations experienced an explosive period of industrialization accompanied by an enormous increase in the use of fossil fuel energy sources and a rapid growth in the manufacture and use of new synthetic chemicals. In response to growing public concern about the pollution and other forms of environmental deterioration resulting from this largely unregulated activity, the U.S. Congress passed the National Environmental Policy Act of 1969. This legislation included a commitment on the part of the government to take an active and aggressive role in protecting the environment. The next year the Environmental Protection Agency (EPA) was established to coordinate and oversee this effort. During the next two decades an unprecedented series of legislative acts and administrative rules were promulgated, placing numerous restrictions on industrial and commercial activities that might result in the pollution, degradation, or contamination of land, air, water, food, and the workplace.

Such forms of regulatory control have always been opposed by the affected industrial corporations and developers as well as by advocates of a free-market policy. More moderate critics of the government's regulatory program recognize that adequate environmental protection will not result from completely voluntary policies. They suggest that a new set of strategies is needed. Arguing that "top down, federal, command and control legislation" is not an appropriate or effective means of preventing ecological degradation, they propose a wide range of alternative tactics, many of which are designed to operate through the economic marketplace. The first significant congressional response to these proposals was the incorporation of tradable pollution emission rights into the 1990 Clean Air Act amendments as a means for achieving the set goals for reducing acid rain–causing sulfur dioxide emissions. More recently, the 1997 international negotiations on controlling global warming in Kyoto, Japan, resulted in a protocol that includes emissions trading as one of the key elements in the plan to limit the atmospheric buildup of greenhouse gases.

Despite past difficulties in obtaining compliance with or enforcing strict statutory pollution limits, the idea of using such market-based strategies as the trading of pollution control credits or the imposition of pollution taxes has won limited acceptance from some major mainstream environmental organizations. Many environmentalists, however, continue to oppose the idea of allowing anyone to pay to pollute, either on moral grounds or because they doubt that these tactics will actually achieve the goal of controlling pollution. Diminishment of the acid rain problem is often cited as an example of how well emission rights trading can work, but in "Dispelling the Myths of the Acid Rain Story," *Environment* (July–August 1998), Don Munton argues that other control measures, such as switching to low-sulfur fuels, deserve much more of the credit for reducing sulfur dioxide emissions.

In "A Low-Cost Way to Control Climate Change," *Issues in Science and Technology* (Spring 1998), Byron Swift argues that the "cap-and-trade" feature of the U.S. Acid Rain Program has been so successful that a similar system for implementing the Kyoto Protocol's emissions trading mandate as a cost-effective means of controlling greenhouse gases should work. In March 2001 the U.S. Senate Committee on Agriculture, Nutrition, and Forestry held a "Hearing on Biomass and Environmental Trading: Opportunities for Agriculture and Forestry," in which witnesses urged Congress to encourage trading for both its economic and its environmental benefits. Richard L. Sandor, chairman and chief executive officer of Environmental Financial Products LLC, said that "200 million tons of CO_2 could be sequestered through soils and forestry in the United States per year. At the most conservative prices of $20–$30 per ton, this could potentially generate $4–$6 billion in additional agricultural income."

In the following selections, Charles W. Schmidt describes the use of economic incentives to motivate corporations to reduce pollution, and he argues that emissions trading schemes represent "the most significant developments" in this area. Brian Tokar has a much more negative assessment of sulfur dioxide pollution credit trading. He argues that such "free-market environmentalism" tactics fail to reduce pollution while turning environmental protection into a commodity that corporate powers can manipulate for private profit.

Charles W. Schmidt  **YES**

The Market for Pollution

Throughout much of its short history, environmental protection in the United States has been guided by a traditional paradigm based on strict regulatory guidelines for reducing emissions and punishments for noncompliance. Experts credit this traditional approach with improvements in air and water quality evident since the U.S. Environmental Protection Agency (EPA) was created more than 30 years ago. Tough environmental standards imposed under programs such as the Clean Water Act and the Clean Air Act filled a regulatory void and forced industries to cut their emissions or face heavy fines. Many of the greatest gains were seen with respect to point sources such as smokestacks and effluent pipes that could be easily monitored. But beyond the avoidance of penalties, industries regulated under those so-called command-and-control programs had little motivation to develop advanced pollution control technologies, which produced little economic gain.

Today, many stakeholders believe a more modern framework based on economic incentives that allow companies to profit from achieving environmental goals will build on the achievements of the past and allow for even greater improvements in environmental protection. Types of incentives vary widely, but they all share one thing in common: they attach a monetary value to the act of reducing pollution. In a January 2001 document titled *The United States Experience with Economic Incentives for Protecting the Environment*, the EPA described several types of incentives, including fees and taxes levied on pollutant releases, tax rebates for environmental technologies, and the trading of air emissions permits on the open market.

Attention is increasingly turning to the use of economic incentives in the wake of President George W. Bush's pledge to make them a foundation of his environmental policy. During the 2000 presidential campaign, Bush said that under his watch government would "set high environmental standards and provide market-based incentives to develop new technologies ... so that Americans could meet and exceed those standards."

Business organizations have responded warmly to the administration's support for incentives. For example, the Business Roundtable, a Washington, D.C.—based nonprofit organization of "CEOs committed to improving public policy," released a statement on 17 May 2001 that "applauds President Bush

From *Environmental Health Perspectives*, vol. 109, no. 8, August 2001. Reproduced with permission from Environmental Health Perspectives.

for incorporating the use of new technologies, as well as incentives that spur technological innovation, as the cornerstone of the administration's national energy policy."

Among the environmental community, the idea that market instruments could be used to control pollution was initially greeted with skepticism and even hostility. But over time, support has risen to a level that Joseph Goffman, a senior attorney with the public interest group Environmental Defense in Washington D.C., describes as "lukewarm to enthusiastic in many cases."

According to Goffman, economic incentives motivate companies to reduce pollution quickly and to exceed environmental standards whenever possible. This is in contrast to command-and-control approaches, which he says stifle innovation while encouraging polluters to do little more than meet minimum requirements. Under a traditional system, the EPA not only sets environmental standards, it often describes how companies should achieve them—a scenario sometimes described as "technology forcing."

Goffman suggests the downside to this approach is that the EPA usually only sets standards that can be met with current technology. This means companies have to wait for the agency to finish a technology review before either the EPA or the states revise a given standard. "With incentive programs," he says, "you don't have this kind of chicken-and-egg mentality. The agency sets a target and leaves the means of compliance up to industry. Companies want to profit from pollution control, so they invest more resources in technology development." Furthermore, Goffman adds, market forces naturally gravitate toward the least-cost option for reducing pollution, while traditional regulatory strategies lock companies into technologies that become progressively less effective, and thus less attractive, over time.

Most experts suggest it's too soon to gauge where and how incentive programs will grow under the Bush administration. This is because a host of key positions at the EPA and other agencies remain unfilled, and policy directions have yet to be fully clarified. However, Bush's commitment to market forces is undiminished, as indicated by comments from White House spokesperson Marcy Viana, who, referring to the president's position on global warming during an interview on 4 June 2001, said, "[He is] committed to reducing greenhouse gas emissions by drawing on the power of the market and the power of technology."

Emissions Trading Schemes

The most significant developments in incentive programs have occurred in the area of emissions trading, through which air pollutants are viewed as tradable commodities, each with its own regional, national, and even international markets. In an emissions trading program, companies that emit less than their assigned limits, or caps, of a pollutant can sell residual allowances on the open market or bank them for future transactions. This gives other, higher-polluting facilities a choice: either buy allowances and continue releasing the same pollutant or clean their own emissions—whichever is cheaper. The only stipulation is that regional environmental quality continue to meet mandated standards.

These so-called cap-and-trade schemes aren't new. The best-known example is the Acid Rain Program established under the Clean Air Act amendments of 1990, which allows electric utilities to trade allowance credits in sulfur dioxide (SO_2). Many experts point to this initiative, which achieved dramatic reductions in SO_2 at lower costs than expected, as an emissions trading success story. The EPA estimates that since the program was formalized in 1995, annual emissions of SO_2 have fallen by 4 million tons, while rainfall acidity in the Northeast has dropped by 25%. Dallas Burtraw, a senior fellow at Resources for the Future in Washington, D.C., says the program works well because it's simple, it sets firm environmental targets, it keeps transaction costs to a minimum, and it's transparent—meaning that information on available allowances and credit trades is freely available to the public.

The success of the Acid Rain Program has fueled the development of similar initiatives within the private sector. Undeterred by President Bush's rejection of the Kyoto Protocol, a diverse group of 34 major companies called the Chicago Climate Exchange (CCX) recently announced an emissions trading scheme for carbon dioxide and other greenhouse gases. Boasting high-profile members such as BP, Ford Motor Company, DuPont, and International Paper, this effort aims to reduce greenhouse gas emissions to 5% below 1999 levels by 2005. The CCX's role will be similar to that of an organized commodity exchange—it will establish the requisite technical infrastructure, common standards, and a computerized platform through which participants can trade in emissions reductions.

Richard Sandor, project leader at the CCX, points to the following hypothetical trade as an example of how the system will work: Two companies, a manufacturer with advanced pollution control technology and a power plant with older controls, agree to cut their combined emissions of greenhouse gases by three tons each for a total of six tons. Taking advantage of its superior technology, the manufacturer can cut its own emissions by five tons at minimal cost while the power plant can only reduce its own emissions cost-effectively by one ton. But by purchasing the rights to the additional two tons from the manufacturer, the power plant pays for another company to reduce greenhouse gases on its behalf. In this win—win situation, the manufacturer takes in revenues for reducing pollution while the power plant avoids higher costs by passing off its emissions reductions agreement to another source.

According to Sandor, the CCX will facilitate trades among seven midwestern states that together comprise the fourth-largest trading bloc in the world. The CCX also plans to include Brazil as a member, indicating the organization hopes to achieve an international presence. Says Sandor, "We've had a fantastic response from industry. We expect to be in the design phase for 12 months and to begin trading by 2002."

The states have also gotten into the game. In Southern California, a cap-and-trade program known as the Regional Clean Air Incentives Market, or RECLAIM, is being used to control SO_2 and nitrogen oxide (NO_x) air emissions from 360 industrial facilities, including power plants, in Los Angeles and the San Bernardino Valley. A coalition known as the Ozone Transport Commission, comprising the environmental agencies from 13 northeastern and midwestern states and the federal EPA, has developed a cap-and-trade program for NO_x. And elsewhere, in Chicago, a cap-and-trade program for volatile organic compounds was established by the Illinois EPA in early 2000.

The states have, for the most part, had a measure of success with these programs. The Ozone Transport Commission announced on 10 May 2001 that NO_x emissions for 1999 and 2000 were less than half those reported in 1990, before the cap-and-trade system was implemented. California's RECLAIM system has been in operation since 1993 but is just now beginning to demonstrate results. The reason for the delay, says Sam Atwood, spokesperson for the Diamond Bar–based South Coast Air Quality Management District, which coordinates RECLAIM, is that state-mandated "allocations" (a state term that defines the emissions that can be traded under the cap) for SO_2 and NO_x have only recently been set at levels below actual emissions released by industry. For several years after the program was initiated, facilities regulated under RECLAIM were allowed to emit SO_2 and NO_x at unusually high levels to cushion the economic shock of a recession that took place during the early 1990s. "By dropping the allocation levels below real emissions, we're just starting to cross over to the point where the incentive begins to kick in," says Atwood. "This is when we expect to see voluntary improvements in technology."

The Question of Mobile Sources

In a recent and somewhat controversial trend, emissions trading schemes have begun incorporating mobile sources, such as cars and trucks. Under this approach, stationary sources such as factories can obtain emission credits from regulators by paying to have old, highly polluting vehicles taken off the road. For example, RECLAIM recently issued a rule allowing stationary sources to receive mobile source credits by replacing diesel-fueled heavy-duty vehicles with cleaner-running alternatives.

Burtraw suggests this practice provides a major opportunity for cost savings. "It can be a lot less expensive to reduce emissions from mobile sources than stationary sources," he explains. But he concedes that adding mobile sources to the mix doesn't come without its own unique set of challenges. "People are all too willing to bring in an old lemon that barely runs so they can collect $500 from a utility company," he says. In a case like this, the emissions reduction is negligible because the car isn't driveable anyway.

Goffman says programs that include mobile sources need to incorporate safeguards to prevent this kind of abuse. The challenges exist, he says, but solutions are available if the systems are well designed at the outset. The South Coast Air Quality Management District, for example, only agrees to pay credits for cars that could continue running for three years or more.

Trading Issues

Despite a generally positive response from the stakeholder community, emissions trading still raises a number of important concerns. Perhaps the greatest worry is that it might lead to "hot spots," or areas of high pollutant exposure.

A company that cuts its emissions in half might help reduce average air pollution concentrations in a particular region, but this means little to those who live close to an older facility that buys credits rather than upgrading its pollution control technology.

John Walke, director of clean air projects with the National Resources Defense Council in Washington, D.C., suggests that environmental justice problems could arise if the dirtier facilities are located close to poor communities. "There are a lot of fundamental issues that need to be addressed with these systems," he says. "One is the extent to which pollution sources may be heavily localized in a particular area. It's important to consider how much pollution the neighboring communities are already saddled with."

And what about facilities located upwind of residential communities? Should they be allowed to purchase air pollution credits if downwind populations don't experience the benefit of cleaner emissions? Experts suggest the answer is no, and that hot spots can be avoided with effective planning. Suellen Keiner, director of the Center for the Economy and the Environment at the National Academy of Public Administration, a public interest group based in Washington, D.C., says potential solutions include discouraging trades across long distances and on-site review of credit uses to protect against hot spots.

Another incentive category that tends to trouble environmentalists is "open market" emissions trading, which is a scheme developed by the EPA in 1995. Unlike cap-and-trade programs, neither the overall sectors nor the individual trading sources regulated under an open market trading system are subject to a cap. Rather, any source that finds that its actual rate of emissions is below permitted levels for even a short time is eligible for credit that it can save for later or sell to another source. A chief concern is that under these schemes industry sets the standard for emissions allowances—not the regulatory agency. This is critical, given widespread agreement among stakeholders that health-protective standards should be set by the government on behalf of the public, while the means of compliance is left to the regulated community.

Burtraw says monitoring emissions under an open market system is particularly challenging. "Unlike cap-and-trade programs, which are often targeted toward large stationary sources that can be monitored at the stack, open trading is geared toward smaller sources, for example dry cleaners," he explains. "It's difficult and expensive to monitor actual emissions from these sources, so they tend to be estimated based on economic activity and the use of a given technology. On paper, open market trading seems promising, but in practice monitoring is often poor, and emissions inventories are weak."

Responding to New Jersey's announcement of an open market trading system for NO_x, approved by the EPA in July 2001, Environmental Defense called on the agency to withhold additional pending approvals in states including Michigan, New Hampshire, and Illinois. Also critical of open market trading is the Washington, D.C.—based organization Public Employees for Environmental Responsibility. This group, which says it represents anonymous EPA employees who fear the repercussions of speaking out publicly, issued a white paper in June 2000 called *Trading Thin Air* in which they claim that state and federal agencies don't have the ability to monitor these programs. According to the paper, open market trading could "cripple enforcement of the Clean Air Act against stationary sources of pollution."

Despite the uproar, many experts believe open market systems will improve over time. "I do have a healthy dose of skepticism about open market trading," says Burtraw. "It isn't based on sound policy and shouldn't be used on a wide scale. But I also see it as a way to include in trading programs a variety of smaller sources of emissions for which there do not exist emission inventories. At best, open market trading should be viewed as a transitional stepping stone to some better-developed institution that will emerge in the future."

Outlook for the Future

When applied to the nation as a whole, the EPA suggests in its April 2001 report that "the potential savings from widespread use of economic incentives ... could be almost one-fourth of the approximately $200 billion per year currently spent on environmental pollution control in the United States." In applying these tools, the EPA recommends that regulators consider their use in the context of political acceptability, potential for stimulating technological improvements, and enforceability. A number of important questions need to be considered. How many sources are there for each pollutant? Does a unit of pollution from each source have the same health and ecologic impact regardless of where it's released? Who's being affected by the pollution, and will the program reduce these impacts?

A key point raised by Burtraw is that incentives are a tool—not a solution. "You can compare incentives to a hammer," he says. "You can use a hammer to build a house, or you can use it to pull out the nails. This is the big issue we're facing now—if we use the incentives to back away from emissions reductions, then we're using the hammer to pull out the nails. But if we use incentives to aggressively pursue emissions reduction in the most cost-effective way, then we're building a stronger house for the future."

NO

Brian Tokar

Trading Away the Earth: Pollution Credits and the Perils of "Free Market Environmentalism"

The Republican takeover of Congress has unleashed an unprecedented assault on all forms of environmental regulation. From the Endangered Species Act to the Clean Water Act and the Superfund for toxic waste cleanup, laws that may need to be strengthened and expanded to meet the environmental challenges of the next century are instead being targeted for complete evisceration.

For some activists, this is a time to renew the grassroots focus of environmental activism, even to adopt a more aggressively anti-corporate approach that exposes the political and ideological agendas underlying the current backlash. But for many, the current impasse suggests that the movement must adapt to the dominant ideological currents of the time. Some environmentalists have thus shifted their focus toward voluntary programs, economic incentives and the mechanisms of the "free market" as means to advance the cause of environmental protection. Among the most controversial, and widespread, of these proposals are tradeable credits for the right to emit pollutants. These became enshrined in national legislation in 1990 with President George Bush's amendments to the 1970 Clean Air Act.

Even in 1990, "free market environmentalism" was not a new phenomenon. In the closing years of the 1980s, an odd alliance had developed among corporate public relations departments, conservative think tanks such as the American Enterprise Institute, Bill Clinton's Democratic Leadership Council (DLC), and mainstream environmental groups such as the Environmental Defense Fund. The market-oriented environmental policies promoted by this eclectic coalition have received little public attention, but have nonetheless significantly influenced debates over national policy.

Glossy catalogs of "environmental products," television commercials featuring environmental themes, and high profile initiatives to give corporate officials a "greener" image are the hallmarks of corporate environmentalism in the 1990s. But the new market environmentalism goes much further than these showcase efforts. It represents a wholesale effort to recast environmental protection based on a model of commercial transactions within the marketplace. "A

From Brian Tokar, "Trading Away the Earth: Pollution Credits and the Perils of 'Free Market Environmentalism,'" *Dollars & Sense* (March/April 1996). Copyright © 1996 by Economic Affairs Bureau, Inc. Reprinted by permission. *Dollars & Sense* is a progressive economics magazine published six times a year.

new environmentalism has emerged," writes economist Robert Stavins, who has been associated with both the Environmental Defense Fund and the DLC's Progressive Policy Institute, "that embraces ... market-oriented environmental protection policies."

Today, aided by the anti-regulatory climate in Congress, market schemes such as trading pollution credits are granting corporations new ways to circumvent environmental concerns, even as the same firms try to pose as champions of the environment. While tradeable credits are sometimes presented as a solution to environmental problems, in reality they do nothing to reduce pollution—at best they help businesses reduce the costs of complying with limits on toxic emissions. Ultimately, such schemes abdicate control over critical environmental decisions to the very same corporations that are responsible for the greatest environmental abuses.

How It Works, and Doesn't

A close look at the scheme for nationwide emissions trading reveals a particular cleverness; for true believers in the invisible hand of the market, it may seem positively ingenious. Here is how it works: The 1990 Clean Air Act amendments were designed to halt the spread of acid rain, which has threatened lakes, rivers and forests across the country. The amendments required a reduction in the total sulfur dioxide emissions from fossil fuel burning power plants, from 19 to just under 9 million tons per year by the year 2000. These facilities were targeted as the largest contributors to acid rain, and participation by other industries remains optional. To achieve this relatively modest goal for pollution reduction, utilities were granted transferable allowances to emit sulfur dioxide in proportion to their current emissions. For the first time, the ability of companies to buy and sell the "right" to pollute was enshrined in U.S. law.

Any facility that continued to pollute more than its allocated amount (roughly half of its 1990 rate) would then have to buy allowances from someone who is polluting less. The 110 most polluting facilities (mostly coal burners) were given five years to comply, while all the others would have until the year 2000. Emissions allowances were expected to begin selling for around $500 per ton of sulfur dioxide, and have a theoretical ceiling of $2000 per ton, which is the legal penalty for violating the new rules. Companies that could reduce emissions for less than their credits are worth would be able to sell them at a profit, while those that lag behind would have to keep buying credits at a steadily rising price. For example, before pollution trading every company had to comply with environmental regulations, even if it cost one firm twice as much as another to do so. Under the new system, a firm could instead choose to exceed the mandated levels, purchasing credits from the second firm instead of implementing costly controls. This exchange would save money, but in principle yield the same overall level of pollution as if both companies had complied equally. Thus, it is argued, market forces will assure that the most cost-effective means of reducing acid rain will be implemented first, saving the economy billions of dollars in "excess" pollution control costs.

Defenders of the Bush plan claimed that the ability to profit from pollution credits would encourage companies to invest more in new environmental technologies than before. Innovation in environmental technology, they argued, was being stifled by regulations mandating specific pollution control methods. With the added flexibility of tradeable credits, companies could postpone costly controls—through the purchase of some other company's credits—until new technologies became available. Proponents argued that, as pollution standards are tightened over time, the credits would become more valuable and their owners could reap large profits while fighting pollution.

Yet the program also included many pages of rules for extensions and substitutions. The plan eliminated requirements for backup systems on smokestack scrubbers, and then eased the rules for estimating how much pollution is emitted when monitoring systems fail. With reduced emissions now a marketable commodity, the range of possible abuses may grow considerably, as utilities will have a direct financial incentive to manipulate reporting of their emissions to improve their position in the pollution credits market.

Once the EPA actually began auctioning pollution credits in 1993, it became clear that virtually nothing was going according to their projections. The first pollution credits sold for between $122 and $310, significantly less than the agency's estimated minimum price, and by 1995, bids at the EPA's annual auction of sulfur dioxide allowances averaged around $130 per ton of emissions. As an artificial mechanism superimposed on existing regulatory structures, emissions allowances have failed to reflect the true cost of pollution controls. So, as the value of the credits has fallen, it has become increasingly attractive to buy credits rather than invest in pollution controls. And, in problem areas air quality can continue to decline, as companies in some parts of the country simply buy their way out of pollution reductions.

At least one company has tried to cash in on the confusion by assembling packages of "multi-year streams of pollution rights" specifically designed to defer or supplant purchases of new pollution control technologies. "What a scrubber really is, is a decision to buy a 30-year stream of allowances," John B. Henry of Clean Air Capital Markets told the *New York Times*, with impeccable financial logic. "If the price of allowances declines in future years," paraphrased the *Times*, "the scrubber would look like a bad buy."

Where pollution credits have been traded between companies, the results have often run counter to the program's stated intentions. One of the first highly publicized deals was a sale of credits by the Long Island Lighting Company to an unidentified company located in the Midwest, where much of the pollution that causes acid rain originates. This raised concerns that places suffering from the effects of acid rain were shifting "pollution rights" to the very region it was coming from. One of the first companies to bid for additional credits, the Illinois Power Company, canceled construction of a $350 million scrubber system in the city of Decatur, Illinois. "Our compliance plan is based almost totally on purchase of credits," an Illinois Power spokesperson told the *Wall Street Journal*. The comparison with more traditional forms of commodity trading came full circle in 1991, when the government announced that the entire system for trading and auctioning emissions allowances would be administered by the Chicago Board of Trade, long famous for its ever-frantic markets in everything from grain futures and pork bellies to foreign currencies.

Some companies have chosen not to engage in trading pollution credits, proceeding with pollution control projects, such as the installation of new scrubbers, that were planned before the credits became available. Others have switched to low-sulfur coal and increased their use of natural gas. If the 1990 Clean Air Act amendments are to be credited for any overall improvement in the air quality, it is clearly the result of these efforts and not the market in tradeable allowances.

Yet while some firms opt not to purchase the credits, others, most notably North Carolina-based Duke Power, are aggressively buying allowances. At the 1995 EPA auction, Duke Power alone bought 35% of the short-term "spot" allowances for sulfur dioxide emissions, and 60% of the long-term allowances redeemable in the years 2001 and 2002. Seven companies, including five utilities and two brokerage firms, bought 97% of the short term allowances that were auctioned in 1995, and 92% of the longer-term allowances, which are redeemable in 2001 and 2002. This gives these companies significant leverage over the future shape of the allowances market.

The remaining credits were purchased by a wide variety of people and organizations, including some who sincerely wished to take pollution allowances out of circulation. Students at several law schools raised hundreds of dollars, and a group at the Glens Falls Middle School on Long Island raised $3,171 to purchase 21 allowances, equivalent to 21 tons of sulfur dioxide emissions over the course of a year. Unfortunately, this represented less than a tenth of one percent of the allowances auctioned off in 1995.

Some of these trends were predicted at the outset. "With a tradeable permit system, technological improvement will normally result in lower control costs and falling permit prices, rather than declining emissions levels," wrote Robert Stavins and Brad Whitehead (a Cleveland-based management consultant with ties to the Rockefeller Foundation) in a 1992 policy paper published by the Progressive Policy Institute. Despite their belief that market-based environmental policies "lead automatically to the cost-effective allocation of the pollution control burden among firms," they are quite willing to concede that a tradeable permit system will not in itself reduce pollution. As the actual pollution levels still need to be set by some form of regulatory mandate, the market in tradeable allowances merely gives some companies greater leverage over how pollution standards are to be implemented.

Without admitting the underlying irrationality of a futures market in pollution, Stavins and Whitehead do acknowledge (albeit in a footnote to an Appendix) that the system can quite easily be compromised by large companies' "strategic behavior." Control of 10% of the market, they suggest, might be enough to allow firms to engage in "price-setting behavior," a goal apparently sought by companies such as Duke Power. To the rest of us, it should be clear that if pollution credits are like any other commodity that can be bought, sold and traded, then the largest "players" will have substantial control over the entire "game." Emissions trading becomes yet another way to assure that large corporate interests will remain free to threaten public health and ecological survival in their unchallenged pursuit of profit.

Trading the Future

Mainstream groups like the Environmental Defense Fund (EDF) continue to throw their full support behind the trading of emissions allowances, including the establishment of a futures market in Chicago. EDF senior economist Daniel Dudek described the trading of acid rain emissions as a "scale model" for a much more ambitious plan to trade emissions of carbon dioxide and other gases responsible for global warming. This plan was unveiled shortly after the passage of the 1990 Clean Air Act amendments, and was endorsed by then-Senator Al Gore as a way to "rationalize investments" in alternatives to carbon dioxide-producing activities.

International emissions trading gained further support via a U.N. Conference on Trade and Development study issued in 1992. The report was coauthored by Kidder and Peabody executive and Chicago Board of Trade director Richard Sandor, who told the *Wall Street Journal*, "Air and water are simply no longer the 'free goods' that economists once assumed. They must be redefined as property rights so that they can be efficiently allocated."

Radical ecologists have long decried the inherent tendency of capitalism to turn everything into a commodity; here we have a rare instance in which the system fully reveals its intentions. There is little doubt that an international market in "pollution rights" would widen existing inequalities among nations. Even within the United States, a single large investor in pollution credits would be able to control the future development of many different industries. Expanded to an international scale, the potential for unaccountable manipulation of industrial policy by a few corporations would easily compound the disruptions already caused by often reckless international traders in stocks, bonds and currencies.

However, as long as public regulation of industry remains under attack, tradeable credits and other such schemes will continue to be promoted as market-savvy alternatives. Along with an acceptance of pollution as "a by-product of modern civilization that can be regulated and reduced, but not eliminated," to quote another Progressive Policy Institute paper, self-proclaimed environmentalists will call for an end to "widespread antagonism toward corporations and a suspicion that anything supported by business was bad for the environment." Market solutions are offered as the only alternative to the "inefficient," "centralized," "command-and-control" regulations of the past, in language closely mirroring the rhetoric of Cold War anti-communism.

While specific technology-based standards can be criticized as inflexible and sometimes even archaic, critics choose to forget that in many cases, they were instituted by Congress as a safeguard against the widespread abuses of the Reagan-era EPA. During the Reagan years, "flexible" regulations opened the door to widely criticized—and often illegal—bending of the rules for the benefit of politically favored corporations, leading to the resignation of EPA administrator Anne Gorsuch Burford and a brief jail sentence for one of her more vocal legal assistants.

The anti-regulatory fervor of the present Congress is bringing a variety of other market-oriented proposals to the fore. Some are genuinely offered to further environmental protection, while others are far more cynical attempts to replace public regulations with virtual blank checks for polluters. Some have proposed a direct charge for pollution, modeled after the comprehensive pollution taxes that have proved popular in Western Europe. Writers as diverse as Supreme Court Justice Stephen Breyer, American Enterprise Institute economist Robert Hahn and environmental business guru Paul Hawken have defended pollution taxes as an ideal market-oriented approach to controlling pollution. Indeed, unlike tradeable credits, taxes might help reduce pollution beyond regulatory levels, as they encourage firms to control emissions as much as possible. With credits, there is no reduction in pollution below the threshold established in legislation. (If many companies were to opt for substantial new emissions controls, the market would soon be glutted and the allowances would rapidly become valueless.) And taxes would work best if combined with vigilant grassroots activism that makes industries accountable to the communities in which they operate. However, given the rapid dismissal of Bill Clinton's early plan for an energy tax, it is most likely that any pollution tax proposal would be immediately dismissed by Congressional ideologues as an outrageous new government intervention into the marketplace.

Air pollution is not the only environmental problem that free marketeers are proposing to solve with the invisible hand. Pro-development interests in Congress have floated various schemes to replace the Endangered Species Act with a system of voluntary incentives, conservation easements and other schemes through which landowners would be compensated by the government to protect critical habitat. While these proposals are being debated in Congress, the Clinton administration has quietly changed the rules for administering the Act in a manner that encourages voluntary compliance and offers some of the very same loopholes that anti-environmental advocates have sought. This, too, is being offered in the name of cooperation and "market environmentalism."

Debates over the management of publicly-owned lands have inspired far more outlandish "free market" schemes. "Nearly all environmental problems are rooted in society's failure to adequately define property rights for some resource," economist Randal O'Toole has written, suggesting a need for "property rights for owls and salmon" developed to "protect them from pollution." O'Toole initially gained the attention of environmentalists in the Pacific Northwest for his detailed studies of the inequities of the U.S. Forest Service's long-term subsidy programs for logging on public lands. Now he has proposed dividing the National Forest system into individual units, each governed by its users and operated on a for-profit basis, with a portion of user fees allocated for such needs as the protection of biological diversity. Environmental values, from clean water to recreation to scenic views, should simply be allocated their proper value in the marketplace, it is argued, and allowed to out-compete unsustainable resource extraction. Other market advocates have suggested far more sweeping transfers of federal lands to the states, an idea seen by many in the West as a first step toward complete privatization.

Market enthusiasts like O'Toole repeatedly overlook the fact that ecological values are far more subjective than the market value of timber and minerals removed from public lands. Efforts to quantify these values are based on various sociological methods, market analysis and psychological studies. People are asked how much they would pay to protect a resource, or how much money they would accept to live without it, and their answers are compared with the prices of everything from wilderness expeditions to vacation homes. Results vary widely depending on how questions are asked, how knowledgeable respondents are, and what assumptions are made in the analysis. Environmentalists are rightfully appalled by such efforts as a recent Resources for the Future study designed to calculate the value of human lives lost due to future toxic exposures. Outlandish absurdities like property rights for owls arouse similar skepticism.

The proliferation of such proposals—and their increasing credibility in Washington—suggest the need for a renewed debate over the relationship between ecological values and those of the free market. For many environmental economists, the processes of capitalism, with a little fine tuning, can be made to serve the needs of environmental protection. For many activists, however, there is a fundamental contradiction between the interconnected nature of ecological processes and an economic system which not only reduces everything to isolated commodities, but seeks to manipulate those commodities to further the single, immutable goal of maximizing individual gain. An ecological economy may need to more closely mirror natural processes in their stability, diversity, long time frame, and the prevalence of cooperative, symbiotic interactions over the more extreme forms of competition that thoroughly dominate today's economy. Ultimately, communities of people need to reestablish social control over economic markets and relationships, restoring an economy which, rather than being seen as the engine of social progress, is instead, in the words of economic historian Karl Polanyi, entirely "submerged in social relationships."

Whatever economic model one proposes for the long-term future, it is clear that the current phase of corporate consolidation is threatening the integrity of the earth's living ecosystems—and communities of people who depend on those ecosystems—as never before. There is little room for consideration of ecological integrity in a global economy where a few ambitious currency traders can trigger the collapse of a nation's currency, its food supply, or a centuries-old forest ecosystem before anyone can even begin to discuss the consequences. In this kind of world, replacing our society's meager attempts to restrain and regulate corporate excesses with market mechanisms can only further the degradation of the natural world and threaten the health and well-being of all the earth's inhabitants.

POSTSCRIPT

Can Pollution Rights Trading Effectively Control Environmental Problems?

Does pollution rights trading give major corporate polluters too much power to control and manipulate the market for emission credits? This is one of the key issues that continues to inspire developing countries to withhold their endorsement of the greenhouse gas emissions trading provisions of the Kyoto Protocol. The evidence that Tokar cites, which is primarily based on short-term experience with trading in sulfur dioxide pollution credits, does not appear to fully justify the broad generalizations he makes about the inherent perils in market-based regulatory plans. Recent assessments of the Acid Rain Program by the EPA and such organizations as the Environmental Defense Fund are more positive. So is the corporate world: In "Economic Man, Cleaner Planet," *The Economist* (September 29, 2001), it is asserted that economic incentives have proved very useful and that "market forces are only just beginning to make inroads into green policymaking." In March 2002 *Pipeline & Gas Journal* reported that "despite uncertainty surrounding U.S. and international environmental policies, companies in a wide range of industries—especially those in the energy field—are increasingly using emission reduction credits as a way to meet the challenges of cutting greenhouse gas emissions." See also Cait Murphy, "Hog Wild for Pollution Trading," *Fortune* (September 2, 2002). Kevin A. Baumert, James F. Perkaus, and Nancy Kete, in "Great Expectations: Can International Emissions Trading Deliver an Equitable Climate Regime?" *Climate Policy* (June 2003), contend that international trading schemes may be more difficult to manage and financially riskier.

The position of those who are ideologically opposed to pollution rights is concisely stated in Michael J. Sandel's op-ed piece "It's Immoral to Buy the Right to Pollute," *The New York Times* (December 15, 1997). In "Selling Air Pollution," *Reason* (May 1996), Brian Doherty supports the concept of pollution rights trading but argues that the kind of emission cap imposed in the case of sulfur dioxide is an inappropriate constraint on what he believes should be a completely free-market program. Richard A. Kerr, in "Acid Rain Control: Success on the Cheap," *Science* (November 6, 1998), contends that emissions trading has greatly reduced acid rain and that the annual cost has been about a tenth of the $10 billion initially forecast. According to Barry D. Solomon and Russell Lee, in "Emissions Trading Systems and Environmental Justice," *Environment* (October 2000), "a significant part of the opposition to emissions trading programs is a perception that they do little to reduce environmental injustice and can even make it worse." However, Byron Swift, in "Allowance Trading and Potential Hot Spots— Good News From the Acid Rain Program," *Environment Reporter* (May 12, 2000), argues that the success of the EPA's emission trading program has not led to the creation of pollution "hot spots" as feared by some critics.

ISSUE 7

Do Environmentalists Overstate Their Case?

YES: Ronald Bailey, from "Debunking Green Myths," *Reason* (February 2002)

NO: David Pimentel, from "Skeptical of the Skeptical Environmentalist," *Skeptic* (vol. 9, no. 2, 2002)

ISSUE SUMMARY

YES: Environmental journalist Ronald Bailey argues that the natural environment is not in trouble, despite the arguments of many environmentalists that it is. He holds that the greatest danger facing the environment is not human activity but "ideological environmentalism, with its hostility to economic growth and technological progress."

NO: David Pimentel, a professor of insect ecology and agricultural sciences, argues that those who contend that the environment is not threatened are using data selectively and that the supply of basic resources to support human life is declining rapidly.

For over two centuries, seemingly everyone who has claimed to see environmental disaster in the offing has been challenged. In 1798, for example, English parson Thomas Malthus thought population must inevitably outstrip the ability of the environment to produce food. When the crisis he foretold did not come about, he was ridiculed; indeed, his failure has been held up ever since as a main reason why we need not be concerned about the consequences of population growth, urbanization, industrialization, and other human activities.

Yet the environmentalists have continued to find things to be concerned about. Rachel Carson (1907–1964) is famous for realizing the dangers of pesticides and other chemicals that we release to the environment, which she reported in her best-seller *Silent Spring* (Houghton Mifflin, 1962). Ecologists Paul Ehrlich and Garrett Hardin have reiterated Malthus's concern about population. A Massachusetts Institute of Technology team lead by Donella Meadows and Dennis Meadows used computer models to analyze population, development, and pollution trends and forecast a crisis of resource depletion and economic collapse before 2050 (see *The Limits to Growth* [Universe Books, 1972]). In 1992 the study was repeated with improved computer models, and even more pessimistic conclusions were reached (see *Beyond the Limits: Confronting Global Collapse, Envisioning a Sustainable Future* [Chelsea Green, 1992]). In 1980 the U.S. government published *The Global 2000 Report to the President*, which projected increased environmental degradation, loss of resources, and a widening gap between the rich and the poor.

No one likes such conclusions. Nor does anyone like the implications for what must be done: limit industrial development and population growth. Conservatives object to proposals to regulate industrial development, for only unchecked industry can generate the wealth needed to solve problems, and the free market can be trusted to produce solutions for all problems that truly need solutions. They also object to proposals to limit family size. Liberals object that restricting development will harm the poor much more than it will the rich. Some also object that the true problem is modern capitalism, which emphasizes short-term economic payoffs over longer-term benefits.

The growing sense that we do indeed face environmental crises lies behind the long series of international conferences arranged by the United Nations, from 1968's Biosphere Conference through 1972's Conference on the Human Environment and 1992's Earth Summit (or the Conference on Environment and Development) to 2002's World Summit on Sustainable Development. The concept of sustainable development became prominent after 1992 and has now taken center stage. Yet the debate is hardly over. Analysts such as Niles Eldredge (*Life in the Balance: Humanity and the Biodiversity Crisis* [Princeton University Press, 1998]) have argued that development in the traditional sense cannot be sustained in a world whose resources are finite; sustainable development must mean development without growth in industrial activity or population.

One prominent contrary voice was that of economist Julian L. Simon (1932–1998), who argued that environmental problems could only be short-term problems; increased population and the free market would ensure an ever-improving standard of living and an ever-healthier environment (see *The Ultimate Resource* [Princeton University Press, 1981]). In 1998 Danish statistician and political scientist Bjorn Lomborg joined the fray with *The Skeptical Environmentalist: Measuring the Real State of the World* (Cambridge University Press, 2001). In it, he accuses environmentalists of distorting the truth in a litany of disaster. The truth, he says, is that "mankind's lot has actually improved in terms of practically every measurable indicator."

Is Lomborg right? The following selections represent two of the many reviews that have discussed this question. In the first selection, Ronald Bailey argues that Lomborg is indeed correct. Despite the claims of environmentalists, he contends, the natural environment is not in trouble from human activity but is, in fact, more threatened by ideological environmentalism. In the second selection, David Pimentel argues that Lomborg misrepresents the truth by selecting only data that support his case. He maintains that human activities do threaten the environment and that mankind's lot is at the mercy of a rapidly declining supply of basic resources.

111

Ronald Bailey  **YES**

Debunking Green Myths

Modern environmentalism, born of the radical movements of the 1960s, has often made recourse to science to press its claims that the world is going to hell in a handbasket. But this environmentalist has never really been a matter of objectively describing the world and calling for the particular social policies that the description implies.

Environmentalism is an ideology, very much like Marxism, which pretended to base its social critique on a "scientific" theory of economic relations. Like Marxists, environmentalists have had to force the facts to fit their theory. Environmentalism is an ideology in crisis: The massive, accumulating contradictions between its pretensions and the actual state of the world can no longer be easily explained away.

The publication of *The Skeptical Environmentalist*, a magnificent and important book by a former member of Greenpeace, deals a major blow to that ideology by superbly documenting a response to environmental doomsaying. The author, Bjorn Lomborg, is an associate professor of statistics at the University of Aarhus in Denmark. On a trip to the United States a few years ago, Lomborg picked up a copy of *Wired* that included an article about the late "doomslayer" Julian Simon.

Simon, a professor of business administration at the University of Maryland, claimed that by most measures, the lot of humanity is improving and the world's natural environment was not critically imperiled. Lomborg, thinking it would be an amusing and instructive exercise to debunk a "right-wing" anti-environmentalist American, assigned his students the project of finding the "real" data that would contradict Simon's outrageous claims.

Lomborg and his students discovered that Simon was essentially right, and that the most famous environmental alarmists (Stanford biologist Paul Ehrlich, Worldwatch Institute founder Lester Brown, former Vice President Al Gore, *Silent Spring* author Rachel Carson) and the leading environmentalist lobbying groups (Greenpeace, the World Wildlife Fund, Friends of the Earth) were wrong. It turns out that the natural environment is in good shape, and the prospects of humanity are actually quite good.

⌑

Lomborg begins with "the Litany" of environmentalist doom, writing: "We are all familiar with the Litany.... Our resources are running out. The population is ever growing, leaving less and less to eat. The air and water are becoming ever more polluted. The planet's species are becoming extinct in vast numbers.... The world's ecosystem is breaking down.... We all know the Litany and have heard it so often that yet another repetition is, well, almost reassuring." Lomborg notes that there is just one problem with the Litany: "It does not seem to be backed up by the available evidence."

Lomborg then proceeds to demolish the Litany. He shows how, time and again, ideological environmentalists misuse, distort, and ignore the vast reams of data that contradict their dour visions. In the course of *The Skeptical Environmentalist*, Lomborg demonstrates that the environmentalist lobby is just that, a collection of interest groups that must hype doom in order to survive monetarily and politically.

Lomborg notes, "As the industry and farming organizations have an obvious interest in portraying the environment as just-fine and no-need-to-do anything, the environmental organizations also have a clear interest in telling us that the environment is in a bad state, and that we need to act now. And the worse they can make this state appear, the easier it is for them to convince us we need to spend more money on the environment rather than on hospitals, kindergartens, etc. Of course, if we were equally skeptical of both sorts of organization there would be less of a problem. But since we tend to treat environmental organizations with much less skepticism, this might cause a grave bias in our understanding of the state of the world." Lomborg's book amply shows that our understanding of the state of the world is indeed biased.

⌑

So what is the real state of humanity and the planet?

Human life expectancy in the developing world has more than doubled in the past century, from 31 years to 65. Since 1960, the average amount of food per person in the developing countries has increased by 38 percent, and although world population has doubled, the percentage of malnourished poor people has fallen globally from 35 percent to 18 percent, and will likely fall further over the next decade, to 12 percent. In real terms, food costs a third of what it did in the 1960s. Lomborg points out that increasing food production trends show no sign of slackening in the future.

What about air pollution? Completely uncontroversial data show that concentrations of sulfur dioxide are down 80 percent in the U.S. since 1962, carbon monoxide levels are down 75 percent since 1970, nitrogen oxides are down 38 percent since 1975, and ground level ozone is down 30 percent since 1977. These trends are mirrored in all developed countries.

Lomborg shows that claims of rapid deforestation are vastly exaggerated. One United Nations Food and Agriculture survey found that globally, forest cover has been reduced by a minuscule 0.44 percent since 1961. The World Wildlife Fund claims that two-thirds of the world's forests have been lost since the dawn of agriculture; the reality is that the world still has 80 percent of its forests. What about the Brazilian rainforests? Eighty-six percent remain uncut, and the rate of clearing is falling. Lomborg also debunks the widely circulated claim that the world will soon lose up to half of its species. In fact, the best evidence indicates that 0.7 percent of species might be lost in the next 50 years if nothing is done. And of course, it is unlikely that nothing will be done.

Finally, Lomborg shows that global warming caused by burning fossil fuels is unlikely to be a catastrophe. Why? First, because actual measured temperatures aren't increasing nearly as fast as the computer climate models say they should be—in fact, any increase is likely to be at the low end of the predictions, and no one thinks that would be a disaster. Second, even in the unlikely event that temperatures were to increase substantially, it will be far less costly and more environmentally sound to adapt to the changes rather than institute draconian cuts in fossil fuel use. The best calculations show that adapting to global warming would cost $5 trillion over the next century. By comparison, substantially cutting back on fossil fuel emissions in the manner suggested by the Kyoto Protocol would cost between $107 and $274 trillion over the same period. (Keep in mind that the current yearly U.S. gross domestic product is $10 trillion.) Such costs would mean that people living in developing countries would lose over 75 percent of their expected increases in income over the next century. That would be not only a human tragedy, but an environmental one as well, since poor people generally have little time for environmental concerns.

Where does Lomborg fall short? He clearly understands that increasing prosperity is the key to improving human and environmental health, but he often takes for granted the institutions of property and markets that make progress and prosperity possible. His analysis, as good as it is, fails to identify the chief cause of most environmental problems. In most cases, imperiled resources such as fisheries and airsheds are in open-access commons where the incentive is for people to take as much as possible of the resource before someone else beats them to it. Since they don't own the resource, they have no incentive to protect and conserve it.

Clearly, regulation has worked to improve the state of many open-access commons in developed countries such as the U.S. Our air and streams are much cleaner than they were 30 years ago, in large part due to things like installing catalytic converters on automobiles and building more municipal sewage treatment plants. Yet there is good evidence that assigning private property rights to these resources would have resulted in a faster and cheaper cleanup.

Lomborg's analysis would have been even stronger had he more directly taken on ideological environmentalism's bias against markets. But perhaps that is asking for too much in an already superb book.

"Things are *better* now," writes Lomborg, "but they are still not *good* enough." He's right. Only continued economic growth will enable the 800 million people who are still malnourished to get the food they need; only continued economic growth will let the 1.2 billion who don't have access to clean water and sanitation obtain those amenities. It turns out that ideological environmentalism, with its hostility to economic growth and technological progress, is the biggest threat to the natural environment and to the hopes of the poorest people in the world for achieving better lives.

"The very message of the book," Lomborg concludes, is that "children born today—in both the industrialized world and the developing countries—will live longer and be healthier, they will get more food, a better education, a higher standard of living, more leisure time and far more possibilities—without the global environment being destroyed. And that is a beautiful world."

NO

David Pimentel

Skeptical of the Skeptical Environmentalist

Bjorn Lomborg discusses a wide range of topics in his book and implies, through his title, that he will inform readers exactly what the real state of world is. In this effort, he criticizes countless world economists, agriculturists, water specialists, and environmentalists, and furthermore, accuses them of misquoting and/or organizing published data to mislead the public concerning the status of world population, food supplies, malnutrition, disease, and pollution. Lomborg bases his optimistic opinion on his selective use of data. Some of Lomborg's assertions will be examined in this review, and where differing information is presented, extensive documentation will be provided.

Lomborg reports that "we now have more food per person than we used to." In contrast, the Food and Agricultural Organization (FAO) of the United Nations reports that food per capita has been declining since 1984, based on available cereal grains (Figure 1). Cereal grains make up about 80% of the world's food. Although grain yields per hectare (abbreviated ha) in both developed and developing countries are still increasing, these increases are slowing while the world population continues to escalate. Specifically from 1950 to 1980, U.S. grains yields increased at about 3% per year, but after 1980 the rate of increase for corn and other grains has declined to only about 1% (Figure 2).

Obviously fertile cropland is an essential resource for the production of foods but Lomborg has chosen not to address this subject directly. Currently, the U.S. has available nearly 0.5 ha of prime cropland per capita, but it will not have this much land if the population continues to grow at its current rapid rate. Worldwide the average cropland available for food production is only 0.25 ha per person. Each person added to the U.S. population requires nearly 0.4 ha (1 acre) of land for urbanization and transportation. One example of the impact of population growth and development is occurring in California where an average of 156,000 ha of agricultural land is being lost each year. At this rate it will not be long before California ceases to be the number one state in U.S. agricultural production.

In addition to the quantity of agricultural land, soil quality and fertility is vital for food production. The productivity of the soil is reduced when it is eroded by rainfall and wind. Soil erosion is not a problem, according to

Figure 1

Cereal Grain Production Per Capita in the World From 1961 to 1999

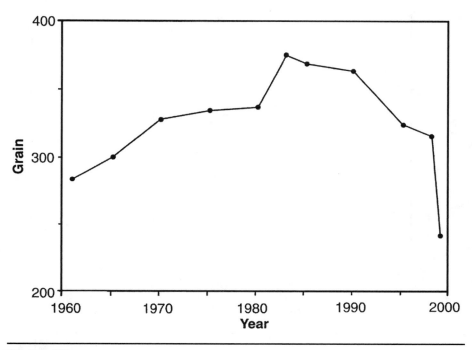

FAO, 1961–1999

Lomborg, especially in the U.S. where soil erosion has declined during the past decade. Yes, as Lomborg states, instead of losing an average of 17 metric tons per hectare per year on cropland, the U.S. cropland is now losing an average of 13 t/ha/yr. However, this average loss is 13 times the sustainability rate of soil replacement. Exceptions occur, as during the 1995–96 winter in Kansas, when it was relatively dry and windy, and some agricultural lands lost as much as 65 t/ha of productive soil. This loss is 65 times the natural soil replacement in agriculture.

Worldwide soil erosion is more damaging than in the United States. For instance, the India soil is being lost at 30 to 40 times its sustainability. Rate of soil loss in Africa is increasing not only because of livestock overgrazing but also because of the burning of crop residues due to the shortages of wood fuel. During the summer of 2000, NASA published a satellite image of a cloud of soil from Africa being blown across the Atlantic Ocean, further attesting to the massive soil erosion problem in Africa. Worldwide evidence concerning soil loss is substantiated and it is difficult to ignore its effect on sustainable agricultural production.

Figure 2

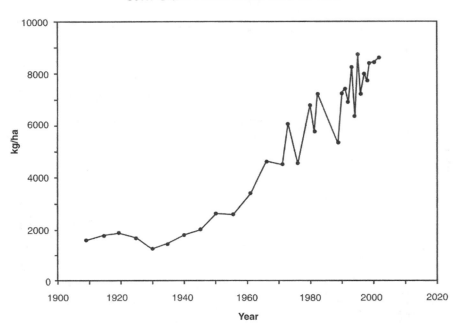

Corn Grain Yields From 1910 to 1999

USDA, 1910–2000

Contrary to Lomborg's belief, crop yields cannot continue to increase in response to the increased applications of more fertilizers and pesticides. In fact, field tests have demonstrated that applying excessive amounts of nitrogen fertilizer stresses the crop plants, resulting in declining yields. The optimum amount of nitrogen for corn, one of the crops that require heavy use of nitrogen, is approximately 120 kg/ha.

Although U.S. farmers frequently apply significantly more nitrogen fertilizer than 120 kg/ha, the extra is a waste and pollutant. The corn crop can only utilize about one-third of the nitrogen applied, while the remainder leaches either into the ground or surface waters. This pollution of aquatic ecosystems in agricultural areas results in the high levels of nitrogen and pesticides occurring in many U.S. water bodies. For example, nitrogen fertilizer has found its way into 97% of the well-water supplies in some regions, like North Carolina. The concentrations of nitrate are above the U.S. Environmental Protection Agency drinking-water standard of 10 milligrams per liter (nitrogen) and are a toxic threat to young children and young livestock. In the last 30 years, the nitrate content has tripled in the Gulf of Mexico, where it is reducing the Gulf fishery.

In an undocumented statement Lomborg reports that pesticides cause very little cancer. Further, he provides no explanation as to why human and other nontarget species are not exposed to pesticides when crops are treated. There is abundant medical and scientific evidence that confirms that pesticides cause significant numbers of cancers in the U.S. and throughout the world. Lomborg also neglects to report that some herbicides stimulate the production of toxic chemicals in some plants, and that these toxicants can cause cancer.

In keeping with Lomborg's view that agriculture and the food supply are improving, he states that "fewer people are starving." Lomborg criticizes the validity of the two World Health Organization [WHO] reports that confirm more than 3 billion people are malnourished. This is the largest number and proportion of malnourished people ever in history! Apparently Lomborg rejects the WHO data because they do not support his basic thesis. Instead, Lomborg argues that only people who suffer from calorie shortages are malnourished, and ignores the fact that humans die from deficiencies of protein, iron, iodine, and vitamin A, B, C, and D.

Further confirming a decline in food supply, the FAO reports that there has been a three-fold decline in the consumption of fish in the human diet during the past seven years. This decline in fish per capita is caused by overfishing, pollution, and the impact of a rapidly growing world population that must share the diminishing fish supply.

In discussing the status of water supply and sanitation services, Lomborg is correct in stating that these services were improved in the developed world during the 19th century, but he ignores the available scientific data when he suggests that these trends have been "replicated in the developing world" during the 20th century. Countless reports confirm that developing countries discharge most of their untreated urban sewage directly into surface waters. For example, of India's 3,119 towns and cities, only eight have full waste water treatment facilities. Furthermore, 114 Indian cities dump untreated sewage and partially cremated bodies directly into the sacred Ganges River. Downstream the untreated water is used for drinking, bathing, and washing. In view of the poor sanitation, it is no wonder that water borne infectious diseases account for 80% of all infections worldwide and 90% of all infections in developing countries.

Contrary to Lomborg's view, most infectious diseases are increasing worldwide. The increase is due not only to population growth but also because of increasing environmental pollution. Food-borne infections are increasing rapidly worldwide and in the United States. For example, during 2000 in the U.S. there were 76 million human food-borne infections with 5,000 associated deaths. Many of these infections are associated with the increasing contamination of food and water by livestock wastes in the United States.

In addition, a large number of malnourished people are highly susceptible to infectious diseases, like tuberculosis (TB), malaria, schistosomiasis, and AIDS. For example, the number of people infected with tuberculosis in the

U.S. and the world is escalating, in part because medicine has not kept up with the new forms of TB. Currently, according to the World Health Organization, more than 2 billion people in the world are infected with TB, with nearly 2 million people dying each year from it.

Consistent with Lomborg's thesis that world natural resources are abundant, he reports that the U.S. Energy Information Agency for the period 2000 to 2020 projects an almost steady oil price over the next two decades at about $22 per barrel. This optimistic projection was crossed late in 2000 when oil rose to $30 or more per barrel in the United States and the world. The best estimates today project that world oil reserves will last approximately 50 years, based on current production rates.

Figure 3

Number of Hectares in Forests Worldwide (x 1 million ha) From 1961 to 1994

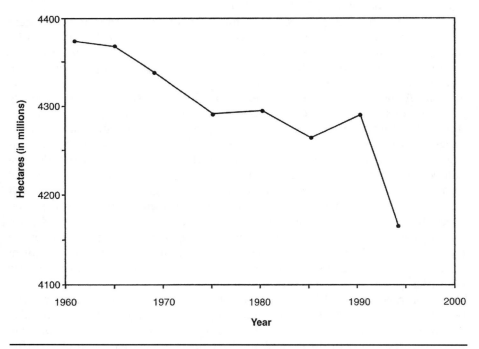

FAOSTAT Database, consulted September 3, 2001

Lomborg takes the World Wildlife Fund (WWF) to task for their estimates on the loss of world forests during the past decade and their emphasis on resulting ecological impacts and loss of biodiversity. Whether the loss of forests is slow, as Lomborg suggests, or rapid as WWF reports, there is no question that forests are disappearing worldwide (Figure 3). Forests not only provide valuable products but they harbor a vast diversity of species of plants, animals and microbes. Progress in medicine, agriculture, genetic engineering, and environmental quality depend on maintaining the species diversity in the world.

This reviewer takes issue with Lomborg's underlying thesis that the size and growth of the human population is not a major problem. The difference between Lomborg's figure that 76 million humans were added to the world population in 2000, or the 80 million reported by the Population Reference Bureau, is not the issue, thought the magnitude of both projections is of serious concern. Lomborg neglects to explain that the major problem with world population growth is the young age structure that now exists. Even if the world adopted a policy of only two children per couple tomorrow, the world population would continue to increase for more than 70 years before stabilizing at more than 12 billion people. As an agricultural scientist and ecologist, I wish I could share Lomborg's optimistic views, but my investigations and those of countless scientists lead me to a more conservative outlook. The supply of basic resources, like fertile cropland, water, energy, and an unpolluted atmosphere that support human life is declining rapidly, as nearly a quarter million people are daily added to the Earth. We all desire a high standard of living for each person on Earth, but with every person added, the supply of resources must be divided and shared. Current losses and degradation of natural resources suggest concern and a need for planning for future generations of humans. Based on our current understanding of the real state of the world and environment, there is need for conservation and protection of vital world resources.

POSTSCRIPT

Do Environmentalists Overstate Their Case?

One of the basic issues at work in the debate between "Malthusians" (who believe that the environment can only support a limited number of people and amount of industrial activity) and "cornucopians" (who believe that there are no such limits) is whether or not past trends can be extrapolated reliably into the future. Cornucopians see little problem in such extrapolation: If the per capita food supply has continued to increase for the last half century, it will continue to do so ad infinitum. Ecologists point out that natural environments have a finite supply of resources (water, soil nutrients, and so on). In nature, population growth follows an S-shaped curve with a steep rise phase followed by a leveling-off at what is known as the "carrying capacity." In "Anticipating Environmental 'Surprise,'" in Lester R. Brown et al., *State of the World 2000* (W. W. Norton, 2000), Chris Brown stresses that straight-line extrapolation cannot be trusted in the real world, because real-world trends are not linear; straight-line trends level off, and they may even peak and fall.

Bjorn Lomborg's *The Skeptical Environmentalist*, with its essentially optimistic message that conditions will continue to get better and better and that we do not need to worry about environmental issues, received very positive responses from the press. However, the scientific community reacted very differently. The January 2002 issue of *Scientific American*, for example, features a series of articles by leading scientists under the heading "Misleading Math About the Earth." Stephen Schneider deals with Lomborg's comments on global warming, John P. Holdren deals with energy, John Bongaarts addresses population, and Thomas Lovejoy discusses biodiversity. All four document the numerous ways in which they feel that Lomborg distorts the truth. Richard B. Norgaard, in "Optimists, Pessimists, and Science," *Bioscience* (March 2002), calls the *Scientific American* article series "devastating" and develops an explanation for Lomborg's approach in the very different ways that economics and science approach the world. Russ Baker, in "The Lomborg File: When the Press Is Lured by a Contrarian's Tale," *Columbia Journalism Review* (March/April 2002), bemoans the uncritical reaction of the nonscientific media, quoting Winston Churchill's famous remark "A lie gets halfway around the world before the truth has a chance to get its pants on."

In "The Skeptical Environmentalist Replies," *Scientific American* (May 2002), Lomborg rebuts his detractors. However, John Rennie, editor in chief of *Scientific American*, charges that Lomborg fails to address the scientists' points and continues to be selective about data.

On the other hand, a review in the May/June 2002 issue of *Foreign Affairs* calls the responses of the scientists "highly critical, even petulant" and says that Lomborg's own "occasionally tendentious examination of sources pales in comparison with the errors of Lomborg's targets." In "Green With Ideology," *Reason* (May 2002), Ronald Bailey argues that the scientists' responses amply demonstrate the truth of his point that environmentalism is an ideology.

Since Lomborg is Danish, the Danish Research Agency's Committee on Scientific Dishonesty responded to the controversy by launching an inquiry. In January 2003, the committee found that he had indeed been selective in his use of data. Lomborg complained to the science ministry, and in December 2003, the ministry declared that the committee had overstepped on the grounds that its role is to rule on issues of fraud, not on failure to follow "good scientific practice." Lomborg and his proponents have taken this as vindication, but the ruling did not address the validity of Lomborg's book. See Lone Frank, "Charges Don't Stick to *The Skeptical Environmentalist,*" *Science* (January 2, 2004).

The Arctic National Wildlife Refuge: A Special Report

This site offers a cogent review of the debate over exploiting the Arctic National Wildlife Refuge (ANWR) for oil.

http://arcticcircle.uconn.edu/ANWR/anwrindex.html

National Wilderness Preservation System

Operated by representatives of the Arthur Carhart National Wilderness Training Center, the Aldo Leopold Wilderness Research Institute, and the Wilderness Institute at the University of Montana's School of Forestry, the National Wilderness Preservation System provides information, news, and links related to wilderness. A database of information on all 644 wilderness areas is included.

http://www.wilderness.net/nwps/default.cfm

Intergovernmental Panel on Climate Change

The Intergovernmental Panel on Climate Change (IPCC) was formed by the World Meteorological Organization (WMO) and the United Nations Environment Programme (UNEP) to assess the scientific, technical, and socio-economic information relevant for the understanding of the risk of human-induced climate change.

http://www.ipcc.ch/

Climate Change

The United Nations Environment Programme maintains this site as a central source for substantive work and information resources with regard to climate change.

http://climatechange.unep.net/

New Source Review

The U.S. Environmental Protection Agency (EPA) has provided a wealth of information about New Source Review and the proposed changes to the rules.

http://www.epa.gov/air/nsr-review/

PART 3

Energy Issues

*H*umans cannot live and society cannot exist without producing environmental impacts. The reason is very simple: Humans cannot live and society cannot exist without using resources (e.g., soil, water, ore, wood, space, plants, animals, oil, sunlight), and those resources come from the environment. Many of these resources (e.g., wood, oil, coal, water, wind, sunlight, uranium) have to do with energy. The environmental impacts come from what must be done to obtain these resources and what must be done to dispose of the wastes generated in the process of obtaining and using them. The issues that arise are whether or not and how these resources should be obtained, whether or not and how these wastes should be dealt with, and whether or not alternative answers to these questions may be preferable to the answers that experts think they already have.

The five issues in this section are not the only issues related to energy, but they serve to demonstrate the vigor of the debate that they engender.

- Should the Arctic National Wildlife Refuge Be Opened to Oil Drilling?
- Should Society Act Now to Forestall Global Warming?
- Will Hydrogen End Our Fossil-Fuel Addiction?
- Should Existing Power Plants Be Required to Install State-of-the-Art Pollution Controls?
- Is It Time to Revive Nuclear Power?

ISSUE 8

Should the Arctic National Wildlife Refuge Be Opened to Oil Drilling?

YES: Dwight R. Lee, from "To Drill or Not to Drill: Let the Environmentalists Decide," *The Independent Review* (Fall 2001)

NO: Katherine Balpataky, from "Protectors of the Herd," *Canadian Wildlife* (Fall 2003)

ISSUE SUMMARY

YES: Professor of economics Dwight R. Lee argues that the economic and other benefits of Arctic National Wildlife Refuge (ANWR) oil are so great that even environmentalists should agree to permit drilling—and they probably would if they stood to benefit directly.

NO: Katherine Balpataky argues that cost-benefit analyses do not support the case for drilling in the ANWR and that the damage done by drilling both to the environment and to the traditional values of the indigenous people, the Gwich'in, cannot be tolerated.

$\mathbf{T}$ he birth of environmental consciousness in the United States was marked by two strong, opposing views. Late in the nineteenth century, John Muir (1838–1914) called for the preservation of natural wilderness, untouched by human activities. At about the same time, Gifford Pinchot (1865–1946) became a strong voice for conservation (not to be confused with preservation; Gifford's conservation allowed the use of nature but in such a way that it was not destroyed; his aim was "the greatest good of the greatest number in the long run"). Both views agree that nature has value; however, they disagree on the form of that value. The preservationist says that nature has value in its own right and has a right to be left alone, neither developed with houses and roads nor exploited with farms, dams, mines, and oil wells. The conservationist says that nature's value lies chiefly in the benefits it provides to human beings.

The first national parks date back to the 1870s. Parks and the national forests are managed for "multiple use" on the premise that wildlife protection, recreation, timber cutting, and even oil drilling and mining can coexist. The first "primitive areas," where all development is barred, were created by the U.S. Forest Service in the 1920s. However, pressure from commercial interests (the timber and mining industries, among others) led to the reclassification of many such areas and their opening to exploitation. In 1964 the Federal Wilderness Act provided a mechanism for designating "wilderness" areas, defined as areas "where the earth and its community of life are untrammeled by man, where man himself is a visitor who does not remain." Since then it has become clear that pesticides and other man-made chemicals are found everywhere on earth, drifting on winds and ocean currents and traveling in migrant birds even to areas without obvious human presence. Humans might not be present in these places, but their effects are. And commercial interests are just as interested in the wealth that may be extracted from these areas as they ever were. There is continual pressure to expand commercial use of national forests and parks and to open wilderness areas to exploitation.

The Arctic National Wildlife Refuge (ANWR) provides a good illustration. It is not a "wilderness" area, for it was designated a wildlife preserve in 1960 and enlarged and renamed in 1980 with the proviso that its coastal plain be evaluated for its potential value in terms of oil and gas production. In 1987 the Department of the Interior recommended that the coastal plain be opened for oil and gas exploration. In 1995 Congress approved doing so, but President Bill Clinton vetoed the legislation. In 2001, after California experienced electrical blackouts, President George W. Bush declared that opening the ANWR to oil exploitation was essential to national energy security but could not muster enough votes in Congress to make it happen. In 2003, an attempt to link the need for Arctic oil to the war in Iraq failed in the Senate. Early in 2004, the Bush administration proposed once more to open the ANWR to oil drilling, but the outcome was by no means certain. See "Bush Budget Calls for Oil Drilling in Alaska Refuge" Reuters (February 4, 2004) (available at http://www.planetark.com/dailynewsstory.cfm/newsid/23687/story.htm).

Strict preservationists still remain, but the debate over protecting wilderness areas generally centers on economic arguments. In the following selections, Dwight R. Lee argues that the economic and other benefits of Arctic National Wildlife Refuge oil are so great that drilling should be permitted. Katherine Balpataky argues that drilling in the ANWR is not cost effective and that to the indigenous people, the Gwich'in, the ANWR is a rich homeland that will only be destroyed by exploitation.

Dwight R. Lee

 **YES**

To Drill or Not to Drill

High prices of gasoline and heating oil have made drilling for oil in Alaska's Arctic National Wildlife Refuge (ANWR) an important issue. ANWR is the largest of Alaska's sixteen national wildlife refuges, containing 19.6 million acres. It also contains significant deposits of petroleum. The question is, Should oil companies be allowed to drill for that petroleum?

The case for drilling is straightforward. Alaskan oil would help to reduce U.S. dependence on foreign sources subject to disruptions caused by the volatile politics of the Middle East. Also, most of the infrastructure necessary for transporting the oil from nearby Prudhoe Bay to major U.S. markets is already in place. Furthermore, because of the experience gained at Prudhoe Bay, much has already been learned about how to mitigate the risks of recovering oil in the Arctic environment.

No one denies the environmental risks of drilling for oil in ANWR. No matter how careful the oil companies are, accidents that damage the environment at least temporarily might happen. Environmental groups consider such risks unacceptable; they argue that the value of the wilderness and natural beauty that would be spoiled by drilling in ANWR far exceeds the value of the oil that would be recovered. For example, the National Audubon Society characterizes opening ANWR to oil drilling as a threat "that will destroy the integrity" of the refuge (see statement at www.audubon.org/campaign/ refuge).

So, which is more valuable, drilling for oil in ANWR or protecting it as an untouched wilderness and wildlife refuge? Are the benefits of the additional oil really less than the costs of bearing the environmental risks of recovering that oil? Obviously, answering this question with great confidence is difficult because the answer depends on subjective values. Just how do we compare the convenience value of using more petroleum with the almost spiritual value of maintaining the "integrity" of a remote and pristine wilderness area? Although such comparisons are difficult, we should recognize that they can be made. Indeed, we make them all the time.

We constantly make decisions that sacrifice environmental values for what many consider more mundane values, such as comfort, convenience, and material well-being. There is nothing wrong with making such sacrifices because up to some point the additional benefits we realize from sacrificing a

This article is reprinted with permission of the publisher from *The Independent Review: A Journal of Political Economy* (Fall 2001, vol. VI, no. 2, pp. 217–226). © Copyright 2003, The Independent. info@independent.org; www.independent.org.

little more environmental "integrity" are worth more than the necessary sacrifice. Ideally, we would somehow acquire the information necessary to determine where that point is and then motivate people with different perspectives and preferences to respond appropriately to that information.

Achieving this ideal is not as utopian as it might seem; in fact, such an achievement has been reached in situations very similar to the one at issue in ANWR. In this article, I discuss cases in which the appropriate sacrifice of wilderness protection for petroleum production has been responsibly determined and harmoniously implemented. Based on this discussion, I conclude that we should let the Audubon Society decide whether to allow drilling in ANWR. That conclusion may seem to recommend a foregone decision on the issue because the society has already said that drilling for oil in ANWR is unacceptable. But actions speak louder than words, and under certain conditions I am willing to accept the actions of environmental groups such as the Audubon Society as the best evidence of how they truly prefer to answer the question, To drill or not to drill in ANWR?

Private Property Changes One's Perspective

What a difference private property makes when it comes to managing multiuse resources. When people make decisions about the use of property they own, they take into account many more alternatives than they do when advocating decisions about the use of property owned by others. This straightforward principle explains why environmental groups' statements about oil drilling in ANWR (and in other publicly owned areas) and their actions in wildlife areas they own are two very different things.

For example, the Audubon Society owns the Rainey Wildlife Sanctuary, a 26,000-acre preserve in Louisiana that provides a home for fish, shrimp, crab, deer, ducks, and wading birds, and is a resting and feeding stopover for more than 100,000 migrating snow geese each year. By all accounts, it is a beautiful wilderness area and provides exactly the type of wildlife habitat that the Audubon Society seeks to preserve. But, as elsewhere in our world of scarcity, the use of the Rainey Sanctuary as a wildlife preserve competes with other valuable uses.

Besides being ideally suited for wildlife, the sanctuary contains commercially valuable reserves of natural gas and oil, which attracted the attention of energy companies when they were discovered in the 1940s. Clearly, the interests served by fossil fuels do not have high priority for the Audubon Society. No doubt, the society regards additional petroleum use as a social problem rather than a social benefit. Of course, most people have different priorities: they place a much higher value on keeping down the cost of energy than they do on bird-watching and on protecting what many regard as little more than mosquito-breeding swamps. One might suppose that members of the Audubon Society have no reason to consider such "anti-environmental" values when deciding how to use their own land. Because the society owns the Rainey Sanctuary, it can ignore interests antithetical to its own and refuse to allow drilling. Yet, precisely because the society owns the land, it has been willing to accommodate the interests of those whose priorities are different and has allowed thirty-seven wells to pump gas and oil from the Rainey Sanctuary. In return, it has received royalties of more than $25 million.

One should not conclude that the Audubon Society has acted hypocritically by putting crass monetary considerations above its stated concerns for protecting wilderness and wildlife. In a wider context, one sees that because of its ownership of the Rainey Sanctuary, the Audubon Society is part of an extensive network of market communication and cooperation that allows it to do a better job of promoting its objectives by helping others promote theirs. Consumers communicate the value they receive from additional gas and oil to petroleum companies through the prices they willingly pay for those products, and this communication is transmitted to owners of oil-producing land through the prices the companies are willing to pay to drill on that land. Money really does "talk" when it takes the form of market prices. The money offered for drilling rights in the Rainey Sanctuary can be viewed as the most effective way for millions of people to tell the Audubon Society how much they value the gas and oil its property can provide.

By responding to the price communication from consumers and by allowing the drilling, the Audubon Society has not sacrificed its environmental values in some debased lust for lucre. Instead, allowing the drilling has served to reaffirm and promote those values in a way that helps others, many of whom have different values, achieve their own purposes. Because of private ownership, the valuations of others for the oil and gas in the Rainey Sanctuary create an opportunity for the Audubon Society to purchase additional sanctuaries to be preserved as habitats for the wildlife it values. So the society has a strong incentive to consider the benefits as well as the costs of drilling on its property. Certainly, environmental risks exist, and the society considers them, but if also responsibly weighs the costs of those risks against the benefits as measured by the income derived from drilling. Obviously, the Audubon Society appraises the benefits from drilling as greater than the costs, and it acts in accordance with that appraisal.

Cooperation Between Bird-Watchers and Hot-Rodders

The advantage of private ownership is not just that it allows people with different interests to interact in mutually beneficial ways. It also creates harmony between those whose interests would otherwise be antagonistic. For example, most members of the Audubon Society surely see the large sport utility vehicles and high-powered cars encouraged by abundant petroleum supplies as environmentally harmful. That perception, along with the environmental risks associated with oil recovery, helps explain why the Audubon Society vehemently opposes drilling for oil in the ANWR as well as in the continental shelves in the Atlantic, the Pacific, and the Gulf of Mexico. Although oil companies promise to take extraordinary precautions to prevent oil spills when drilling in these areas, the Audubon Society's position is no off-shore drilling, none. One might expect to find Audubon Society members completely unsympathetic with hot-rodding enthusiasts, NASCAR racing fans, and drivers of Chevy Suburbans. Yet, as we have seen, by allowing drilling for gas and oil in the Rainey Sanctuary, the society is accommodating the interests of those with gas-guzzling lifestyles, risking the "integrity" of its prized wildlife sanctuary to make more gasoline available to those whose energy consumption it verbally condemns as excessive.

The incentives provided by private property and market prices not only motivate the Audubon Society to cooperate with NASCAR racing fans, but also motivate those racing enthusiasts to cooperate with the Audubon Society. Imagine the reaction you would get if you went to a stock-car race and tried to convince the spectators to skip the race and go bird-watching instead. Be prepared for some beer bottles tossed your way. Yet by purchasing tickets to their favorite sport, racing fans contribute to the purchase of gasoline that allows the Audubon Society to obtain additional wildlife habitat and to promote bird-watching. Many members of the Audubon Society may feel contempt for racing fans, and most racing fans may laugh at bird-watchers, but because of private property and market prices, they nevertheless act to promote one another's interests.

The Audubon Society is not the only environmental group that, because of the incentives of private ownership, promotes its environmental objectives by serving the interests of those with different objectives. The Nature Conservancy accepts land and monetary contributions for the purpose of maintaining natural areas for wildlife habitat and ecological preservation. It currently owns thousands of acres and has a well-deserved reputation for preventing development in environmentally sensitive areas. Because it owns the land, it has also a strong incentive to use that land wisely to achieve its objectives, which sometimes means recognizing the value of developing the land.

For example, soon after the Wisconsin chapter received title to 40 acres of beach-front land on St. Croix in the Virgin Islands, it was offered a much larger parcel of land in northern Wisconsin in exchange for its beach land. The Wisconsin chapter made this trade (with some covenants on development of the beach land) because owning the Wisconsin land allowed it to protect an entire watershed containing endangered plants that it considered of greater environmental value than what was sacrificed by allowing the beach to be developed.

Thanks to a gift from the Mobil Oil Company, the Nature Conservancy of Texas owns the Galveston Bay Prairie Preserve in Texas City, a 2,263-acre refuge that is home to the Attwater's prairie chicken, a highly endangered species (once numbering almost a million, its population had fallen to fewer than ten by the early 1990s). The conservancy has entered into an agreement with Galveston Bay Resources of Houston and Aspects Resources, LLC, of Denver to drill for oil and natural gas in the preserve. Clearly some risks attend oil drilling in the habitat of a fragile endangered species, and the conservancy has considered them, but it considers the gains sufficient to justify bearing the risks. According to Ray Johnson, East County program manager for the Nature Conservancy of Texas. "We believe this could provide a tremendous opportunity to raise funds to acquire additional habitat for the Attwater's prairie chicken, one of the most threatened birds in North America." Obviously the primary concern is to protect the endangered species, but the demand for gas and oil is helping achieve that objective. Johnson is quick to point out, "We have taken every precaution to minimize the impact of the drilling on the prairie chickens and to ensure their continued health and safety."

Back to ANWR

Without private ownership, the incentive to take a balanced and accommodating view toward competing land-use values disappears. So, it is hardly surprising that the Audubon Society and other major environmental groups categorically oppose drilling in ANWR. Because ANWR is publicly owned, the environmental groups have no incentive to take into account the benefits of drilling. The Audubon Society does not capture any of the benefits if drilling is allowed, as it does at the Rainey Sanctuary; in ANWR, it sacrifices nothing if drilling is prevented. In opposing drilling in ANWR, despite the fact that the precautions to be taken there would be greater than those required of companies operating in the Rainey Sanctuary, the Audubon Society is completely unaccountable for the sacrificed value of the recoverable petroleum.

Obviously, my recommendation to "let the environmentalists decide" whether to allow oil to be recovered from ANWR makes no sense if they are not accountable for any of the costs (sacrificed benefits) of preventing drilling. I am confident, however, that environmentalists would immediately see the advantages of drilling in ANWR if they were responsible for both the costs and the benefits of that drilling. As a thought experiment about how incentives work, imagine that a consortium of environmental organizations is given veto power over drilling, but is also given a portion (say, 10 percent) of what energy companies are willing to pay for the right to recover oil in ANWR. These organizations could capture tens of millions of dollars by giving their permission to drill. Suddenly the opportunity to realize important environmental objectives by favorably considering the benefits others gain from more energy consumption would come into sharp focus. The environmentalists might easily conclude that although ANWR is an "environmental treasure," other environmental treasures in other parts of the country (or the world) are even more valuable; moreover, with just a portion of the petroleum value of the ANWR, efforts might be made to reduce the risks to other natural habitats, more than compensating for the risks to the Arctic wilderness associated with recovering that value.

Some people who are deeply concerned with protecting the environment see the concentration on "saving" ANWR from any development as misguided even without a vested claim on the oil wealth it contains. For example, according to Craig Medred, the outdoor writer for the *Anchorage Daily News* and a self-described "development-phobic wilderness lover,"

> That people would fight to keep the scar of clearcut logging from the spectacular and productive rain-forests of Southeast Alaska is easily understandable to a shopper in Seattle or a farmer in Nebraska. That people would argue against sinking a few holes through the surface of a frozen

wasteland, however, can prove more than a little baffling even to development-phobic, wilderness lovers like me. Truth be known, I'd trade the preservation rights to any 100 acres on the [ANWR] slope for similar rights to any acre of central California wetlands.... It would seem of far more environmental concern that Alaska's ducks and geese have a place to winter in overcrowded, overdeveloped California than that California's ducks and geese have a place to breed each summer in uncrowded and undeveloped Alaska.

— (1996, Cl)

Even a small share of the petroleum wealth in ANWR would dramatically reverse the trade-off Medred is willing to make because it would allow environmental groups to afford easily a hundred acres of central California wetlands in exchange for what they would receive for each acre of ANWR released to drilling.

We need not agree with Medred's characterization of the ANWR as "a frozen wasteland" to suspect that environmentalists are overstating the environmental amenities that drilling would put at risk. With the incentives provided by private property, environmental groups would quickly reevaluate the costs of drilling in wilderness refuges and soften their rhetoric about how drilling would "destroy the integrity" of these places. Such hyperbolic rhetoric is to be expected when drilling is being considered on public land because environmentalists can go to the bank with it. It is easier to get contributions by depicting decisions about oil drilling on public land as righteous crusades against evil corporations out to destroy our priceless environment for short-run profit than it is to work toward minimizing drilling costs to accommodate better the interests of others. Environmentalists are concerned about protecting wildlife and wilderness areas in which they have ownership interest, but the debate over any threat from drilling and development in those areas is far more productive and less acrimonious than in the case of ANWR and other publicly owned wilderness areas.

The evidence is overwhelming that the risks of oil drilling to the arctic environment are far less than commonly claimed. The experience gained in Prudhoe Bay has both demonstrated and increased the oil companies' ability to recover oil while leaving a "light footprint" on arctic tundra and wildlife. Oil-recovery operations are now sited on gravel pads providing foundations that protect the underlying permafrost. Instead of using pits to contain the residual mud and other waste from drilling, techniques are now available for pumping the waste back into the well in ways that help maintain well pressure and reduce the risks of spills on the tundra. Improvements in arctic road construction have eliminated the need for the gravel access roads used in the development of the Prudhoe Bay oil fields. Roads are now made from ocean water pumped onto the tundra, where it freezes to form a road surface. Such roads melt without a trace during the short summers. The oversize rubber tires used on the roads further minimize any impact on the land.

Improvements in technology now permit horizontal drilling to recover oil that is far from directly below the wellhead. This technique reduces further the already small amount of land directly affected by drilling operations. Of the more than 19 million acres contained in ANWR, almost 18 million acres have been set aside by Congress—somewhat more than 8 million as wilderness and 9.5 million as wildlife refuge. Oil companies estimate that only 2,000 acres would be needed to develop the coastal plain.

This carefully conducted and closely confined activity hardly sounds like a sufficient threat to justify the rhetoric of a righteous crusade to prevent the destruction of ANWR, so the environmentalists warn of a detrimental effect on arctic wildlife that cannot be gauged by the limited acreage directly affected. Given the experience at Prudhoe Bay, however, such warnings are difficult to take seriously. The oil companies have gone to great lengths and spent tens of millions of dollars to reduce any harm to the fish, fowl, and mammals that live and breed on Alaska's North Slope. The protections they have provided for wildlife at Prudhoe Bay have been every bit as serious and effective as those the Audubon Society and the Nature Conservancy find acceptable in the Rainey Sanctuary and the Galveston Bay Prairie Preserve. As the numbers of various wildlife species show, many have thrived better since the drilling than they did before.

Before drilling began at Prudhoe Bay, a good deal of concern was expressed about its effect on caribou herds. As with many wildlife species, the population of the caribou on Alaska's North Slope fluctuates (often substantially) from year to year for completely natural reasons, so it is difficult to determine with confidence the effect of development on the caribou population. It is noteworthy, however, that the caribou population in the area around Prudhoe Bay has increased greatly since that oil field was developed, from approximately 3,000 to a high of some 23,400.... Some argue that the increase has occurred because the caribou's natural predators have avoided the area—some of these predators are shot, whereas the caribou are not. But even if this argument explains some or even all of the increase in the population, the increase still casts doubt on claims that the drilling threatens the caribou. Nor has it been shown that the viability of any other species has been genuinely threatened by oil drilling at Prudhoe Bay.

Caribou Versus Humans

Although consistency in government policy may be too much to hope for, it is interesting to contrast the federal government's refusal to open ANWR with some of its other oil-related policies. While opposing drilling in ANWR, ostensibly because we should not put caribou and other Alaskan wildlife at risk for the sake of getting more petroleum, we are exposing humans to far greater risks because of federal policies motivated by concern over petroleum supplies.

For example, the United States maintains a military presence in the Middle East in large part because of the petroleum reserves there. It is doubtful that the U.S. government would have mounted a large military action and sacrificed American lives to prevent Iraq from taking over the tiny sheikdom of Kuwait except to allay the threat to a major oil supplier. Nor would the United States have lost the nineteen military personnel in the barracks blown up in Saudi Arabia in 1996 or the seventeen killed onboard the USS Cole in a Yemeni harbor in

2000. I am not arguing against maintaining a military presence in the Middle East, but if it is worthwhile to sacrifice Americans' lives to protect oil supplies in the Middle East, is it not worthwhile to take a small (perhaps nonexistent) risk of sacrificing the lives of a few caribou to recover oil in Alaska?

Domestic energy policy also entails the sacrifice of human lives for oil. To save gasoline, the federal government imposes Corporate Average Fuel Economy (CAFE) standards on automobile producers. These standards now require all new cars to average 27.5 miles per gallon and new light trucks to average 20.5 miles per gallon. The one thing that is not controversial about the CAFE standards is that they cost lives by inducing manufacturers to reduce the weight of vehicles. Even Ralph Nader has acknowledged that "larger cars are safer—there is more bulk to protect the occupant." An interesting question is, How many lives might be saved by using more (ANWR) oil and driving heavier cars rather than using less oil and driving lighter, more dangerous cars?

It has been estimated that increasing the average weight of passenger cars by 100 pounds would reduce U.S. highway fatalities by 200 a year. By determining how much additional gas would be consumed each year if all passenger cars were 100 pounds heavier, and then estimating how much gas might be recovered from ANWR oil, we can arrive at a rough estimate of how many human lives potentially might be saved by that oil. To make this estimate, I first used data for the technical specifications of fifty-four randomly selected 2001 model passenger cars to obtain a simple regression of car weight on miles per gallon. This regression equation indicates that every additional 100 pounds decreases mileage by 0.85 miles per gallon. So 200 lives a year could be saved by relaxing the CAFE standards to allow a 0.85 miles per gallon reduction in the average mileage of passenger cars. How much gasoline would be required to compensate for this decrease of average mileage? Some 135 million passenger cars are currently in use, being driven roughly 10,000 miles per year on average (1994–95 data from U.S. Bureau of the Census 1997, 843). Assuming these vehicles travel 24 miles per gallon on average, the annual consumption of gasoline by passenger cars is 56.25 billion gallons (= 135 million × 10,000/24). If instead of an average of 24 miles per gallon the average were reduced to 23.15 miles per gallon, the annual consumption of gasoline by passenger cars would be 58.32 billion gallons (= 135 million × 10,000/23.15). So, 200 lives could be saved annually by an extra 2.07 billion gallons of gas. It is estimated that ANWR contains from 3 to 16 billion barrels of recoverable petroleum. Let us take the midpoint in this estimated range, or 9.5 billion barrels. Given that on average each barrel of petroleum is refined into 19.5 gallons of gasoline, the ANWR oil could be turned into 185.25 billion additional gallons of gas, or enough to save 200 lives a year for almost ninety years (185.25/2.07 = 89.5). Hence, in total almost 18,000 lives could be saved by opening up ANWR to drilling and using the fuel made available to compensate for increasing the weight of passenger cars.

I claim no great precision for this estimate. There may be less petroleum in ANWR than the midpoint estimate indicates, and the study I have relied on may have overestimated the number of lives saved by heavier passenger cars. Still, any reasonable estimate will lead to the conclusion that preventing the recovery of ANWR oil and its use in heavier passenger cars entails the loss of thousands of lives on the highways. Are we willing to bear such a cost in order to avoid the risks, if any, to ANWR and its caribou?

Conclusion

I am not recommending that ANWR actually be given to some consortium of environmental groups. In thinking about whether to drill for oil in ANWR, however, it is instructive to consider seriously what such a group would do if it owned ANWR and therefore bore the costs as well as enjoyed the benefits of preventing drilling. Those costs are measured by what people are willing to pay for the additional comfort, convenience, and safety that could be derived from the use of ANWR oil. Unfortunately, without the price communication that is possible only by means of private property and voluntary exchange, we cannot be sure what those costs are or how private owners would evaluate either the costs or the benefits of preventing drilling in ANWR. However, the willingness of environmental groups such as the Audubon Society and the Nature Conservancy to allow drilling for oil an environmentally sensitive land they own suggests strongly that their adamant verbal opposition to drilling in ANWR is a poor reflection of what they would do if they owned even a small fraction of the ANWR territory containing oil.

Katherine Balpataky

Protectors of the Herd

When Norma Kassi was a young girl, she spent hours sitting by a small lake near Old Crow Flats, 73 kilometres north of the Arctic Circle, listening to her grandfather recount stories of her people and the wildlife that inhabited the lands around them. She remembers watching in awe as flocks of blackpoll warblers, arctic terns, and other migrants arrived each spring to mate on the Yukon's most valuable wetlands. It was this place and through her grandfather's teachings that her appreciation for the natural world took form. "I was taught the science of our environment—from the smallest of insects and plankton in the waters to the biggest of mammals and predators—and how they govern us," says Kassi.

She also learned about the integral connection between her people and the Porcupine caribou—a herd of 130,000 named for its semi-annual migration to the coastal plains of the Arctic National Wildlife Refuge (ANWR) during which it crosses the Porcupine River. For thousands of generations, the Vuntut Gwich'in peoples of northeastern Alaska and northwestern Canada have depended on the Porcupine herd. Their traditional knowledge states that every *vadzaih* (caribou) has a bit of *ezi* (human heart) in it and every human has a bit of vadzaih heart. Thus the life of the caribou and the Gwich'in Nation are historically intertwined.

Yet Kassi's grandfather understood that one day their world would change. "[He] told me many times, 'You have light skin, and one day you will be speaking to many white people. You will speak their language,'" she says. "And I guess that has been my destiny so far."

❦

Since the discovery of oil at Prudhoe Bay in 1967, ANWR has been at the apex of an ongoing battle that pits oil corporations against native peoples and conservationists. Those in favour of opening up the coastal plains for development state that it will create more than 170,000 jobs and that there are 16 billion barrels of oil waiting to be tapped. The oil—deep beneath the surface of ANWR—has tempted every new American president and member of congress. Time and time again new bills and budget acts with provisions for drilling have been put forward.

Many cost-benefit studies have shown, however, that the resources required to extract the oil are too great to make drilling worthwhile. Other studies, such as a report released in March by the National Research Council, show that the effects of oil development on the North Slope—an area that creeps into the refuge—have been accumulating for more than three decades. The report states that migration patterns of bowheads, reproductive success of birds, and geographical distribution of the Porcupine caribou herd have all been affected.

To Kassi, the report is a warning that her family, her culture, and everything she has ever known is in jeopardy. "What is sacred about these lands is that we live near the calving grounds. During times of famine, we travel those areas, but we never step into the sacred calving grounds," says Kassi. If they were to open up the Arctic National Wildlife Refuge, it would be [to us] like going into a hospital nursery and tearing that apart. We will do everything in our power to stop that, because it means our life."

Kassi has spent the past 10 years working with groups such as the World Wildlife Fund, the Alaska Wilderness League, and the Caribou Commons Project (CCP) to raise awareness about the importance of the refuge. The CCP produces multimedia presentations that combine Gwich'in stories handed down from generation to generation with footage by photographer Ken Madsen and sounds of the refuge set to music by Matthew Lien in order to educate the public about the incredible biodiversity of the arctic plains and the slick politics that threaten it. "The music, Norma speaking, and Ken's photos really give you an emotional attachment—and that's what's missing from this debate," says Peter Mather, Arctic Coordinator for CCP. "The oil industry and the government like to reduce the issue to a simple black-and-white argument. But this place—once you see it or feel it, understand it—it goes beyond facts," he says.

As we sit in a small coffee shop in Sainte-Anne de Bellevue, on Montreal's west island, I sense that Kassi is tired. After a full day at McGill University on a panel discussing problems related to indigenous health, the environment, and the lack of resources and trained workers in remote reserves, her fatigue is understandable. But when she explains how her work frequently takes her away from her 82-year-old mother and her four sons—'notably' her youngest, Yudii, who is only 11—I realize it is the constant battles that have worn her down. Her work—all unpaid—requires much travel. Moreover, with each victory comes another critical fork in the road. Each piece of legislation is another potential crisis requiring the fast mobilization of voters to lobby government in order to sway the vote.

Listening to Kassi tell me about the land on which she lives, I am captivated by her manner of storytelling. Lowering her head, she seems to leave the room, as if she is drawing on some greater flow of knowledge that circles far beyond the hustle and bustle of this modern café.

"She's very spiritual about it," says Lien. "Kassi goes somewhere and speaks from a place—you can tell there is a power there, she is tapping into her ancestors and her culture. It's more than just one person speaking."

Although CCP was the brainchild of Lien and Madsen, Kassi brings a spiritual component to the focus and the strength of their work. Since 1992, the two men have been producing visual concerts inspired by their expeditions into the wilderness. And in 1997, they decided to make ANWR the sole focus of their work. While they were busy recruiting sound and light technicians, publicists, musicians, and other collaborators, the Gwich'in Nation was also making plans to address the situation.

For the first time in 100 years, the chiefs called a meeting and more than 500 Gwich'in people from 88 villages attended. The elders elected six representatives whose life mission would be to share their traditional knowledge with the rest of the world and to make it known that an intrusion into the caribou's birthing grounds was also a human-rights issue. Kassi was one of the six. As partial fulfillment of her commitment, she became an active member of any non-governmental organization that would support her, including CCP. "Norma represents a culture that is the refuge. There is no separating the two. Their spirituality, identity, and sense of past, present, and future are tied to the caribou," says Lien. "Norma is able to put that into words."

The CCP set its show in motion, flying to cities such as Victoria, Vancouver, Winnipeg, Saskatoon, and Washington. The trips were intended to cover as much ground as possible, especially to areas where government representatives were in favour of drilling.

Then in the spring of 2002, Madsen and Lien came up with the idea of a trek to Washington, D.C. The original inspiration was to start a *Forrest Gump* phenomenon—to walk from ANWR to Washington, collecting followers along the way. But they realized that a large leg of the trip would be through areas inaccessible to media. Instead, Madsen, Lien, and U.S. Project Coordinator Tim Leach opted to start three separate cycling journeys—spanning a total of more than 12,800 kilometres—from Sarasota, Seattle, and Kansas City to Washington. The trek was modified to expedite travel through small communities, enabling the trio to present slide shows along the way. They arrived in Washington in mid-November, just after the U.S. Senate elections.

The bicycle trek garnered a lot of media attention and captured the interest of thousands of people who otherwise might not have attended the slide presentations. At the same time, Kassi, along with two other Gwich'in activists, won the prestigious Goldman award—the world's largest prize program honouring grassroots environmentalists—adding further momentum to the tour.

Despite all these efforts, many oil-development supporters in Washington have remained firm in their convictions. The greatest slap in the face came after Gale Norton, Secretary of the Interior, visited ANWR and heard presentations from schoolchildren who told her why this land is special to them. On Norton's return, she told the American public that ANWR is a barren wasteland. Moreover, President [George W.] Bush on one occasion encouraged people to visit ANWR to see it for themselves—presumably so that visitors would determine that it's not worth all the fuss. Ironically, comments such as those of Norton and Bush have helped conservation efforts by bringing them into the spotlight.

"You have people like Gale Norton, Secretary of the Interior, saying that it's just a great white nothingness. It's an outright lie," says Lien. "When the caribou are coming in over the mountain passes to congregate, you're witnessing this peculiar miracle—like the hand of a mysterious, invisible clock that sweeps across the land. At the same time, the tundra is finally losing the last of its snow coating, and within a couple of days, this explosion of wildflowers takes off across the tundra and the mountainscapes that have never been glaciated ... Then, in 10 days or so, 40,000 calves are born, and all of a sudden this place is just teeming with all this life."

Lien channels his fire by using samples of Bush's see-it-for-yourself speech in songs about the refuge. The group continues to create new material, because previous experience has taught them that victories for the refuge are usually short-lived.

Last March [2003], for example, the United States Senate voted 52 (including 8 Republicans) to 48 to keep the refuge closed to energy exploration. But less than six months later, the Bush administration told Congress that opening up ANWR for drilling would be back on the agenda in its new energy bill. The news sparked criticism from all sides—Gwich'in people, conservation organizations, and Senate Democrats, who promised to filibuster any legislation that would give oil companies access to the Alaskan wilderness. Then, almost as quickly as the ANWR media storm began, it vanished when in late October, the House and Senate Republicans failed to reach a compromise on energy tax credits and incentives, effectively shelving the energy bill for the time being.

So where does that leave ANWR? "[Clearly] for us to win, we need to get 60 votes within the Senate to protect this area," says Peter Mather. "I do see an end to this issue, but I don't know which way it's going to go. This issue has been coming up for 15 years." And with an issue like ANWR, you need lose only one battle and the war is lost. However, such people as Kassi, Mather, Madsen, and Lien will not let that happen quietly. Like the bulls that lead the Porcupine caribou to the sacred ground of regeneration, they will continue marching along the only path they know.

POSTSCRIPT

Should the Arctic National Wildlife Refuge Be Opened to Oil Drilling?

Those who see in nature only values that can be expressed in human terms are well represented by Jonah Goldberg, who, in "Ugh, Wilderness! The Horror of 'ANWR,' the American Elite's Favorite Hellhole," *National Review* (August 6, 2001), describes the ANWR as so bleak and desolate that development can only improve it. On the other hand, Adam Kolton, testifying before the House Committee on Resources on July 11, 2001, in opposition to the National Energy Security Act of 2001 (NESA), presented the coastal plain as "the site of one of our continent's most awe-inspiring wildlife spectacles" and, thus, deserving of protection from exploitation. Kennan Ward, in *The Last Wilderness: Arctic National Wildlife Refuge* (Wildlight Press, 2001), describes a realm where human impact is still minimal and wilderness endures. John G. Mitchell, in "Oil Field or Sanctuary?" *National Geographic* (August 2001), is more balanced in his appraisal but sides with Amory B. Lovins and L. Hunter Lovins, "Fool's Gold in Alaska" *Foreign Affairs*, (July/August 2001), concluding that better alternatives to developing the ANWR exist.

The House of Representatives approved the NESA in August 2001. The bill then stalled in the Senate, with pro-drilling senators attempting to woo votes with such measures as promising to use oil revenues to pay pension benefits for steelworkers. Their efforts failed in April 2002, when the bill was defeated and a competing energy bill took the lead. This alternative bill, introduced in December 2001 and sponsored by Senate Majority Leader Tom Daschle (Dsouth Dakota) and Senator Jeff Bingaman (D-New Mexico), does not allow for oil exploration in the ANWR.

In "ANWR Oil: An Alternative to War Over Oil," *American Enterprise* (June 2002), Walter J. Hickle, former U.S. secretary of the interior and twice the governor of Alaska, writes, "[T]he issue is not going to go away. Given our continuing precarious dependence on overseas oil suppliers ranging from Saddam Hussein to the Saudis to Venezuela's Castro-clone Hugo Chavez, sensible Americans will continue to press Congress in the months and years ahead to unlock America's great Arctic energy storehouse."

Similar debate has centered on mineral exploitation in the American Southwest. President Clinton created the Grand Staircase–Escalante National Monument by executive order to protect an important part of Utah's remaining wilderness, but opposition remains. See T. H. Watkins, *The Redrock Chronicles: Saving Wild Utah* (Johns Hopkins University Press, 2000). For a survey of the wilderness system created by the 1964 Wilderness Act, see John G. Mitchell and Peter Essick, "Wilderness: America's Land Apart," *National Geographic* (November 1998). Indeed, the Bush administration has continued to propose opening the ANWR to oil drilling, most recently in the 2005 budget proposal put before Congress early in 2004. In addition, in January 2004, Interior Secretary Gale Norton approved a plan to open a large portion of Alaska's North Slope, just west of the ANWR, to oil drilling. Given the rapid rise in gasoline prices in spring 2004, these plans may prevail.

ISSUE 9

Should Society Act Now to Forestall Global Warming?

YES: George Marshall and Mark Lynas, from "Why We Don't Give a Damn," *New Statesman* (December 2003)

NO: Stephen Goode, from "Singer Cool on Global Warming," *Insight on the News* (April 27, 2004)

ISSUE SUMMARY

YES: George Marshall and Mark Lynas argue that despite a remarkable level of agreement that the threat of global warming is real, human psychology keeps us "in denial." But survival demands that we escape denial and seek more positive action.

NO: Long-time anti-global warming spokesman Fred Singer argues in an interview by Stephen Goode that global warming just is not happening in any significant way and if it were, it would—judging from the past—be good for humanity.

Scientists have known for more than a century that carbon dioxide and other "greenhouse gases" (including water vapor, methane, and chlorofluorocarbons) help prevent heat from escaping the earth's atmosphere. In fact, it is this "greenhouse effect" that keeps the earth warm enough to support life. Yet there can be too much of a good thing. Ever since the dawn of the industrial age, humans have been burning vast quantities of fossil fuels, releasing the carbon they contain as carbon dioxide. Because of this, some estimate that by the year 2050, the amount of carbon dioxide in the air will be double what it was in 1850. By 1982 an increase was apparent. Less than a decade later, many researchers were saying that the climate had already begun to warm. Now there is a strong consensus that the global climate is warming and will continue to warm. There is less agreement on just how much it will warm or what the impact of the warming will be on human (and other) life. See Spencer R. Weart, "The Discovery of the Risk of Global Warming," *Physics Today* (January 1997).

The debate has been heated. The June 1992 issue of *The Bulletin of the Atomic Scientists* carries two articles on the possible consequences of the greenhouse effect. In "Global Warming: The Worst Case," Jeremy Leggett says that although there are enormous uncertainties, a warmer climate will release more carbon dioxide, which will warm the climate even further. As a result, soil will grow drier, forest fires will occur more frequently, plant pests will thrive, and methane trapped in the world's seabeds will be released and will increase global warming much further—in effect, there will be a "runaway greenhouse effect." Leggett also hints at the possibility that polar ice caps will melt and raise sea levels by hundreds of feet.

Taking the opposing view, in "Warming Theories Need Warning Label," S. Fred Singer emphasizes the uncertainties in the projections of global warming and their dependence on the accuracy of the computer models that generate them, and he argues that improvements in the models have consistently shrunk the size of the predicted change. There will be no catastrophe, he argues, and money spent to ward off the climate warming would be better spent on "so many pressing—and real—problems in need of resources."

Global warming, says the UN Environment Programme, will do some $300 billion in damage each year to the world economy by 2050. In March 2001 President George W. Bush announced that the United States would not take steps to reduce greenhouse emissions—called for by the international treaty negotiated in 1997 in Kyoto, Japan—because such reductions would harm the American economy (the U.S. Senate has not ratified the Kyoto treaty). Since the Intergovernmental Panel on Climate Change (IPCC) had just released its third report saying that past forecasts were, in essence, too conservative, Bush's stance provoked immense outcry.

According to the IPCC (see *Climate Change 2001* [IPCC, 2001], available at http://www.ipcc.ch/), climate warming is already apparent and will get worse than previous forecasts had suggested. Sea level will rise, ice cover will shrink, rainfall patterns will change, and human activities—particularly emissions of carbon dioxide—are to blame. Writers, such as Stephen H. Schneider and Kristin Kuntz-Duriseti ("Facing Global Warming," *The World & I* [June 2001]), pull no punches: "Nearly all knowledgeable scientists agree that some global warming is inevitable, that major warming is quite possible, and that for the bulk of humanity the net effects are more likely to be negative than positive. This will hold true particularly if global warming is allowed to increase beyond a few degrees, which is likely to occur by the middle of this century if no policies are undertaken to mitigate emissions."

In the following selections, environmentalist writers George Marshall and Mark Lynas argue that despite a remarkable level of agreement that the threat of global warming is real, the human tendency to respond first to short-term, simple threats prevents us from acting. Unfortunately, the threat is so great that without action, human survival may be at stake. Professor Fred Singer argues that although humanity is producing greenhouse gases, there is little evidence that global warming is happening in any significant way and if it were, it would—judging from the past—be good for humanity. Those who promote the "global warming scare" are environmentalist ideologues.

**George Marshall and
Mark Lynas**

 **YES**

Why We Don't Give a Damn

With [the] year's United Nations climate jamboree about to get under way in Milan, it's the season for politicians from around the world to express their heartfelt concerns about global warming. Every scientific institution and national government in the world now endorses the conclusions of the UN's Intergovernmental Panel on Climate Change (IPCC) that global warming is a major threat to the planet's future. Few international issues generate so much agreement.

Yet with the Kyoto Protocol still in limbo thanks to US and Russian intransigence, the conference is taking place in a political no man's land. The international process that began in 1992 at the first Earth Summit has yet to bear significant fruit. Despite plentiful proposals for windfarms, solar panels and hydrogen cells—enough to fill many glossy brochures—the grim reality is that the use of fossil fuels increases relentlessly, and with it the atmospheric concentration of greenhouse gases. So why are we proving so utterly incapable of facing up to the challenge?

First, let us remind ourselves of the magnitude of the threat. Global warming is already well under way: even if all greenhouse gas emissions stopped tomorrow, we would see a rise in planetary temperatures of 1.1°C, twice the warming experienced over the past century, and enough to wipe out most of the world's tropical coral reefs as well as a good proportion of mountain glaciers. Bad as that is, it is still an unrealistically optimistic scenario. It is projected that greenhouse gas emissions will go on rising for decades; the IPCC predicts a global temperature rise of between 1.4° and 5.8° by 2100. At the lower end of this scale, large areas of agriculturally productive land will be destroyed; entire countries will disappear through rapid sea-level rise; and entire regions in the arid subtropics will become uninhabitable.

The financial impact of this, according to Munich Re, the world's largest reinsurer, will run at more than $300 [billion] a year by 2050, while the IPCC estimates that the cost to Europe of climate change at the "moderate" end of its predictions will be $280 [billion] a year.

Some free-market sceptics argue that such costs can be regarded as a containable tax on economic growth. But while rich countries benefit from the growth, the "tax" falls most heavily on the poorest peoples. And according to Munich Re, the cost of climate change is growing two to three times faster than the global economy that pays for it.

From *The New Statesman*, December 1, 2003, pp. 18–20. Reproduced with permission from New Statesman, Ltd.

Greater risks lurk at the upper ends of the IPCC predictions. A global warming episode 250 million years ago wiped out 95 per cent of all species. It took a rise in average global temperatures of only 6° to trigger this catastrophe, which palaeontologists call "the post-apocalyptic greenhouse". The IPCC's current worst-case scenario is 5.8°. One can scarcely imagine a more sombre warning.

The implication is clear: if we do not take immediate action to slash greenhouse gas emissions, we will in effect condemn our children—and all generations that follow—to a permanently impoverished and more threatening world dominated by extreme weather and ecological collapse.

Yet as if in a parallel universe, plans continue to be made for business as usual, with rapid economic growth projected to continue unabated, still largely driven by fossil-fuel energy: oil consumption will increase by 50 per cent over the next two decades. Some calculations show emissions of countries from the south alone breaking through the safe "corridor" (within which we could avoid major climate impacts) in as little as a decade.

These dangerous trends continue almost unchallenged. Why? Because we appear to be experiencing a disastrous form of collective denial, more typically found among societies suffering major institutional human rights abuses—such as apartheid South Africa or Nazi Germany—where individuals may understand the reality of the problems, but refuse to accept the implications. In his book *States of Denial*, the sociologist Stanley Cohen terms this condition "implicatory denial" and identifies it as a natural defence that humans tend to adopt when faced with a morally unthinkable situation. It has resulted in, to borrow another term from psychology, "cognitive dissonance" among opinion-formers and the public. Nearly everyone professes to care about global warming while simultaneously continuing with set patterns of behaviour that make the problem worse.

[British Prime Minister] Tony Blair illustrates this well. In Johannesburg [2002], he told delegates to the second Earth Summit: "We know that if climate change is not stopped, all parts of the world will suffer. Some will even be destroyed. It remains unquestionably the most urgent environmental challenge." At the same time, his government does nothing to reverse the growth in road traffic, plans an expansion of airports and promotes development of oil supplies overseas. Moreover, Blair has just helped to deliver the second-largest reserves of oil on the planet into the hands of the most dangerous climate denier of all, the US. Sir John Houghton, an eminent climate scientist, expressed it thus in the *Guardian* recently: "I have no hesitation in describing [climate change] as a 'weapon of mass destruction'."

In showing such a profound disconnection between what he says and what he does, Blair is not demonstrating insanity. His position is all too human. Asked in opinion polls, 85 per cent of the British public say they are concerned about climate change. Yet domestic energy consumption still rises by 2 per cent per year, cars get bigger, and people boast of their holidays to ever-more-distant resorts. Blair, like the rest of us, is in denial.

Even progressive movements and groups have shown only patchy concern. Unions and the socialist left as a whole are suspicious of measures that might affect employment and growth. In the US, unions joined the Christian right in opposing the Kyoto Protocol, while in the UK, development and aid organisations have maintained a baffling silence in the face of a threat that will wipe out most, if not all, of the benefits of their work. Among the major groups, only Christian Aid has called openly for stronger political action on climate change.

Just as oddly, those who devote their lives to studying the future manage to miss what is in front of their noses. In *Our Final Century*, a book that examines worrying scenarios for the coming hundred years, Martin Rees, the Cambridge cosmologist, absent-mindedly devotes a mere five and a half pages to climate change, the rest to bio-warfare, genetics and rampaging nanobots. Colin Tudge, in his excellent treatise on global agriculture, concludes his three pages on climate by metaphorically throwing up his hands and hoping for the best. Acknowledging that its effects could be "devastating", he labels global warming "the joker in the pack". But it is not the joker, it's the trump card that could alone negate the rest of his prescriptions for sustainable agriculture.

We have come to dominate the planet through our exceptional ability to anticipate, plan and adapt. Despite an innate selfishness, we have time and again been goaded into action by appeals to our sense of nationhood, responsibility to our children, or our ideas about historical destiny. People willingly lay down their lives to defend cultural identities and religious beliefs. Nor, once a threat is perceived, are we resistant to paying a heavy financial price. Every year, trillions of dollars are spent worldwide on weapons to defend nations against threats that cannot be quantified and are often extremely remote. Even the Y2K computer panic mobilised a $320 [billion] investment in compliance, and persuaded people to stockpile food and flee the cities.

Why, then, are we paralysed in the face of the climate crisis? The answer lies in our evolutionary heritage: we defend ourselves against specific predators and rival tribes of humans. We are "hard-wired" to mobilise rapidly in response to clear and immediate dangers. But as threats become less certain, or causally complex, it becomes harder to find the urgency to tackle them.

Climate change, unfortunately, matches our evolutionary weaknesses. Not only is it complex, ambiguous and inter-generational, but it is largely self-inflicted. This neutralises our natural tendency to identify as threats rival social groups—whether they be asylum-seekers or rival foreign empires. Clearly, there are degrees of responsibility—the British produce 50 times the quantity of emissions of Bangladeshis, for example. Yet it is impossible to establish direct linkage between one person's sports utility vehicle and another's crop failure. It is hard to blame someone else for a problem we are all causing, hence the almost universal efforts to make global warming fit familiar perpetrator-victim polarities. The south blames the north, cyclists blame drivers, activists blame oil companies, and almost everyone blames George Bush. It's tough to admit that Bush is a victim, too—his children and grandchildren will grow up in the same unstable and devastated world.

WHO'S WHO AMONG THE CLIMATE-CHANGE DENIERS

Bjørn Lomborg, a statistician from Denmark, came to media prominence in 2001 with the launch of his book *The Skeptical Environmentalist*. He appears convincing by aggregating voluminous references without subjecting himself to the rigours of the scientific process. He accepts that climate change is happening, but applies a crude and selective cost-benefit analysis to argue that the cheapest option is to maintain economic growth and adapt to the impacts. He was the guest of honour and award-winner this year at a dinner of the Competitive Enterprise Institute, a far-right US think-tank to which ExxonMobil has donated $1m since 1998.

Richard Lindzen, professor of meteorology at the Massachusetts Institute of Technology, is the only sceptic with credentials in the relevant area of climate science. His work focuses on atmospheric water vapour, which he claims will act through cloud formation to prevent excessive global warming. There is little evidence to support this hypothesis, which has gained no support from the wider scientific community. He has been a paid consultant to oil and coal interests in the US, and has compared the environmental movement to the Nazis.

Willie Soon and *Sallie Baliunas*, astronomers at the Harvard-Smithsonian Centre for Astrophysics, co-wrote a paper this year challenging the accepted scientific wisdom that the planet is now hotter than it has been for at least a thousand years. The White House and Republican senators loved the message, which supports their denials about human-induced climate change. It transpired that the paper was partly funded by the American Petroleum Institute, and that Soon and Baliunas are scientific advisers to the Marshall Institute, another far-right US think-tank. Three editors at *Climate Research*, which published the paper, resigned when prevented from printing a repudiation.

Philip Stott is Britain's leading climate-change denier and has built a career on criticising environmentalists. Professor emeritus of biogeography at the University of London, he has no climate-science qualifications. A skilled communicator who has written for the *Times* and *New Scientist*, he describes global warming as a "lie". On an advisory board of the Scientific Alliance, an anti-environmentalist campaign group that denies climate change; opposes organic agriculture and promotes genetically modified foods and nuclear power.

Julian Morris, director of the International Policy Network, is also research fellow at the Institute of Economic Affairs, for which he co-wrote a report called *Global Warming: apocalypse or hot air?* He is often in the media, undermining the case for Kyoto. The policy network's "partners" around the world include Tech Central Station (funded by ExxonMobil, General Motors and McDonald's) and the Cambridge-based European Science and Environment Forum, an anti-environmentalist group originally set up for the Philip Morris tobacco company by a PR firm. Philip Morris often accuses environmentalists of inventing the global warming "myth" in order to generate cash.

⋄⊚⋄

The complex causality of climate change also plays particularly strongly to the natural human tendency to diffuse responsibility. This is the "passive bystander effect", after the frequently observed phenomenon that violent crimes can be committed in a crowded street without anyone intervening.

This is not a moral failure; it is simply that everyone is waiting for someone else to act first; the more people there are on the scene, the less individual responsibility we feel. In the case of climate change, we are all simultaneously bystanders, perpetrators and victims. These internal conflicts cripple our ability to act, and are only amplified by the vast denial of others. We doubt the reliability of our own instincts, and our power to make any difference.

More profoundly, we simply find it impossible to imagine the globally warmed future. Again, there are good reasons: throughout history, humans have looked to the past to guide future behaviour. From the wisdom of social elders to the courts, we seek precedents. But there is no historical parallel for what is happening. This is the very essence of our denial: while we accept the evidence for climate change intellectually, we reject it emotionally. We find ourselves unable to believe it really, truly exists.

So what options do we have? One vision of the future sees little more than a nightmare of ecological despoliation, mass starvation and perpetual war. "The mass of mankind," writes John Gray in *Straw Dogs*, "is ruled not by its intermittent moral sensations, still less by self-interest, but by the needs of the moment. It seems fated to wreck the balance of life on earth—and thereby to be the agent of its own destruction." If Gray is right, then people will delay taking action until the effects of climate change are severe. Even then, our strongest impulse may be to adapt—tackling droughts with dams, floods with dykes and hurricanes with storm shelters. A fuller response may be triggered only if climate change is converted into a more common struggle between competing "tribes", such as direct conflicts over emissions or, more likely, wars over diminishing environmental resources.

But humans can change behaviour in anticipation of rewards or punishments. The world's religions are founded on this principle. We could transform our lifestyles, but only if we recognise and confront the psychological barriers to major behavioural change. A big shift in world-view is essential, and time is running short.

The social herd instinct may yet be our salvation. Malcolm Gladwell of the *New Yorker* argues in his book *The Tipping Point* that all it takes for an idea to "tip" from the margin to the mainstream is a certain alignment of social factors. The passive bystander effect stops operating as soon as sufficient people break ranks and become involved. It may become "normal" to eschew cars, to shop locally and to consume renewable energy only. This outcome feels remote, but it is up to all of us to escape denial and despair, and seek something more positive. Ultimately, this something is not wealth or power, or even moral purpose: it is survival.

Stephen Goode

Singer Cool on Global Warming

INSIGHT: When did you first get interested in the question of global warming as an example of bad science?

Fred Singer: My interest in the global-warming scare began about 1988 with the testimony of Jim Hansen (then head of NASA's Goddard Institute for Space Studies) before Sen. Al Gore in a Senate hearing. I looked at his testimony and discovered some holes in it. I published a piece in the *Wall Street Journal* pointing out the weak points in the argument.

Q: What are some of the weak points about the global-warming argument?

A: The fact that they don't properly take into account the effects of clouds in the atmosphere. Clouds will cool the climate rather than warm the climate. When you try to warm the ocean, I argued—and the argument is still sound—you evaporate more water and create more clouds and this reduces the amount of solar radiation. What you have is a kind of negative feedback which keeps the temperature from rising very much.

Q: Why is the disagreement so wide between those who see global warming happening right now and those who don't? What is a nonscientist to make of such a disagreement?

A: Let me explain the origin of this scientific disagreement. There are two kinds of scientists. Let's assume for the moment that both of them are honest. In the first group there are quite a few who argue as follows:

They say "Carbon dioxide in the atmosphere is increasing." It is. Second, they say, "Carbon dioxide is a greenhouse gas." It is. They then say, "Because carbon dioxide is on the increase and it is a greenhouse gas, therefore the climate must be warming. The [mathematical] models support this assumption," they say, "and the models show the climate is warming; therefore evidence that goes contrary to this we will ignore. We will only look at supporting evidence."

That's how they are. The other group, of which I am one, says, "This is all true, but as far as we can tell, the climate is not warming as it should be if the greenhouse theory is correct. In fact, the warming is a great deal less than what the models predict. Therefore, something is wrong with the models."

I belong to the latter school, as I say, and what we do is analyze the data. Just now we have a new result. It's been known for a long time that the weather satellites do not show any warming, but the first group tends to neglect this information. They argue that the weather satellites have only been around for 25 years and that's too short a time to tell. It's a specious argument. Or they say there's something wrong with the weather satellites, though they haven't been able to show that there's anything at all wrong with them.

So now we find that not only the weather satellites but also weather balloons, which measure temperature in a completely different way than the satellites, give the same results as the satellites.

Q: The data collected by weather balloons also say there is no global warming?

A: Yes. So now we have a situation in which most of the evidence is showing there is essentially no warming. The first group of scientists is aware of this information, but they tend to ignore it. They say, "Something's wrong with it because it doesn't support our hypothesis, so we will push it aside."

The second group of scientists, of which I am one, says, "There must be something wrong with the first group's models because they don't agree with what we observe and measure." So what you have is one group of people who believe in models or theory and the other group who believe in what they are measuring in the atmosphere! That's the major science issue in a nutshell.

Q: These two groups of scientists also have vast differences when it comes to policies that should be developed to deal with the increase of carbon dioxide in the atmosphere, don't they?

A: Well, yes. As far as policy goes, the first group of scientists says, "Even if we don't see any warming, nonetheless, assuming the theory is right, there should be a warming given the increase in carbon dioxide. And we had better do something about it!" It's called the precautionary principle. As the culture puts it, "Better safe than sorry."

But the first group of scientists does not ask, "How much does it cost to be safe?" They don't ask—and this is very important—"What does safety mean?"

Put another way, when you buy an insurance policy you look at the cost of the premium and you look at the risk. You don't buy insurance policies against being hit by a meteorite. The risk is very small.

Q: Won't one of the arguments the first group of scientists put forth be that we should slow our use of energy, conserve it, and in the process save the environment?

A: If the policy were cost-free, I would say, "Sure, why not?" So, for example, if people say, "Well, we should conserve energy," I would say, "Yes, of course. It's cost-free and conservation not only saves you energy, it even saves you money, and for that reason you should be doing it irrespective of a warming."

But I would add, "When you say, 'We have to do away with fossil fuels and use wind energy exclusively or solar energy,' well ... I would then say, 'That's very expensive and it doesn't even work very well.'" So there is a basic policy difference between the two groups of scientists. The first group believes in the precautionary principle. And the second group, to use another slogan from the culture, believes, "Look before you leap!"

Q: "Look before you leap" means let's not adopt large government programs to deal with a problem that the evidence says isn't taking place but which theory and mathematical models say must take place?

A: If we don't see anything happening despite the fact that carbon dioxide is increasing, then maybe something else is happening and the effect of the increase will be minimal. I won't say an effect won't be there, but that maybe it is minimal—or not even enough to be detectable. If it's not detectable, it means it probably can't do you any harm.

There's an additional argument, which is this: Supposing it did warm up, is that good or bad? You cannot automatically assume it is bad, because we've had warming in the past and coolings. Climate is always changing. Every time the climate has been warm, it's been good for mankind, and every time it has been cold it has been bad.

Q: How is a nonscientist to deal with these questions? How can a layperson look at the science and decide for himself or herself which side to be convinced by?

A: I think that the overall way of handling it is to look at the indices of human well-being. One is longevity. If people are now living longer and healthier lives than they used to—and this is certainly true—then things must be improving. So you have to conclude that air pollution, climate change, radiation, chemicals and whatever else you want to think about within the environment are not doing us in to a greater degree than before.

That's one way of looking at it. The other, more detailed [way] is to look at the individual items that are being held up as dangerous. Again, for example, air pollution. Air pollution assuredly can be unhealthy. In present-day China it is horrible, truly awful. But according to the EPA [Environmental Protection Agency], air pollution virtually has disappeared from the United States. Today we have fewer particulates, less sulphur, fewer ozone events and so on. The air is cleaner and better, according to the EPA. I don't question that. It's EPA's data, and, when you think about it, it would be in EPA's interest to show that this is not so. It would be in the EPA's interest to show that air pollution is a serious problem and maybe even getting worse. But in fact, the outdoor air has become so clean that probably the greater health hazard is indoor air. Most of us spend 80 percent or so of our lives indoors, so in a sense outdoor air pollution is almost irrelevant.

Q: Do we politicize science now more than we used to?

A: I think yes. I remember when Earth Day first was proclaimed in 1970; that's when the heavy politicizing started.

Q: What's your impression of science education in this country?

A: It goes up and down. It peaked after *Sputnik* in science and engineering, and it's been slowly going down. We're lagging behind, as I read it, many other countries. We're well down in the middle, lagging behind India and Japan.

Q: Does good science education help make people immune to being convinced by bad science, and isn't solid science training essential?

A: That's true. In fact, when I speak out about climate change and global warming, the greatest amount of support I get is from people who know something about the subject. They don't have to be specialists, but they have to be able to read and absorb data when I show them a graph—to understand what it means.

Q: What about the Bush administration's space program? Should we be getting back to, and deeper into, space exploration?

A: Should we be spending money at all on science? On astronomy and other scientific fields that have no practical payoff in the short term? Black holes are interesting. Discovering new planets is interesting. But where's the practical payoff for those from whom the money is taken to pay for such programs?

Even so, let us assume that space exploration is important. Then the question is, how best to do it. I have always pointed out that some things are more important than others, which means some things are of less importance.

Among the things that are less important is putting a base on the moon. I don't see any good reason to put a permanent base on the moon. It's not just the expense involved, but the fact that a moon base would delay or make impossible other things we should be doing.

Supposing you get a half-dozen people to sit in an enclosure on the moon, so what? To me, a base on the moon is just another space station, and we've already proved that people can survive in space. We've known that for a long time, so we're not learning anything new.

Q: What could we be doing that would be more beneficial to science?

A: We should be going to Mars. Not with a base, but a short exploratory visit. Not to the surface of Mars, because that's difficult and costly and would take forever. But to Demos, a moon of Mars, and from that moon conduct an unmanned exploration of the planet.

Q: What do you think of the Bush administration's attitude toward science in general?

A: The administration is conducting continually a climate-research program to the tune of about $2 billion a year. If I were doing it, I would spend a lot less and try to focus on what the really important issues are. But it's turned out to be a great support project for scientists, not only for physical scientists but also for the social scientists who study the social, philosophical and theological implications of climate change. Everyone is getting in on this because they can get money from the program.

Q: Any other problems with the administration when it comes to science?

A: The Bush administration has quite properly said we're not going to go along with the Kyoto Protocol. They're not going to do all those crazy things demanded by the protocol, such as rationing energy and making energy even more expensive and causing ourselves economic harm. But, on the other hand, the administration is acting like this is a real problem, as though the problems the protocol was supposed to address are real. So they have a great big research program on hydrogen cars and so on, or sequestering carbon dioxide.

It makes no sense. It tells people, "This is a problem after all." Why would you want to sequester carbon dioxide? To do so implies carbon dioxide is bad—when it's not bad, it's good. We should have more carbon dioxide in the atmosphere. It's good for plants. It makes them grow faster.

Q: What are your views on energy?

A: The best we have now are coal, oil, and gas—and these will be with us a long time, long enough until they become too expensive, meaning scarce. But we have other sources of energy. We have nuclear energy, for example, nuclear energy which works. One of the real curious things about this whole debate is that the people who are concerned about global climate change are also the people who are opposed to advancing nuclear energy. The very same people.

Never mind that nuclear energy would do the job that needs to be done. It would produce energy without any carbon dioxide, so it's the obvious answer. But they don't want anything to do with it, so you see they can't be serious. It shows how ideological they are.

POSTSCRIPT

Should Society Act Now to Forestall Global Warming?

The United Nations Conference on Environment and Development in Rio de Janeiro, Brazil, took place in 1992. High on the agenda was the problem of global warming, but despite widespread concern and calls for reductions in carbon dioxide releases, the United States refused to consider rigid deadlines or set quotas. The uncertainties seemed too great, and some thought the economic costs of cutting back on carbon dioxide might be greater than the costs of letting the climate warm.

The nations that signed the UN Framework Convention on Climate Change in Rio de Janeiro in 1992 met again in Kyoto, Japan, in December 1997 to set carbon emissions limits for the industrial nations. The United States agreed to reduce its annual greenhouse gas emissions 7 percent below the 1990 level between 2008 and 2012. In November 1998 they met in Buenos Aires, Argentina, to work out practical details (see Christopher Flavin, "Last Tango in Buenos Aires," *World Watch* [November/December 1998]). Unfortunately, developing countries, where carbon emissions are growing most rapidly, face few restrictions, and political opposition in developed nations—especially in the United States—remains strong. Ross Gelbspan, in "Rx for a Planetary Fever," *American Prospect* (May 8, 2000), blames much of that opposition on "big oil and big coal [which] have relentlessly obstructed the best-faith efforts of government negotiators." Nor do some portions of the industry seem interested in acting on their own. In May 2003 Exxon Mobil rejected proposals that it address global warming and develop renewable energy. CEO Lee Raymond, who had previously denounced the Kyoto Protocol, said the company does not "make social statements at the expense of shareholder return."

The opposition remains visible despite the latest IPCC report. Critics stress uncertainties in the data and the potential economic impacts of attempting to reduce carbon dioxide emissions. See Richard A. Kerr, "Rising Global Temperature, Rising Uncertainty," *Science* (April 13, 2001). Some feel that climate change may well be less severe than expected and also beneficial overall to agriculture and human well-being. See Patrick J. Michaels and Robert C. Balling, Jr., *The Satanic Gases: Clearing the Air About Global Warming* (Cato Institute, 2000).

There is also opposition based on the view that the methods of reducing greenhouse gas emissions called for in the Kyoto treaty are, at root, unworkable. See Frank N. Laird, "Just Say No to Greenhouse Gas Emissions Targets," *Issues in Science and Technology* (Winter 2000–2001). However, researchers have proposed a number of innovative ways to keep from adding carbon dioxide to the atmosphere. See Howard Herzog, Baldue Eliasson, and Olav Kaarstad, "Capturing Greenhouse Gases," *Scientific American* (February 2000). Fred Krupp, president of Environmental Defense, in "Global Warming and the USA," *Vital Speeches of the Day* (April 15, 2003), recommends a market-based method to finding and developing innovative approaches. Thomas J. Wilbanks, et al., in "Possible Responses to Global Climate Change: Integrating Mitigation and Adaptation," *Environment* (June 2003), note that many mitigation techniques are under study around the world but that people will also have to adapt to a warming world.

In June 2002 the U.S. Environmental Protection Agency (EPA) issued its *U.S. Climate Action Report—2002* (available at http://www.epa.gov/globalwarming/publications/car/index.html) to the United Nations. In it, the EPA admits for the first time that global warming is real and that human activities are most likely to blame. President George W. Bush immediately dismissed the report as "put out by the bureaucracy" and said he still opposes the Kyoto Protocol. He insists that more research is necessary before anyone can even begin to plan a proper response, which prompted Ian Frazier, in "As the World Burns," *Mother Jones* (March/April 2003), tongue slightly in cheek, to call him "a man with a plan—about planning to plan." A plan for that additional research was announced later in 2003; see David Malakoff, "New Climate Science Plan Garners Split Opinions," *Science* (August 1, 2003). Meanwhile, the evidence for climatic effects continues to mount; see Matthew Sturm, Donald K. Perovich, and Mark C. Serreze, "Meltdown in the North," *Scientific American* (October 2003). Experts recognize the uncertainties in the data and analyses but agree that climate change and its impacts "could be quite disruptive"; see Thomas R. Karl and Kevin E. Trenberth, "Modern Global Climate Change," *Science* (December 5, 2003).

Seth Dunn, in *Reading the Weathervane: Climate Policy from Rio to Johannesburg*, Worldwatch Paper 160 (August 2002), urges swift implementation of the Kyoto Protocol as "the best way to achieve global action on climate change." On the other hand, Richard B. Stewart and Jonathan B. Wiener, in "Practical Climate Change Policy," *Issues in Science and Technology* (Winter 2004), declare, "It's time for a new, more pragmatic approach," meaning a new treaty with more emphasis on costs and benefits. James Kasting of Pennsylvania State University and James Walker of the University of Michigan warn that if one looks a little further into the future than the next century, the prospects look even more alarming. By the 2200s, the amount of carbon dioxide in the atmosphere could be 7.6 times the preindustrial level; with draconian restrictions, it could be held to a fourfold increase. Global warming may therefore turn out to be much worse in the long run than anyone is predicting now, they say. See Thomas R. Karl, Neville Nichols, and Jonathan Gregory, "The Coming Climate," *Scientific American* (May 1997). See also "Bangladesh: The Next Atlantis?" *Environment* (June 2003), which reports on recent modeling studies warning that if the IPCC projections are correct, over half of Bangladesh could be under water for most of each year by 2100.

ISSUE 10

Will Hydrogen End Our Fossil-Fuel Addiction?

YES: Jeremy Rifkin, from "Hydrogen: Empowering the People," *The Nation* (December 23, 2002)

NO: Henry Payne and Diane Katz, from "Gas and Gasbags... Or, the Open Road and Its Enemies," *National Review* (March 25, 2002)

ISSUE SUMMARY

YES: Social activist Jeremy Rifkin maintains that fossil fuels are approaching the end of their usefulness and that hydrogen fuel holds the potential not only to replace them but also to reshape society.

NO: Writer Henry Payne and director of science, environment, and technology policy at the Mackinac Center for Public Policy Diane Katz argue that hydrogen can only be made widely available if society invests heavily in nuclear power. Market mechanisms will keep fossil fuels in play for years to come.

The 1973 oil crisis heightened awareness that the world—even if it was not yet running out of oil—was extraordinarily dependent on that fossil fuel (and therefore on supplier nations) for transportation, home heating, and electricity generation. Since the supply of oil and other fossil fuels is clearly finite, some people worried that there would come a time when demand could not be satisfied—and our dependence would leave us helpless. At the same time, we became acutely aware of the many unfortunate side-effects of fossil fuels, including air pollution, strip mines, oil spills, and more.

The 1970s saw the modern environmental movement gain momentum. The first Earth Day was in 1970. Numerous government steps were taken to deal with air pollution, water pollution, and other environmental problems. In response to the oil crisis, a great deal of public money went into developing alternative energy supplies. The emphasis was on "renewable" energy, meaning conservation, wind, solar, and fuels such as hydrogen gas (which when burned with pure oxygen produces only water vapor as exhaust). However, when the crisis passed and oil supplies were once more ample (albeit it did cost more to fill a gasoline tank), most public funding for alternative-energy research and demonstration projects vanished. What work continued was at the hands of a few enthusiasts and those corporations that saw future opportunities. In 1991, Roger Billings, who had converted cars to run on hydrogen, developed the use of metal hydrides for hydrogen storage and founded corporations to develop and market hydrogen technology, self-published *The Hydrogen World View* (a new edition appeared in 2000); his dream was a future when hydrogen would be the universal fuel, as widely employed for transportation and home heating (among other uses) as oil is today. That dream was not his alone. For instance, in 2001, the WorldWatch Institute published Seth Dunn's *Hydrogen Futures: Toward a Sustainable Energy System*. In 2002, MIT Press published Peter Hoffman's *Tomorrow's Energy: Hydrogen, Fuel Cells, and the Prospects for a Cleaner Planet*. On the corporate side, fossil fuel companies have long been major investors in alternative energy systems; in just the last few years, Shell, BP/Amoco, and ChevronTexaco have invested large amounts of money in renewables, hydrogen, photovoltaics, and fuel cells.

What drives the continuing interest in hydrogen and other alternative or renewable energy systems is the continuing problems associated with fossil fuels (and the discovery of new problems such as global warming), concern about dependence and potential political instability, and the growing realization that the availability of petroleum will peak in the near future. See Colin J. Campbell, "Depletion and Denial: The Final Years of Oil," *USA Today Magazine* (November 2000), Charles C. Mann, "Getting Over Oil," *Technology Review* (January/February 2002), and Tim Appenzeller, "The End of Cheap Oil, " *National Geographic* (June 2004).

Will that interest come to anything? There are, after all, a number of other ways to meet the need. Coal can be converted into oil and gasoline (though the air pollution and global warming problems will remain). Cars can be made more efficient (and mileage efficiency is much greater than it was in the 1970s despite the popularity of SUVs). Cars can be designed to use natural gas or battery power; "hybrid" cars use combinations of gasoline and electricity, and some are already on the market. See Jennifer Hattam, "Righteous Road Trip," *Sierra* (May/June 2002).

The hydrogen enthusiasts remain. In the selections that follow, Jeremy Rifkin argues that as oil supplies decline, hydrogen can fill the gap with many fewer side effects. In addition, because hydrogen can be produced by small-scale operations, it will take energy supplies out of the hands of major corporations and favor the development of a decentralized, environmentally benign economy. Henry Payne and Diane Katz do not agree that hydrogen can be produced locally; they argue that it can only be made widely available if society invests heavily in nuclear power. They conclude that oil and other fossil fuels will remain in use for years to come, with market mechanisms ensuring steady supply.

Jeremy Rifkin

Hydrogen: Empowering the People

While the fossil-fuel era enters its sunset years, a new energy regime is being born that has the potential to remake civilization along radically new lines—hydrogen. Hydrogen is the most basic and ubiquitous element in the universe. It never runs out and produces no harmful CO_2 emissions when burned; the only byproducts are heat and pure water. That is why it's been called "the forever fuel."

Hydrogen has the potential to end the world's reliance on oil. Switching to hydrogen and creating a decentralized power grid would also be the best assurance against terrorist attacks aimed at disrupting the national power grid and energy infrastructure. Moreover, hydrogen power will dramatically reduce carbon dioxide emissions and mitigate the effects of global warming. In the long run, the hydrogen-powered economy will fundamentally change the very nature of our market, political and social institutions, just as coal and steam power did at the beginning of the Industrial Revolution.

Hydrogen must be extracted from natural sources. Today, nearly half the hydrogen produced in the world is derived from natural gas via a steam-reforming process. The natural gas reacts with steam in a catalytic converter. The process strips away the hydrogen atoms, leaving carbon dioxide as the byproduct.

There is, however, another way to produce hydrogen without using fossil fuels in the process. Renewable sources of energy—wind, photovoltaic, hydro, geothermal and biomass—can be harnessed to produce electricity. The electricity, in turn, can be used, in a process called electrolysis, to split water into hydrogen and oxygen. The hydrogen can then be stored and used, when needed, in a fuel cell to generate electricity for power, heat and light.

Why generate electricity twice, first to produce electricity for the process of electrolysis and then to produce power, heat and light by way of a fuel cell? The reason is that electricity doesn't store. So, if the sun isn't shining or the wind isn't blowing or the water isn't flowing, electricity can't be generated and economic activity grinds to a halt. Hydrogen provides a way to store renewable sources of energy and insure an ongoing and continuous supply of power.

Hydrogen-powered fuel cells are just now being introduced into the market for home, office and industrial use. The major auto makers have spent

Reprinted with permission from the December 23, 2002 issue of *The Nation*. For subscription information, call 1-800-333-8536. Portions of each week's Nation magazine can be accessed at http://www.thenation.com.

more than $2 billion developing hydrogen-powered cars, buses and trucks, and the first mass-produced vehicles are expected to be on the road in just a few years.

In a hydrogen economy the centralized, top-down flow of energy, controlled by global oil companies and utilities, would become obsolete. Instead, millions of end users would connect their fuel cells into local, regional and national hydrogen energy webs (HEWs), using the same design principles and smart technologies that made the World Wide Web possible. Automobiles with hydrogen cells would be power stations on wheels, each with a generating capacity of 20 kilowatts. Since the average car is parked most of the time, it can be plugged in, during nonuse hours, to the home, office or the main interactive electricity network. Thus, car owners could sell electricity back to the grid. If just 25 percent of all U.S. cars supplied energy to the grid, all the power plants in the country could be eliminated.

Once the HEW is set up, millions of local operators, generating electricity from fuel cells onsite, could produce more power more cheaply than can today's giant power plants. When the end users also become the producers of their energy, the only role remaining for existing electrical utilities is to become "virtual power plants" that manufacture and market fuel cells, bundle energy services and coordinate the flow of energy over the existing power grids.

To realize the promise of decentralized generation of energy, however, the energy grid will have to be redesigned. The problem with the existing power grid is that it was designed to insure a one-way flow of energy from a central source to all the end users. Before the HEW can be fully actualized, changes in the existing power grid will have to be made to facilitate both easy access to the web and a smooth flow of energy services over the web. Connecting thousands, and then millions, of fuel cells to main grids will require sophisticated dispatch and control mechanisms to route energy traffic during peak and nonpeak periods. A new technology developed by the Electric Power Research Institute called FACTS (flexible alternative current transmission system) gives transmission companies the capacity to "deliver measured quantities of power to specified areas of the grid."

Whether hydrogen becomes the people's energy depends, to a large extent, on how it is harnessed in the early stages of development. The global energy and utility companies will make every effort to control access to this new, decentralized energy network just as software, telecommunications and content companies like Microsoft and AOL Time Warner have attempted to control access to the World Wide Web. It is critical that public institutions and nonprofit organizations—local governments, cooperatives, community development corporations, credit unions and the like—become involved early on in establishing distributed-generation associations (DGAs) in every country. Again, the analogy to the World Wide Web is apt. In the new hydrogen energy era, millions of end users will generate their own "content" in the form of hydrogen and electricity. By organizing collectively to control the energy they produce—just as workers in the twentieth century organized into unions to control their labor power—end users can better dictate the terms with commercial suppliers of fuel cells for lease, purchase or other use arrangements and with virtual utility companies, which will manage the decentralized "smart" energy grids. Creating the appropriate partnership between commercial and noncommercial interests will be critical to establishing the legitimacy, effectiveness and long-term viability of the new energy regime.

I have been describing, thus far, the implementation of hydrogen power mainly in industrialized countries, but it could have an even greater impact on emerging nations. The per capita use of energy throughout the developing world is a mere one-fifteenth of the consumption enjoyed in the United States. The global average per capita energy use for all countries is only one-fifth the level of this country. Lack of access to energy, especially electricity, is a key factor in perpetuating poverty around the world. Conversely, access to energy means more economic opportunity. In South Africa, for example, for every 100 households electrified, ten to twenty new businesses are created. Making the shift to a hydrogen energy regime—using renewable resources and technologies to produce the hydrogen—and creating distributed generation energy webs that can connect communities all over the world could lift billions of people out of poverty. As the price of fuel cells and accompanying appliances continues to plummet with innovations and economies of scale, they will become far more broadly available, as was the case with transistor radios, computers and cellular phones. The goal ought to be to provide stationary fuel cells for every neighborhood and village in the developing world.

Renewable energy technologies—wind, photovoltaic, hydro, biomass, etc.—can be installed in villages, enabling them to produce their own electricity and then use it to separate hydrogen from water and store it for subsequent use in fuel cells. In rural areas, where commercial power lines have not yet been extended because they are too expensive, stand-alone fuel cells can provide energy quickly and cheaply.

After enough fuel cells have been leased or purchased, and installed, mini energy grids can connect urban neighborhoods as well as rural villages into expanding energy networks. The HEW can be built organically and spread as the distributed generation becomes more widely used. The larger hydrogen fuel cells have the additional advantage of producing pure drinking water as a byproduct, an important consideration in village communities around the world where access to clean water is often a critical concern.

Were all individuals and communities in the world to become the producers of their own energy, the result would be a dramatic shift in the configuration of power: no longer from the top down but from the bottom up. Local peoples would be less subject to the will of far-off centers of power. Communities would be able to produce many of their own goods and services and consume the fruits of their own labor locally. But, because they would also be connected via the worldwide communications and energy webs, they would be able to share their unique commercial skills, products and services with other communities around the planet. This kind of economic self-sufficiency becomes the starting point for global commercial interdependence, and is a far different economic reality from that of colonial regimes of the past, in which local peoples were made subservient to and dependent on powerful forces from the outside. By redistributing power broadly to everyone, it is possible to establish the conditions for a truly equitable sharing of the earth's bounty. This is the essence of reglobalization from the bottom up.

Two great forces have dominated human affairs over the course of the past two centuries. The American Revolution unleashed a new human aspiration to universalize the radical notion of political democracy. That force continues to gain momentum and will likely spread to the Middle East, China and every corner of the earth before the current century is half over.

A second force was unleashed on the eve of the American Revolution when James Watt patented his steam engine, inaugurating the beginning of the fossil fuel era and an industrial way of life that fundamentally changed the way we work.

The problem is that these two powerful forces have been at odds with each other from the very beginning, making for a deep contradiction in the way we live our lives. While in the political arena we covet greater participation and equal representation, our economic life has been characterized by ever greater concentration of power in ever fewer institutional hands. In large part that is because of the very nature of the fossil-fuel energy regime that we rely on to maintain an industrialized society. Unevenly distributed, difficult to extract, costly to transport, complicated to refine and multifaceted in the forms in which they are used, fossil fuels, from the very beginning, required a highly centralized command-and-control structure to finance exploration and production, and coordinate the flow of energy to end users. The highly centralized fossil-fuel infrastructure inevitably gave rise to commercial enterprises organized along similar lines. Recall that small cottage industries gave way to large-scale factory production in the late nineteenth and early twentieth centuries to take advantage of the capital-intensive costs and economies of scale that went hand in hand with steam power, and later oil and electrification. In the discussion of the emergence of industrial capitalism, little attention has been paid to the fact that the energy regime that emerged determined, to a great extent, the nature of the commercial forms that took shape.

Now, on the cusp of the hydrogen era, we have at least the "possibility" of making energy available in every community of the world—hydrogen exists everywhere on earth—empowering the whole of the human race. By creating an energy regime that is decentralized and potentially universally accessible to everyone, we establish the technological framework for creating a more participatory and sustainable economic life—one that is compatible with the principle of democratic participation in our political life. Making the commercial and political arenas seamless, however, will require a human struggle of truly epic proportions in the coming decades. What is in doubt is not the technological know-how to make it happen but, rather, the collective human will, determination and resolve to transform the great hope of hydrogen into a democratic reality.

NO

**Henry Payne and
Diane Katz**

Gas and Gasbags... Or, the Open Road and Its Enemies

Any crisis in the Middle East inevitably prompts Washington to scapegoat the automobile as a threat to national security. The dust had barely settled on lower Manhattan last fall before calls went forth—from pundits and pols across the spectrum—to relinquish our "gas-guzzlers" in the name of energy independence.

But just as the Cassandras will dominate media coverage of energy, so will Middle Eastern oil continue to fuel America's vehicles for the foreseeable future. Simple economics, geography, and consumer choice all demand it

Since Sept. 11, Washington has mobilized to end our "dangerous addiction" to foreign energy sources. Sens. John Kerry and John McCain are proposing dramatic increases in federal fuel-economy standards. The energy package crafted by majority leader Tom Daschle advocates "biodiesels," and the Natural Resources Defense Council is insisting that we could cut gasoline consumption by 50 percent over ten years—if only the feds would mandate what and where we drove.

Even the "oil men" in the Bush administration have advocated doling out millions in research subsidies for hydrogen fuel cells that supposedly would replace the internal-combustion engine. The project, Energy Secretary Spencer Abraham announced in January, is "rooted in President Bush's call to reduce American reliance on foreign oil."

In fact, the price of oil has declined since Sept. 11, as it consistently has for decades, and with producers scattered all over the world, no single nation or region can stop the flow.

But supporters of a comprehensive energy policy seem undeterred by these realities. "Logic," Robert Samuelson writes in the *Washington Post*, "is no defense against instability. We need to make it harder for [Middle Easterners] to use the oil weapon and take steps to protect ourselves if it is used. Even if we avoid trouble now, the threat will remain."

Past efforts to attain a petroleum-free utopia, however, have largely failed. For example, despite three decades of federal fuel-economy standards, oil imports as a share of U.S. consumption have risen from 35 to 59 percent.

A market-based solution, such as a gas tax, is the most obvious approach to cutting consumption, but even environmentalists concede that proposing one would spell political suicide. Moreover, gas taxes are an expensive solution and come with no guarantee of energy independence. The European Union, for example, taxes gas up to $4 per gallon—and still imports over half its oil.

So instead of enraging consumers at the pump, Washington has largely relied on backdoor taxes.

The regulatory regime known as CAFE (Corporate Average Fuel Economy) was hatched in the wake of the oil-price shocks of the early 1970s, when sedans still made up most of the nation's fleet. Instead of the redesigned smaller, lighter, and less powerful vehicles, however, consumers flocked to minivans, small trucks, and sport utility vehicles, which are held to a lower CAFE standard (20.7 mpg versus the 27.5 mpg required for cars).

Today, both passenger cars and light trucks are more efficient than ever, having improved 114 percent and 56 percent, respectively, since 1974. But gasoline is so cheap, despite continuing Middle Eastern crises, that on average Americans are driving twice as many miles as in years past.

A recent study by H. Sterling Burnett of the National Center for Policy Analysis found that raising CAFE standards by 40 percent—as Kerry and others recommend—would not "reduce future U.S. dependence on foreign oil." CAFE's only function is to keep regulators busy calculating elaborate formulas for determining compliance in which manufacturers then look for loopholes. (CAFE requires that a manufacturer's trucks meet an *average* standard of 20.7 mpg. Thus DaimlerChrysler AG, for example, designates its popular PT Cruiser as a "truck" in order to offset the low mpg of its large SUVs, such as the Dodge Durango.)

Worse, stricter CAFE standards would surely undermine the very economic security that proponents vow to protect. The profits of U.S. automakers—and tens of thousands of UAW jobs—depend on sales of SUVs and light trucks. According to an analysis by Andrew N. Kleit, a professor at Pennsylvania State University, the Kerry CAFE proposal would reduce the profits of GeneralMotors by $3.8 billion, of Ford by $3.4 billion, and of Daimler-Chrysler by $2 billion. Foreign manufacturers, which largely specialize in smaller vehicles, would see a profit *increase* of $4.4 billion.

Evidently hoping to shield automakers from a CAFE assault—and to win PR points for expanded domestic drilling—the Bush administration has embraced the latest alternative-fuel fad: the hydrogen fuel cell.

The Bush plan replaces the Partnership for a New Generation of Vehicles, Al Gore's vain attempt to produce an affordable, emissions-free family sedan capable of 80 mpg by 2004. Over eight years, Washington pumped more than $1.5 billion into the program—in addition to the $1.5 billion sunk into it by the Big Three. In its annual review of the project last August, the National Research Council judged the super-car goals to be inherently "unrealistic."

The Bush plan has drawn broad political support. Former Clinton chief of staff John Podesta cheers, "The next step is hydrogen-powered fuel-cell vehicles. But the only way to get these vehicles out of the lab and onto the road is with incentives and requirements aimed at producing 100,000 vehicles by 2010, 2.5 million by 2020."

But the 100-year dominance of conventional internal-combustion engines over alternatives is no accident. A quick primer on the complexities of hydrogen power helps explain why.

Hydrogen's status as the new darling of the sustainable-energy movement is understandable. Its promise lies first in its performance: Unlike ethanol, it supplies more energy per pound than gasoline. When used to power an automobile, its only emission is water—making it especially attractive to an industry already under pressure from clean-air and global-warming rules. And hydrogen is one of the most plentiful elements on the planet.

The trouble is, hydrogen always comes married to another element—as in methane gas or water.

Most fuel-cell technology today relies on hydrogen extracted from methane, in a process that emits large quantities of greenhouse gases. And as *Car and Driver* magazine's technical analyst, Patrick Bedard, explains, domestic sources of methane are "[t]oo limited to serve any significant demand for automobiles." A study by the Argonne National Laboratory concluded that the U.S. would have to look to foreign sources—primarily in Russia and Iran, and in other Middle East nations.

Goodbye, oil dependence. Hello, methane dependence.

Given these hurdles, attention is turning instead to electrolysis—the extraction of hydrogen from water, which is readily obtainable along America's ample coasts. Electrolysis is, however, the most energy-intensive process of any fuel alternative; studies differ on whether it would consume more carbon-based fuels than the use of hydrogen would save. What is certain, points out Stanford University professor John McCarthy, is that "the advantage of hydrogen, if you have to burn carbon fuels (coal, oil, or gas) to manufacture it, would be negligible."

In other words, McCarthy explains, the unspoken truth about hydrogen is that "it is a synonym for nuclear power."

Leading researchers in the field—including David Scott of the University of Victoria in Canada, Cesare Marchetti of the Internaitonal Institute for Applied Systems Analysis, and Jesse Ausubel of Rockefeller University—say that the only way to produce liquid hydrogen in the mass quantities needed for transportation is with a major investment in nuclear power. Says Scott: "[A]pplying the most elementary numeracy, nuclear fission is the only realistic option."

Ironically, many of the political voices now embracing hydrogen fuel are the same ones that have prevented the construction of a single new U.S. nuclear plant in 25 years. Ausubel has written in *The industrial Physicist* magazine that "understanding how to use nuclear power, and its acceptance, will take a century or more."

For now, the answer is still gasoline. Compared with the technical barriers to developing alternative fuels, there already exist numerous market mechanisms to mitigate potential oil shortages. As suggested by Donald Losman, a National Defense University economist, these include: stockpiling, futures contracts, diversifying the supplier base, and relaxing the restrictions that currently mandate some 13 different fuel blends in 30 cities.

Dramatic improvements in fuel efficiency also could be achieved if Washington allowed automakers to market diesel-powered vehicles. In Germany, for example, Volkswagen mass markets the 80-mpg Lupo, which is powered by a direct-injection diesel engine. But that's anathema to American greens who insist—without evidence—that diesel's particulate emissions are dangerous to public health.

All fuels require trade-offs, of course. But politically correct, misguided energy schemes will not make America more independent. Gasoline remains by far the best deal we have.

POSTSCRIPT

Will Hydrogen End Our Fossil-Fuel Addiction?

Hydrogen as a fuel offers definite benefits. As Joan M. Ogden notes in "Hydrogen: The Fuel of the Future?" *Physics Today* (April 2002), the technology is available, and compared to the alternatives, it "offers the greatest potential environmental and energy-supply benefits." To put hydrogen to use, however, will require massive investments in facilities for generating, storing, and transporting the gas, as well as manufacturing hydrogen-burning engines and fuel cells. Currently, large amounts of hydrogen can easily be generated by "reforming" natural gas or other hydrocarbons. Hydrolysis—splitting hydrogen from water molecules with electricity—is also possible, and in the future this may use electricity from renewable sources, such as wind or nuclear power. The basic technologies are available right now. See Thammy Evans, Peter Light, and Ty Cashman, "Hydrogen—A Little PR," *Whole Earth* (Winter 2001). Daniel Sperling, in "Updating Automotive Research," *Issues in Science and Technology* (Spring 2002), notes, "Fuel cells and hydrogen show huge promise. They may indeed prove to be the Holy Grail, eventually taking vehicles out of the environmental equation." Making that happen, however, will require research, government assistance in building a hydrogen distribution system, and incentives for both industry and car buyers. See Also Matthew L. Wald, "Questions about a Hydrogen Economy," *Scientific American* (May 2004).

Joseph J. Romm, in "The Hype about Hydrogen," *Issues in Science and Technology* (Spring 2004), cautions that replacing fossil fuels with hydrogen is not something that can be done overnight. It is a process that will take decades, and for now, efforts are best bent toward reducing greenhouse gas emissions in other ways. Lacking other long-term options, Sperling and Ogden, in "The Hope for Hydrogen," *Issues in Science and Technology* (Spring 2004), we should be working hard on the transition now. Hydrogen, they insist, "accesses a broad array of energy resources, potentially provides broader and deeper societal benefits than any other option, potentially provides large private benefits, has no natural political or economic enemies, and has a strong industrial proponent in the automotive industry."

Is the hydrogen economy likely to be as decentralized as Rifkin envisions? Jim Motavalli, in "Hijacking Hydrogen," *E Magazine* (January-February 2003), worries that the fossil fuel and nuclear industries will dominate the hydrogen future. The fossil fuel industries wish to use "reforming" to generate hydrogen from coal, and the nuclear industries see hydrolysis as creating demand for nuclear power. Nuclear power, Motavalli says, is particularly favored by the U.S. government's 2001 National Energy Policy.

In January 2003, President George W. Bush proposed $1.2 billion in funding for making hydrogen-powered cars an on-the-road reality. Gregg Easterbrook, in "Why Bush's H-Car Is Just Hot Air," *New Republic* (February 24, 2003), thinks it would make much more sense to address fuel-economy standards; Bush should "leave futurism to the futurists." Peter Schwartz and Doug Randall, in "How Hydrogen Can Save America," *Wired* (April 2003), commend Bush's proposal but say that the proposed funding is not enough. We need, they say, "an Apollo-scale commitment to hydrogen power. The fate of the republic depends on it." Toward that end, Schwartz and Randall list five steps essential to making the hydrogen future real:

- Develop fuel tanks that can store hydrogen safely and in adequate quantity.
- Encourage mass production of fuel cell vehicles.
- Convert the fueling infrastructure to hydrogen.
- Increase hydrogen production.
- Mount a PR campaign.

But are fossil fuels as scarce as industry critics and hydrogen enthusiasts say? They are certainly finite, but it has become apparent that not all fossil fuels are included in most accountings of available supply. A major omission is methane hydrate, a form of natural gas locked into cage-like arrangements of water molecules and found as masses of white ice-like material on the sea-bed. There appear to be vast quantities of methane hydrate on the bottom of the world's seas. If they can be recovered—and there are major difficulties in doing so—they may provide huge amounts of additional fossil fuels. Once liberated, their methane may be burned directly or "reformed" to generate hydrogen, making the nuclear approach less necessary. However, the methane still poses global warming risks. Indeed, if methane hydrate deposits ever gave up their methane naturally, they could change world climate abruptly (just such releases may have been responsible for past global climate warmings). See Erwin Suess, Gerhard Bohrmann, Jens Greinert, and Erwin Lausch, "Flammable Ice," *Scientific American* (November 1999).

Will nuclear power be part of the hydrogen economy? Some scientists and even some environmentalists are now recognizing that properly designed and managed nuclear power plants may have fewer environmental side effects than fossil fuel power plants. See Issue 12.

ISSUE 11

Should Existing Power Plants Be Required to Install State-of-the-Art Pollution Controls?

YES: Eliot Spitzer, from Testimony Before the United States Senate Committee on Environment and Public Works and the Committee on the Judiciary (July 16, 2002)

NO: Jeffrey Holmstead, from Testimony Before the United States Senate Committee on Environment and Public Works and the Committee on the Judiciary (July 16, 2002)

ISSUE SUMMARY

YES: New York attorney general Eliot Spitzer states that removing regulatory requirements for power plant pollution controls will prevent needed improvements in air quality.

NO: Environmental Protection Agency assistant administrator Jeffrey Holmstead argues that removing regulatory requirements for power plant pollution controls in favor of a markets-based approach will improve air quality.

The U.S. government first attempted to address air pollution problems with the 1967 Air Quality Act. In 1970, that act became the much stronger Clean Air Act, which included a requirement that the Environmental Protection Agency (EPA) develop a program to regulate new and modified sources of air pollution, such as new power plants and power plants that had altered their operations. This program eventually became known as New Source Review (NSR). Under it, the EPA has permitted routine maintenance projects without subjecting them to the review and permit process. New power plants and major upgrades to existing power plants, if they would result in a more-than-minimal increase in pollutant emissions, must be reviewed, must obtain a permit, and must incorporate state-of-the-art pollution controls. The aim was to ensure that new power plants be the cleanest possible, that old power plants become progressively cleaner, and that the air breathed by the American people become progressively sweeter. Not surprisingly, since NSR could mean greatly increased expenses for upgrades, the industry objected, fought NSR in the courts, and lobbied Washington for changes. The Bush administration has responded favorably, asking the EPA to review NSR and recommend changes.

The 1990 amendments to the Clean Air Act introduced the idea of setting limits (caps) on the amounts of particular pollutants that could be emitted and permitting companies that emitted less than their limit to trade (sell) the "unused" portion of their limit. This was applied to sulfur dioxide (SO_2) emissions in an effort to solve the acid rain problem, and over the next decade that problem diminished greatly. Some people asserted that the "cap-and-trade" approach was responsible. Others said that changes in technology and a shift to low-sulfur fuels had more to do with success.

In spring 2003, the Subcommittee on Clean Air, Climate Change, and Nuclear Safety of the Senate Committee on Environment and Public Works heard testimony on amending the Clean Air Act. The Bush administration urged its Clear Skies Act of 2003, a program that uses the acid rain experience to justify a similar cap-and-trade approach to controlling power plant emissions of SO_2, nitrogen oxide, and mercury. Senator George Voinovich (R-Ohio) said, "The Clear Skies program will provide power plants with the flexibility to choose among various options for reducing emissions that best fit their specific circumstances while saving over $1 billion annually in compliance costs.... The flexibility of the Clear Skies market-based cap and trade program and the certainty of its emissions reduction targets ... will ensure that the real reductions called for in this bill can be achieved without forcing utilities to fuel switch and without forcing electricity and natural gas prices through the roof. Perhaps most importantly, Clear Skies will help ensure that the least of our brothers and sisters will not be forced to forgo heating their homes—and that our companies will not be forced to move overseas to remain competitive in the global market—due to sky-high electricity and natural gas prices." According to DeWitt John and Lee Paddock, "Clean Air and the Politics of Coal," *Issues in Science and Technology* (Winter 2004), there is "a surprising consensus about how to fix NSR: emissions caps and a trading system. But ... the administration would protect [the powerful] coal [industry], whereas others give precedence to public health."

In the following selections, the EPA's defense of changing New Source Review is presented by Jeffrey Holmstead, who describes those changes and argues that removing regulatory requirements for power plant pollution controls in favor of a markets-based approach will improve air quality. Eliot Spitzer, on the other hand, takes the stance that weakening NSR will prevent needed improvements in air quality.

Eliot Spitzer

 **YES**

Testimony of Eliot Spitzer

Chairman Leahy and Chairman Jeffords, Senator Schumer and Senator Clinton, and distinguished members of the committees: Thank you for convening this hearing and thank you for providing me with the opportunity to testify about the need to maintain and enforce the New Source Review (NSR) provisions of the federal Clean Air Act.

New York State has been hard hit by air pollution from coal-burning power plants. Hundreds of lakes and ponds in the Adirondack and Catskill Mountains have been ravaged by acid rain. Ground level ozone has triggered asthma attacks and other respiratory diseases in every corner of our state, particularly in New York City. In addition, nitrate and sulfate particulates cause respiratory and cardiac illness, lung cancer and thousands of deaths in the regions downwind from polluting plants.

The New Source Review provisions of the Clean Air Act constitute a powerful tool to rein in this harmful pollution. For years, power plants have been exploiting an exemption, added to the Clean Air Act in 1977, which temporarily excused existing power plants from having to install modern pollution control devices. This exemption, however, was not intended to be permanent. Congress understood in 1977—twenty-five years ago—that existing plants could not operate indefinitely without having to undertake expensive life extension projects. At that time, Congress mandated, power plants would have to install state-of-the-art pollution controls. But now, decades later, many of these power plants continue to spew huge quantities of air contaminants and operate with no pollution controls, in blatant violation of the Clean Air Act.

The aim of the Clean Air Act litigation brought by New York, other northeast states, the federal Environmental Protection Agency (EPA) and various environmental organizations is to address these harms by going to their source. In 1999, working in partnership with EPA and other Attorneys General from the northeast, my office identified various power plants that were in violation of the New Source Review requirements. These coal-burning power plants had undergone major multi-million dollar improvements without installing NSR dictated pollution controls. To date, I have filed lawsuits with respect to 17 of these power plants—which are located in Ohio, West Virginia, Virginia and Indiana—under the citizen suit provision of the Clean Air Act.

From U.S. Senate, Committee on Environment and Public Works and Committee on the Judiciary. *New Source Review Program of the Clean Air Act*. Hearing, July 16, 2002. Washington, DC: Government Printing Office, 2002.

Each of these cases has been joined by EPA and other states. The plants involved emit tons of nitrogen oxides and sulfur dioxide every day, harming New York's air quality and damaging its natural resources.

My office also has taken enforcement action against several power plants located in New York State even though they are generally responsible for much less pollution than their counterparts in the Midwestern and southern states. Working with the New York State Department of Environmental Conservation, we have identified 7 power plants that were in violation within New York, and we have filed a lawsuit against the owner of the two largest plants. The Commissioner of the State Department of Environmental Conservation and I are currently in negotiations with the owners of the other five plants.[1]

Unfortunately, however, our efforts to enforce the Clean Air Act have prompted the Bush Administration to propose a set of illegal regulatory changes that would essentially neutralize New Source Review as an enforcement mechanism and deprive the public of the benefits of this laudably far-sighted legislation. The Administration's efforts to dismantle NSR must be defeated, and I will go to court, if necessary, to stop them. I also urge Congress to ensure that the proposed changes do not come to fruition. In the meantime, however, the Administration's retrenchment on clean air already has jeopardized all of the existing NSR cases brought by the states and the federal government, and threatens to thwart any future NSR enforcement efforts.

My testimony today addresses four points. First, I explain how the Administration's proposed changes would, if enacted, illegally contravene the Clean Air Act. I intend to go to court to challenge these illegal changes if the Administration puts them into effect. And I intend to win. Second, I demonstrate that the Administration's plans to gut the NSR provisions are already—before the changes even become effective—jeopardizing our existing enforcement cases and depriving us of the millions of tons in pollution reductions that those cases would yield. Third, I refute both the Administration's claim that the NSR program needs "clarification" and industry's contention that it was "unfairly surprised" by our enforcement cases. Finally, I offer my recommendations as to how Congress should respond to the Administration's assault on the Clean Air Act.

I. The Administration's Proposed Changes Are Illegal

The Administration's proposed changes—so far as we know them through EPA's press statements—are illegal because they purport to amend the Clean Air Act. I will first explain the existing law, as enacted and enforced under the prior Reagan and Bush administrations. I will then review the changes and explain why they are illegal.

A. New Source Review Law and Regulations

In 1977, Congress created the Prevention of Significant Deterioration (PSD) program to ensure that increased pollution from the construction of new emissions sources or the modification of existing emission sources would be minimized, and to ensure that construction activities would be consistent with air quality planning requirements. This program only applied to areas of the country where the air quality met or exceeded the national ambient air quality standards. The non-attainment New Source Review program, also created in 1977, contains virtually identical requirements applicable to facilities in non-attainment areas. (I refer to both programs together as the NSR program.)

Generally, the NSR program requires such sources to obtain permits from the permitting authority *before* the sources undertake construction projects if those projects will result in an increase in pollution above a *de minimis* amount. In addition, the NSR regulations usually require that sources install state-of-the-art controls to limit or eliminate pollution. Congress required and fully expected that those older existing sources would either incorporate the required controls as they underwent "modifications," or would instead be allowed to "die" and be replaced with new, state-of-the-art units that fully complied with pollution control requirements.

The Clean Air Act defines "modification" as any physical change or change in the method of operation that increases the amount of an air pollutant emitted by the source. 42 U.S.C. § 7411(a). Courts for many years have interpreted the Clean Air Act term "modification" broadly. *Alabama Power Co. v. Costle*, 636 F.2d 323, 400 (D.C. Cir. 1979) (the term "'modification' is nowhere limited to physical changes exceeding a certain magnitude"); *Wisconsin Electric Power Co. v. Reilly*, 893 F.2d 901, 905 (7th Cir. 1990) ("*WEPCO*") ("[e]ven at first blush, the potential reach of these modification provisions is apparent: the most trivial activities—the replacement of leaky pipes, for example—may trigger the modification provisions if the change results in an increase in the emissions of a facility.") The *WEPCO* court noted that Congress did not intend to provide "indefinite immunity [to grandfathered facilities] from the provisions of [the Clean Air Act]," *id.* at 909, and that "courts considering the modification provisions of [the Clean Air Act] have assumed that '*any* physical change' means precisely that." *Id.* at 908 (emphasis added) (citations omitted).

EPA recognized, however, that interpreting "modification" to include literally "any physical change" could become administratively unworkable ("the definition of physical or operational change in Section 111(a)(4) could, standing alone, encompass the most mundane activities at an industrial facility (even the repair or replacement of a single leaky pipe, or a change in the way that pipe is utilized)"). 57 Fed. Reg. 32,314, 32,316 (July 21, 1992). To exclude these trivial activities from the scope of the NSR provisions, EPA regulations have exempted routine maintenance, repair, and replacement from the definition of modification since 1977. 40 C.F.R. ' 52.21(b)(2)(iii).

EPA historically has analyzed and applied the "routine maintenance" exemption to modification by using a common sense test that assesses four primary factors, the (1) nature and extent, (2) purpose, (3) frequency, and (4) cost of the proposed work. *See, e.g.*, Memorandum from Don R. Clay, EPA Acting Assistant Administrator for Air and Radiation, to David A. Kee, Air and Radiation Division, EPA Region V (Sept. 9, 1988). This approach was upheld by the

U.S. Court of Appeals for the Seventh Circuit in *WEPCO*, a case brought under the first President Bush. Our cases follow these standards.

Although Congress did not authorize EPA to create this "routine mainte-nance" exemption, the Court of Appeals for the D.C. Circuit ruled, in a chal-lenge to the exemption in the PSD regulations for minor emission increases, recognized that EPA may exempt *de minimis* activity from the scope of the modification provisions. *Alabama Power Co. v. Costle*, 636 F.2d at 360-61. *See also Natural Resources Defense Council v. Costle*, 568 F.2d 1369 (D.C. Cir. 1977) (similar holding regarding the Clean Water Act). Thus, as long as it is con-strued narrowly, the routine maintenance exemption is legal.

Another change EPA made over a decade ago was to limit the scope of the modification provisions to those modifications that generate a significant increase in pollution. This requirement is essential when one considers the justifications offered by the present Administration for its NSR "reforms." In announcing the NSR changes, EPA has claimed repeatedly that NSR require-ments have deterred emissions-reducing projects. In offering this justifica-tion, EPA appears to have bought into one of the power industry's favorite arguments against the NSR program—that the program somehow prevents companies from making efficiency improvements that would benefit the environment. However, efficiency improvements that are environmentally beneficial and reduce emissions do not trigger NSR: if emissions decrease—or even increase only slightly— existing NSR requirements are inapplicable.[2]

B. The Bush Administration's Proposals

The Bush Administration proposed changes would sanction plant modifica-tions that are far from *de minimis*. For example, EPA proposes to allow large facilities to operate under a single plant-wide emissions cap (plant-wide appli-cability limit or PAL) for a period of 10–15 years. Unlike what some who sup-port plant-wide caps would require—that the caps decline over time—the Administration would allow the caps to remain high. Emissions at such a plant would remain the same throughout the 10–15 year period, regardless of changes in air quality, technology, or air quality standards. Because the plant's emissions are set for the duration of the PAL, states likely would be prohibited from imposing emission reduction requirements beyond what the PAL required, regardless of air quality needs.

Similarly, EPA proposes that any unit that has installed "Best Available Control Technology" (BACT) or BACT equivalent since 1990 would not be required to undergo NSR review for a period of 10–15 years, unless "allowable" emissions increase. Again, this limit on review of the source's emissions fails to consider evolving air quality needs, and may prevent a state from imposing more stringent emission reduction requirements, even if air quality consider-ations would justify such measures. Congress's clear intention to have the Clean Air Act stimulate technology improvement will be frustrated.

EPA also proposes several significant revisions in the method by which NSR-triggering emissions increases are calculated. For example, EPA proposes that the baseline for measuring emissions (for facilities other than power plants) become the highest emission level achieved over any two year period during the last ten years. By allowing a source to use a baseline that extends back 10 years, EPA is proposing to permit inflation of the source's baseline, because many regulations in the last ten years have forced sources to reduce emissions. These required emission reductions, however, may not be reflected in the source's baseline generated under the Administration's proposal. Thus, a source would actually be allowed to *increase* emissions from current levels without any attendant pollution control upgrade.

The most alarming revision proposed by EPA is the wholesale expansion of the Routine Repair and Maintenance (RRM) exception. Specifically, EPA is proposing to allow companies to treat multi-million dollar once-in-a-lifetime projects as "routine maintenance," even though, as industry documents establish, power plant staff never considered the projects routine. EPA is planning to forgo pollution control requirements for virtually limitless "like-kind" replacements that would restore and perhaps expand an old plant's capacity and dramatically prolong its life. To accomplish this, EPA proposes to include in the definition of RRM projects [those] that are below a specified cost threshold (inflated to reflect facility replacement cost, not original cost), and that involve installation of replacement equipment that serves the same function and does not alter basic design parameters. The cost threshold test fails to consider air quality and places no limit on any emissions increase the project might produce. Thus, significant increases in emissions could occur with no attendant pollution control requirement. Similarly, the equipment replacement exemption could essentially allow a company to rebuild a source without undergoing any governmental review and without meeting pollution control requirements. Significant emission increases could result.

These impacts have severe consequences for the American public and particularly for the states. EPA's proposal would severely blunt one of the states' most important anti-pollution tools, placing the states in an extraordinarily difficult position regarding their responsibilities under the Clean Air Act. It is the states—not EPA, not the federal government—that have the responsibility for insuring that National Ambient Air Quality Standards (NAAQS) are met. 42 U.S.C. §§ 7404; 7410. Under EPA's proposed revisions, the states stand to lose flexibility in determining how best to achieve or maintain air quality because the largest sources of pollution—which generally are the most efficient to control—will essentially be exempted from regulation.

C. States Will Sue to Prevent This Illegal Rollback of Clean Air Protections

I will do all in my power to prevent the Administration from unilaterally gutting the Clean Air Act. The Administration cannot change the law retroactively as it is seeking to do,[3] it cannot change regulations without adequate notice and comment. And, most importantly, the Administration cannot eviscerate the Clean Air Act without getting Congress to pass legislation allowing such a rollback. As explained above, the CAA itself contains no exemption for routine maintenance. Nor does it exempt like-kind replacement activities, no matter how massive or infrequent, from the definition of modification. With the statute so clear, the permissible scope for agency-created exemptions is very narrow. When in the *Alabama Power* case the D.C. Circuit held, following ample Supreme Court and D.C. Circuit precedent, that EPA can exempt *de minimis* activity, it emphasized that EPA could only exempt the most minor of activities so that the program would be workable administratively. Indeed, the court stated in very strong terms that "there exists no general administrative power to create exemptions to statutory requirements based upon the agency's perceptions of costs and benefits." *Alabama Power*, 636 F.2d at 357. The court also held that the power to create exceptions "is not an ability to depart from the statute, but rather a tool to be used implementing the legislative design." *Id*. at 359.

That is not what the Administration proposes to do. The Administration's proposed changes are far from *de minimis*. EPA's changes would have the effect of essentially eliminating the applicability of New Source Review to modifications, contrary to the express language of the statute. EPA's announced changes will confer on existing, dirty power plants indefinite immunity from the requirements of the Clean Air Act, contrary to Congress's clear intention when it enacted the NSR provisions twenty-five years ago. This is illegal and for that reason, I—and I expect to be joined by many other states—intend to sue EPA if it carries out its plans.

II. The Proposed Changes and the Administration's Hostility to NSR Are Already Jeopardizing the Enforcement Cases

If enacted, the Administration's proposed changes would impermissibly undercut existing law and reduce the scope of the Clean Air Act. Simply by signaling its hostility to the NSR program, however, the Administration already has compromised our existing enforcement cases. Indeed, from the day administrations in Washington changed, industry has sought to avail itself of its enhanced bargaining position.

A. The Administration Is Overtly Hostile to NSR

Fifteen months ago, the Administration released [Vice] President Cheney's "National Energy Policy: A Report of the National Energy Policy Development Group." The report directed Attorney General Ashcroft to "review existing enforcement actions regarding NSR to ensure that the enforcement actions are consistent with the Clean Air Act and its regulations." That directive immediately undercut the Department of Justice's lawyers; yet, on January 15, 2002, DOJ concluded that the NSR cases were legally sound.

The Vice President also directed the EPA "in consultation with the Secretary of Energy and other relevant agencies, to review NSR regulations, including administrative interpretations and implementation, and report to the President within 90 days on the impact of the regulations on investment in new utility and refinery generation capacity, energy efficiency, and environmental protection." Over a year later, EPA finally announced its illegal, wholesale administrative rollback of NSR.

In its press statements, EPA claims to be simply "clarifying" the existing regulations and maintains that its proposed rewriting of the law will not affect the filed cases. Indeed, on the day of EPA's announcement, Administrator Whitman explained that EPA would continue its enforcement efforts against past violations, "because you can't get away with violating the law just because the law gets changed." See June 14, 2002, *Atlanta Journal and Constitution* article "Air Proposals Irk Environmentalists; Bush Plan a 'Massive Gift' to Energy Industry, Critics Say."

Earlier, on March 27, 2002, the Justice Department's environmental chief, Thomas Sansonetti, said that pursuing NSR cases was one of his top priorities. Quoted in the "Daily Environment Report," Mr. Sansonetti stated: "We're going full steam ahead. We're actively pursuing all cases. When companies refuse to settle, DOJ will take them to trial." He predicted that DOJ would prosecute two or three NSR cases in court in the coming year. He also said that DOJ had budgeted $3 million in the current fiscal year to pursue such cases. I'd like to believe Mr. Sansonetti; his attorneys at the Justice Department have done excellent work on the pending cases and I want to continue our partnership. But his statements were made before EPA announced its retrenchment. Since then, DOJ has been silent as to its future intentions regarding NSR.

B. The Existing NSR Cases Are in Jeopardy

Although we agree with the Administration that any new regulations should not be retroactive, it would be naive to believe that industry will not try to use the "NSR reforms" in court to justify their past conduct. We are already seeing the effects of this Administration's misguided and illegal policy changes: settlements are stalled, judges are wondering about the impact of the reforms on their cases, and industry lawyers are already arguing in court that the cases should not go forward. Whether or not the rollback will affect the existing cases is an issue of first impression for the courts because of the unprecedented nature of EPA's action. Never before has EPA—or Congress, for that matter— undertaken such a clear retreat on environmental protection. Conducting such a rollback while enforcement cases under the old rules are pending is not only unprecedented but was unimaginable, at least before this Administration came to power. Simply put, the existing NSR cases are in jeopardy and we are fooling ourselves if we believe that the federal government will be filing more cases after rewriting the regulations to legalize the conduct at issue.

I would like to focus my comments now on three concrete examples of how the Administration's policies are adversely affecting our pending enforcement cases.

1. Cinergy and VEPCO

On November 16, 2000, my office and the EPA reached a $1.2 billion dollar settlement in principle covering eight coal-fired power plants run by the Virginia Electric Power Company (VEPCO)—one subject to New York's pending lawsuit and seven others that VEPCO brought into the settlement. The settlement would have reduced air pollution by more than 270,000 tons annually. VEPCO was to spend $1.2 billion over 12 years to reduce its sulfur dioxide emissions by 70 percent and its nitrous oxides emissions by 71 percent from pre-existing levels. Further, VEPCO was to pay $5.3 million in penalties to the federal government and an additional $13.9 million to fund environmental benefit projects, with a portion going to New York State. The intent at the time was to finalize the agreement within 60–90 days. Eighteen months later, this agreement remains unexecuted. My staff has spent countless hours in meetings with VEPCO and the federal government, but the regulatory uncertainty has prevented any final agreement. This is a terrible loss for the people of this nation, who expect, and deserve, cleaner air.

Similar delay has beset our effort to reach a final agreement with the Ohio-based utility Cinergy. In December 2000, I joined the federal government and the States of Connecticut and New Jersey in reaching a settlement in principle covering ten of Cinergy's coal-fired power plants (one subject to New York's lawsuit and nine others). We were to see over 300,000 tons in emission reductions, and $30 million in penalties and environmental projects. Like VEPCO, the Cinergy agreement remains in limbo. After tolerating two years of settlement discussions, the Cinergy court has placed the case back on the litigation track. Although DOJ advised the court that it intended to file an amended complaint by July 10, it has not yet done so, raising questions about DOJ's willingness to pursue NSR enforcement cases when its client, EPA, is in the process of changing the rules.

Although Cinergy and VEPCO have continued to express their interest in settlement, their actions speak louder than words. As might be expected, the softening of EPA's regulatory posture has only hardened Cinergy's and VEPCO's positions on the remaining issues to be worked out. I now see no way for these settlements to become final unless the states and DOJ capitulate on the remaining issues, something that I am not prepared to do.

2. Tennessee Valley Authority Case

In 2000, EPA issued a final determination that TVA had violated the NSR requirements of the Act by undertaking enormous and expensive modification projects at several of its power plants. TVA appealed to the Eleventh Circuit, briefs were submitted and oral argument was held this past May. Like many others involved in these cases, I was hopeful that the Eleventh Circuit would issue a quick decision, affirming EPA's determinations. A decision from the Eleventh Circuit would be an extremely important precedent for the other NSR cases.

Instead, in the wake of EPA's recent announcement on NSR "reform," the Eleventh Circuit took the extraordinary step of ordering the parties to mediation. Although we cannot be certain that this order was issued in direct response to the EPA announcement, it is unlikely that the timing of the two events is coincidental.

3. Niagara Mohawk Case

On January 10, 2002, Governor Pataki and I filed a lawsuit in federal court against Niagara Mohawk Power Corporation and NRG (the current owner of the power plants) for violating NSR at two power plants in western New York. The Dunkirk and Huntley coal-burning power plants account for more than 20 percent of the nitrogen oxide emissions and 38 percent of the sulfur dioxide emissions released by all power plants in New York State.

The defendants filed a motion to dismiss all or portions of the case on jurisdictional grounds. Briefing was completed and my attorneys were preparing to argue the case. But shortly after EPA's announcement, the judge called us in to explain how the Administration's announced intention to change the NSR rules would affect the existing case. In its brief on this issue (see Exhibit 2), Niagara Mohawk has described EPA as "reconsidering" its position on NSR and recommended that the Court put the case on hold until EPA takes final action on the NSR changes:

> In order to consider the merits of the case, the Court would ultimately have to decide whether EPA's interpretation of the Act and regulations, as applied by DEC, is reasonable and in accordance with law. The Court cannot properly make that decision until the EPA decides finally what its interpretation is.

> In short, EPA has said that its recommendations involve clarification of existing law and policy, and definition of a regulatory concept (routine maintenance, repair and replacement) that derives from EPA's interpretation of the Clean Air Act. Accordingly, to the extent that EPA's final action follows its recommendations, its action may affect not only the State's request for prospective injunctive relief, but also its request for penalties for alleged past violations.

Niagara Mohawk also contends that even if the new rules were purely prospective, "they would still affect the State's request for injunctive relief." We think this argument is wrong. When a business breaks the law—no matter how much influence it may now have in Washington—the rule of law requires courts to order compliance. However, Niagara Mohawk's argument evidences a practical problem that judges will face if the Administration succeeds in implementing its "reforms." We expect the courts to find with relative ease that the utilities violated the law. But when it comes time to select a remedy, will they require substantial emission reductions even though the Administration's proposed policy would not require such reductions? Will a practical judge require a company to spend millions of dollars on pollution controls for actions that EPA is now saying do not require such controls? Indeed, now can EPA even ask for that relief with a straight face? If any of these cases go to trial, we might see the payment of some fines for past wrongdoing, but we may be deprived of the emission reductions we so desperately need. More money in the state and federal coffers, while welcomed, will not help us reverse the ravages of acid rain and respiratory disease in New York State and elsewhere.

I intend to continue to press forward on this important case. Niagara Mohawk violated the law and we need the remedy of dramatic emission reductions. Unless EPA tries to take away the states' authority to reject the regulatory changes—something I hear may be in the works—New York can continue to implement the law as it has existed for 25 years within New York. But we enjoy no such comfort in our out-of-state cases, where it will be difficult to proceed if EPA pulls the rug out from under us.

III. NSR Needs No "Clarification"

The power industry has always understood the scope of NSR and has never considered the modifications at issue to be routine maintenance. These modifications were large-scale capital projects that required significant advance planning and typically cost millions of dollars; they were intended to fix problems that routine repair or replacement had been unable to address. By contrast, activities considered by industry to be "routine" include relatively mundane actions, such as the day-to-day repair of leaky or broken pipes. In short, the record supplies no basis for the Administration's claims that the law was somehow unclear and that industry was somehow ambushed by our enforcement cases.

A. Industry Officials Originally Distinguished Routine Activities From Upgrades

Industry documents establish that industry officials appreciated the potential applicability of the NSR provisions to their power plant life extension projects. Because of protective orders entered in our various cases, I am unable to quote from most of these documents in my testimony. However, despite the utilities' attempt to cloak their plant life extension projects in secrecy, publicly available industry documents amply demonstrate industry's acknowledgment of the routine maintenance exemption's limited scope. For example, the Babcock and Wilcox company, in its definitive power plant treatise, *Steam, Its Generation and Use*, distinguished some of the very plant life extension activities at issue in our NSR cases from routine maintenance activities as follows: "*Older boilers represent important resources in meeting energy production needs. A strategic approach is required to optimize and extend the life of these units. Initially, routine maintenance is sufficient to maintain high availability. However, as the unit matures and components wear, more significant steps become necessary to extend equipment life.*" *Id.* at 46-1 (Exhibit 3). Our cases involve such "more significant," as opposed to the routine maintenance activities that the plants conduct on a day-to-day basis.

Similarly, the American Electric Power Company (AEP) explained to the Ohio Public Utilities Commission that life extension activities go beyond routine maintenance: "*As time goes on, the cumulative effects of operation affect more components, and affect those components more severely. Finally, the major subsystems and components reach a stage at which 'normal' maintenance and repair become inadequate to support satisfactory continued operation.*" Direct Testimony of Myron Adams, AEP's Manager of Integrated Resource Planning, filed with the Public Utilities Commission of Ohio on July 20, 1994 at 20 (Exhibit 4).

Publicly available information likewise demonstrates the magnitude of the projects we have cited in our cases. For example, modifications performed by TVA include projects costing $57 million, $23 million, and $29 million.

These modifications required that the affected units be shut down for 13 months, 3 months and 6 months respectively. Another TVA project costing $11 million required construction of a railroad track and a monorail to facilitate the replacement of 44% of the 234,000 square feet of total boiler surface area. At Ohio Edison, the NSR violations include installation of an entirely new and redesigned furnace and burner system—the core of any power plant—at the W.H. Sammis plant, as described in the accompanying article (Exhibit 5).

Documents produced by Niagara Mohawk show that the company originally used the term "routine maintenance" to apply to only a narrow category of work done at the plant. (Exhibit 6 A). In another company document, Niagara Mohawk made clear that work done at the plant for the purpose of extending the life of an electric generating unit concerned *components that are not routinely replaced.* (Exhibit 6 B). Indeed, Niagara Mohawk requested that its contractor not include "maintenance" type recommendations in a life extension report for one of the generating units. (Exhibit 6 C).

Industry's complaint that EPA suddenly changed its interpretation of the NSR requirements during the Clinton Administration is similarly contradicted by industry documents dating from the 1980s, which cite particular plant life extension projects as exceeding routine maintenance and therefore triggering the NSR requirements. Thus, in 1984—seven years after the enactment of the NSR requirements—the Electric Power Research Institute (EPRI) held a conference that included the topic of extending the lives of old power plants. The conference literature explicitly recognized that *"a fossil fuel power plant is designed for a 30-year life,"* meaning that all plants existing when the NSR/PSD requirements were enacted would reach the end of their useful lives by 2007. (Exhibit 7). Conference attendees then discussed the life extension activities that would be needed. A Duke Power representative stated that keeping the old plants running *"necessitated us developing a different approach than routine maintenance"* which only keep *"the plant in service until the end of its design life."* (Exhibit 8).

Similarly, at 1985 and 1986 EPRI conferences, industry representatives recognized that life extension activities transcend routine maintenance:

> *If plant life extension serves the balanced interests of stockholders and ratepayers, capital improvements and increased attention to equipment above and beyond routine maintenance may be warranted....*
>
> *It is of primary importance to define the distinction between plant life extension work and routine maintenance.*

(Exhibit 9).

B. Industry Was Fully Aware That Its Activities Were Not Exempt From NSR

Not only did industry recognize that plant life extension activities failed to qualify as "routine maintenance," industry also understood that NSR requirements would likely be applicable. For example, an article entitled *"Regulatory Aspects of Power Plant Life Extension"*—which was presented at a 1985 industry conference—expressly discussed the circumstances under which life extension projects could require NSR permits. (Exhibit 10). As a result, EPRI recommended *"that corporate counsel be consulted as a part of life extension planning activities, particularly for the interpretation of regulatory and environmental issues when such activities are clearly beyond the scope of what might be considered typical maintenance."* (Exhibit 11).

Rather than seeking EPA's guidance, however, industry simply attempted to conceal its activities. For example, a 1984 EPRI workshop on life extension recommended that life extension projects be described as maintenance activities in order to avoid triggering NSR requirements:

> *[T]here are a number of issues which require clarification. Several of these are: What is considered 'routine' repair, replacement, or maintenance for the purpose of qualifying for an exemption to the NSPS modification provisions? Some aspects of life extension projects may not be considered routine repair/maintenance/replacement. To the extent possible these projects should be identified as upgraded maintenance programs....*

> *Life extension projects will result in increased regulatory agency sensitivity to facility retirement dates.... Regulatory agencies may contend that since life extension projects will defer the need for new generation, additional pollution control should be required for the older, higher emitting affected plants.*

> *It may be appropriate to downplay the life extension aspects of these projects (and extended retirement dates) by referring to them as plant restoration (reliability/ availability improvement) projects. To the extent possible, air quality regulatory issues associated with these projects should be dealt with at the state and local level and not elevated to the status of a national environmental issue.*

> *To the extent possible, project elements should be stressed as maintenance related activities to maximize chances for NSPS exemptions. Utility accounting practices play a significant role here.*

(Exhibit 12).

In 1988, EPA issued an applicability determination to the Wisconsin Electric Power Company, or WEPCO, in which EPA determined that WEPCO's multi-million dollar life extension projects were not covered by the routine maintenance exemption. The issuance of the WEPCO interpretation conclusively disabused industry of any notion that it might avoid compliance with NSR requirements. Shortly after EPA issued its WEPCO applicability determination concerning the *life extension* projects at issue there, the Utility Air Regulatory Group (UARG), a leading industry group, advised its members that *"Life Extension is [now] an unpopular term in the wake of WEPCO."* (Exhibit 13, p. 2.). Consistent with other industry missives at the time, the memo further recommended against using *"the term 'life extension' to describe any project."* *Id.*, at 5. The same industry memorandum demonstrates that UARG and its members fully understood EPA's interpretation limiting the routine maintenance exception:

According to UARG, EPA equates 'routine' with 'frequent'.... UARG believes that under present EPA policy, in order to qualify for the routine maintenance exemption, the activity would have to be:

- *frequent,*
- *inexpensive,*
- *able to be accomplished at a scheduled outage,*
- *will not extend the normal economic life of the unit,*
- *be of standard industry design.*

Id., at 4. UARG also advised its members that if the WEPCO applicability determination were upheld by the courts, it *"will set a serious precedent if it is adverse." Id.*, at 5.

After the WEPCO determination, one of Ohio Edison's in house attorneys and one of the lawyers at the law firm representing Ohio Edison wrote an article explaining that, under the EPA interpretation reflected in WEPCO, Ohio Edison's own plant improvements would be subject to NSR, since: *"[a]fter WEPCo, virtually any physical change to an existing facility, even pollution abatement activities and an unpredictable array of repair, replacement, and maintenance projects, can trigger new source control obligations." See* June 18, 1990 letter from David Feltner, Senior Attorney for Ohio Edison, to Ms. Cheryl Romo, with enclosed draft article entitled *"Is There Life Extension After WEPCo?."* (Exhibit 14). (I note that the authors of this article overstate the reach of the NSR requirements by overlooking that the requirements apply only if an emissions increase is projected.) Despite the opinions of its attorneys, Ohio Edison continued to undertake expensive life extension activities at its plants without applying for an NSR permit or otherwise notifying the permitting authorities.[4]

IV. The Role of Congress

Congress need not sit idly while the Administration unilaterally ignores its earlier mandates and jeopardizes public health and the environment. As I've said, I will fight these changes; I urge you to do so as well.

First, while I can go to the courts, you have a greater ability to ensure this rollback does not occur. Any litigation I bring may take years to be resolved. You can act strongly and quickly. I urge you to pass specific legislation, this session, that would expressly prohibit the Administration from proposing or finalizing any new exemptions from NSR, including those that EPA has announced.

Second, I urge you not to be seduced by the Administration's claim that NSR can be replaced by the Administration's so-called "Clear Skies" initiative. That plan is an inadequate substitute for existing law and a wholly unsatisfactory alternative to Senator Jeffords's "Clean Power Act." At the outset, I note that "Clear Skies" is still no more than a press release. Although months have elapsed since the "Clear Skies" replacement for NSR was announced, no plan has even been introduced in Congress. Many of us took note of Administrator Whitman's criticism of the "Clean Power Act," which she dismissed on the grounds that it is unlikely [to] win Congressional approval. I would point out that Senator Jeffords's legislation *has* been introduced, and has passed the Senate Environment and Public Works Committee—so it is at least two steps ahead of "Clear Skies."

Even if the Administration were serious about "Clear Skies," the pollution reductions that program would offer are too little, too late: the caps are too high and would not take effect until the distant future.

To be blunt, the "Clear Skies" caps are based on little more than politics. They do not guarantee compliance with air quality standards. The caps certainly are not based on sound science. Every month, another study shows the need to reduce pollution more aggressively. For example, a recent study finds new links between fine particulate matter (PM) and cancer.

Nor does technical feasibility stand in the way of higher caps. More aggressive SO_2 and NOx cuts are clearly technically feasible even with existing technology. Nor is it a question of rates that consumers must pay for power. The Department of Energy itself determined that the country could cut NOx and SO_2 by 60–80% by 2010 with virtually no rate impact. *See* Energy Information Administration, *Analysis of Strategies for Reducing Multiple Emissions from Power Plants: Sulfur Dioxide, Nitrogen Oxides, and Carbon Dioxide* (December 2000).

The Administration tries to sell its plan by using faulty comparisons to current emissions. Don't be deceived. Even at their end point, the Bush pollution caps would be 50% higher than, for example S.556, the Clean Power Act, or EPA's own initial proposal. This 50% is roughly equivalent to all emissions produced within the State of Ohio, a leading producer of emissions. This difference alone could lead to hundreds, and perhaps thousands, of additional deaths each year. Under the Administration's program, states will find it far more difficult, if not impossible, to attain their mandated air quality standards.

Under the Administration's program, many dirty old plants will remain uncontrolled. In 1977, when it enacted the NSR provisions, Congress clearly expected that all plants would be controlled by 2018—over 40 years after the 1977 amendments made the NSR requirements applicable to plant modifications. However, if all plants were controlled with "best available control technology" by 2018, the SO_2 cap would be below 2 million tons, not 3 million tons as contemplated by "Clear Skies."

Moreover, the "Clear Skies" caps would not be fully phased in until the 2020's. Even EPA's own graphs acknowledge that pollution levels will not reach the cap level by the Administration's announced target dates. While EPA speaks instead of incentives for early reductions, the flip side of early reductions is late compliance. Under the Administration's program, any cuts now can be banked, ton-for-ton, to offset subsequent emissions. We should insist on early reduction *and* caps that are lower and take effect sooner.

Finally, the Administration's claim that the President's plan achieves more reductions than current law is directly contrary to what EPA and the Department of Energy found when they included the emission reductions attributable to full enforcement of the New Source Review provisions. See, e.g., Energy information Administration, *Analysis of Strategies for Reducing Multiple Emissions from Power Plants: Sulfur Dioxide, Nitrogen Oxides, and Car-*

bon Dioxide (December 2000). Furthermore, in its analysis, EPA ignores the emission reductions that will result under current law from other programs, such as the regional haze rule, the mercury Maximum Available Control Technology (MACT) requirements and the new ozone and particulate matter standards. Thus, the Administration is not comparing its proposal to the Clean Air Act as it is now written and as it should be implemented and enforced. Comparing Clear Skies to a Clean Air Act that is ignored or eviscerated is World-Com-style math at best.

I support the "Clean Power Act" because we need swift and significant reductions in sulfur dioxide, nitrous oxides, mercury and carbon emissions. I am especially supportive of including carbon in the four pollutant legislation and commend Senator Jeffords for working so hard on this legislation. The Administration finally admits that global climate change is happening. Unlike the Administration, however, Senator Jeffords has a plan of action. I urge you to pass the Jeffords "Clean Power Act."

Conclusion

Allow me, and others who are serious about environmental law enforcement, to continue to use the Clean Air Act to reduce pollution. That is what Congress intended when it adopted New Source Review twenty-five years ago. Don't allow the most serious attack on the Clean Air Act since it was adopted to succeed. Don't allow the product of 30-plus years of bi-partisan cooperation on clean air to be cast aside.

Notes

1. Attached to my testimony (Exhibit 1) is a list of the twenty-four plants, within and outside of New York, against which we have taken action, along with the amounts of air pollution they emit.
2. That NSR applies only when both a modification is large enough and the emission increase is significant was clearly demonstrated in EPA's May 23, 2000 applicability determination concerning a proposal by the Detroit Edison Company to replace and reconfigure the high pressure section of two steam turbines at its Monroe Power Plant. There, EPA determined that, although the modification was significant enough to trigger the NSR provisions, because the project would not lead to an increase in emissions, it was *not* subject to the pollution control requirements of the PSD program. Applicability Determination, p. 20. Indeed, as Detroit Edison explained to EPA, "because the change would increase efficiency, it would allow increased electricity generation using the same amount of coal, boiler heat input and steam flow while producing the same level of emissions as currently emitted." *Id*. Thus, contrary to the Administration's rhetoric, EPA's existing implementation of the NSR program does not weaken the utility industry's incentive to undertake efficiency programs (or any other projects for that matter) that do not involve increased pollution.

3. To the extent EPA has indicated it will make retroactive changes to the Act, any such changes would be of questionable validity. The D.C. Circuit, which would have exclusive jurisdiction of such changes under 42 U.S.C. § 7607(b)(1), prohibits retroactive application of interpretive rules absent authority delegated by Congress, *see Health Ins. Ass'n of America v. Shalala*, 23 F.3d 412, 423 (D.C. Cir. 1994) ("[I]nterpretive rules, no less than legislative rules, are subject to *Georgetown Hospital's* ban on retroactivity."), and such authority is entirely lacking here.

4. Likewise, a decade ago, one of the attorneys at Porter, Wright, Morris & Arthur, counsel for AEP and Ohio Edison, wrote:

> *The "Routine maintenance, repair, and replacement" exclusion may be available only if: (1) the repair/replacement is immediate after discovery of deterioration; (2) the replaced equipment is standard in the industry and fails frequently; (3) the repair/replacement is inexpensive; and (4) the repair/replacement does not appreciably prolong the life of the unit.*

"What You need to Know About Modifications/Major Modifications" by Robert Meyer at p. 28. (Exhibit 15).

NO

Jeffrey Holmstead

Testimony of Jeffrey Holmstead

Good morning Chairmen and members of the committees. Thank you for the opportunity to talk with you about the New Source Review (NSR) program under the Clean Air Act and the proposed improvements we have announced.

There has been longstanding agreement among virtually all interested parties that the NSR program can and should be improved. For well over ten years, representatives of industry, state and local agencies, and environmental groups have worked closely with EPA to find ways to make the program work better. In 1996, EPA proposed rules to amend several key elements of the program. In 1998, EPA sought additional public input on related issues. Since 1996, EPA has had countless discussions with stakeholders and has invested substantial resources in an effort to develop final revisions to the program. Between the 1996 proposal and January 2001, EPA held two public hearings and more than 50 stakeholder meetings. Environmental groups, industry, and state, local and federal agency representatives participated in these many discussions. Over 600 detailed comments were submitted to EPA between 1992 and 2001.

In 2001, the National Energy Policy Development Group asked EPA to investigate the impact of NSR on investment in new utility and refinery generation capacity, energy efficiency and environmental protection. During this review, the Agency met with more than 100 groups, held four public meetings around the country, and received more than 130,000 written comments. EPA issued a report to President Bush on June 13 in which we concluded that the NSR program does, in fact, adversely affect or discourage some projects at existing facilities that would maintain or improve reliability, efficiency, and safety of existing energy capacity. This report lends strong support to the decade-long effort to improve the NSR program.

We now believe that it is time to finish the task of improving and reforming the NSR program. At the same time that we submitted our report to the President, we published a set of recommended reforms that we intend to make to the NSR program. These reforms are designed to remove barriers to environmentally beneficial projects, provide incentives for companies to install good controls and reduce actual emissions, specify when NSR applies, and streamline and simplify several key NSR provisions.

From U.S. Senate, Committee on Environment and Public Works and Committee on the Judiciary. *New Source Review Program of the Clean Air Act.* Hearing, July 16, 2002. Washington, DC: Government Printing Office, 2002.

We plan to move ahead with this rulemaking effort in the very near future. We look forward to working with you during this important effort.

Background

The NSR program is by no means the primary regulatory tool to address air pollution from existing sources. The Clean Air Act provides authority for several other public health-driven and visibility-related control efforts: for example, the National Ambient Air Quality Standards (NAAQS) Program implemented through enforceable State Implementation Plans, the NOx SIP Call, the Acid Rain Program, the Regional Haze Program, the National Emissions Standards for Hazardous Air Pollutants (NESHAP) program, etc. Thus, while NSR was designed by Congress to focus particularly on sources that are newly constructed or that make major modifications, Congress provided numerous other tools for assuring that emissions from existing sources are adequately controlled.

The NSR provisions of the Clean Air Act combine air quality planning, air pollution technology requirements, and stakeholder participation. NSR is a preconstruction permitting program. If new construction or making a modification will increase emissions by an amount large enough to trigger NSR requirements, then the source must obtain a permit before it can begin construction. To obtain the permit, the owners must meet several requirements, including applying state-of-the-art control technology. States are key partners in the program. Under the Act, States have the primary responsibility for issuing permits, and they can customize their NSR programs within the limits of EPA regulations. EPA's role has been approving State programs and assuring consistency with EPA rules, the State's implementation plan, and the Clean Air Act. EPA also issues permits where there is no approved NSR program, such as on some Tribal lands.

The NSR permit program for major sources has two different components— one for areas with air quality problems, and the other for areas where the air is cleaner. Under the Clean Air Act, geographic areas, such as counties or metropolitan statistical areas, are designated as "attainment" or "nonattainment" for the NAAQS, which are the air quality standards used to protect human health and the environment. Preconstruction permits for sources located in attainment or unclassifiable areas are called Prevention of Significant Deterioration (PSD) permits and those for sources located in nonattainment areas are called nonattainment NSR permits.

A major difference in the two programs is that the control technology requirement is more stringent in nonattainment areas and is called the Lowest Achievable Emission Rate (LAER). In attainment areas, a source must apply Best Available Control Technology (BACT). The statute allows consideration of cost in determining BACT.

Also, in keeping with the goal of progress toward attaining the NAAQS, sources in nonattainment areas must always provide or purchase "offsets"— decreases in emissions which compensate for the increases from the new source or modification. In attainment areas, PSD sources typically do not need to obtain offsets. However, under the PSD provisions, facilities are required to undertake an air quality modeling analysis of the impact of the construction project. If the analysis finds that the project contributes to ambient air pollution that exceeds allowable levels, the facility must take steps to reduce emissions and mitigate this impact. In addition to ensuring compliance with the NAAQS, States track and control emissions of air pollution by calculating the maximum increase in concentration allowed to occur above an established background level—that change in concentration is known as a PSD increment.

Another key requirement is the provision in the PSD program to protect pristine areas like national parks or wilderness areas, also referred to as Class I areas. If a source constructs or modifies in a way that could affect a Class I area, the law allows a federal land manager, for example, a National Park Service superintendent, an opportunity to review the permit and the air quality analysis to assure that relevant factors associated with the protection of national parks and wilderness areas are taken into consideration, and, if necessary, that harmful effects are mitigated.

Current Status of the NSR Program

Let me give you a few statistics about the NSR program to put things in perspective. Estimates based on our most recent data indicate that typically more than 250 facilities apply for a PSD or nonattainment NSR permit annually. The nonattainment NSR and PSD programs are designed to focus on changes to facilities that have a major impact on air quality.

EPA has worked for over 10 years to make changes to the NSR program to provide more flexibility and certainty for industry while ensuring environmental protection. In 1992, EPA issued a regulation addressing issues regarding NSR at electric utility steam generating units making major modifications. This is referred to as the "WEPCO" rule. And in 1996, EPA proposed to make changes to the existing NSR program that would significantly streamline and simplify the program. In 1998, EPA issued a notice of availability where we asked for additional public comment on several issues.

EPA held public hearings and more than 50 stakeholder meetings on the 1996 proposed rules and related issues. Environmental groups, industry, and State, local and Federal agency representatives variously participated in these discussions. Despite widespread acknowledgment of the need for reforms, EPA has not yet finalized these proposed regulations.

In May 2001, the President issued the National Energy Policy. The Policy included numerous recommendations for action, including a recommendation that the EPA Administrator, in consultation with the Secretary of Energy and other relevant agencies, review New Source Review regulations, including administrative interpretation and implementation. The recommendation requested EPA to issue a report to the President on the impact of the regulations on investment in new utility and refinery generation capacity, energy efficiency, and environmental protection.

In June 2001, EPA issued a background paper giving an overview of the NSR program. EPA solicited public comments on the background paper and other information relevant to New Source Review. In developing the final report responding to the National Energy Policy recommendation, EPA met with more than 100 industry, environmental, and consumer groups, and public officials, held public meetings around the country, and evaluated more than 130,000 written comments.

On June 13, 2002, EPA submitted the final report on NSR to President Bush. At that time, EPA also released a set of recommended reforms to the program. With regard to the energy sector, EPA found that the NSR program has not significantly impeded investment in new power plants or refineries. For the utility industry, this is evidenced by significant recent and future planned investment in new power plants. Lack of construction of new greenfield refineries is generally attributed to economic reasons and environmental or other permitting restrictions unrelated to NSR.

With respect to the maintenance and operation of existing utility generation capacity, there is more evidence of adverse impacts from NSR. EPA's review found that uncertainty about the exemption for routine activities has resulted in the delay or cancellation of some projects that would maintain or improve reliability, efficiency and safety of existing energy capacity. Reforms to NSR will remove barriers to pollution prevention projects, energy efficiency improvements, and investments in new technologies and modernization of facilities.

EPA announced that it intends to take a series of actions to improve the NSR program, promote energy efficiency and pollution prevention, and enhance energy security while encouraging emissions reductions.

These improvements include finalizing NSR rule changes that were proposed in 1996 and recommending some new changes to the rules. The 1996 recommendations and subsequent notice of availability were subject to extensive technical review and public comment over the past six years. EPA will conduct notice-and-comment rulemaking for changes not proposed in 1996.

Our actions are completely consistent with the strong public health protection provided by the Clean Air Act. The key provisions of the Clean Air Act include several programs designed to protect human health and the environment from the harmful effects of air pollution and all of them remain in place. Moreover, the changes that we make to the NSR program will be prospective in nature, and EPA will continue to vigorously pursue its current enforcement actions. Accordingly, EPA does not intend for its future rulemaking or proposed changes to be used in, or have any impact on, current litigation.

Summary of Improvements

Congress established the New Source Review Program in order to maintain or improve air quality while still providing for economic growth. The reforms announced last month will improve the program to ensure that it is meeting these goals. These reforms will:

- Provide greater assurance about which activities are covered by the NSR program;
- Remove barriers to environmentally beneficial projects;

- Provide incentives for industries to improve environmental performance when they make changes to their facilities; and
- Maintain provisions of NSR and other Clean Air Act programs that protect air quality.

The following NSR reforms, all of which were originally proposed in 1996, have been subject to extensive technical review and public comment:

$ **Pollution control and prevention projects** To encourage pollution control and prevention, EPA will create a simplified process for companies that undertake environmentally beneficial projects. NSR can discourage investments in certain pollution control and prevention projects, even if they are environmentally beneficial.

$ **Plantwide Applicability Limits (PALs)** To provide facilities with greater flexibility to modernize their operations without increasing air pollution, a facility would agree to operate within strict site-wide emissions caps called PALs. PALs provide clarity, certainty and superior environmental protection.

$ **Clean unit provision** To encourage the installation of state-of-the-art air pollution controls, EPA will give plants that install "clean units" operational flexibility if they continue to operate within permitted limits. Clean units must have an NSR permit or other regulatory limit that requires the use of the best air pollution control technologies.

$ **Calculating emissions increases and establishing actual emissions baseline** Currently, the NSR program estimates emissions increases based upon what a plant would emit if operated 24 hours a day, year-round. This can make it difficult to make certain modest changes in a facility without triggering NSR, even if those changes will not actually increase emissions. This commonsense reform will require an evaluation of how much a facility will actually emit after the proposed change. Also, to more accurately measure actual emissions, account for variations in business cycles, and clarify what may be a "more representative" period, facilities will be allowed to use any consecutive 24-month period in the previous decade as a baseline, as long as all current control requirements are taken into account.

EPA also intends to propose three new reforms that will go through the full rulemaking process, including public comment, before they are finalized. These include:

C **Routine maintenance, repair and replacement** To increase environmental protection and promote the implementation of routine repair and replacement projects, EPA will propose a new definition of "routine" repairs. NSR excludes repairs and maintenance activities that are "routine," but a multifactored case-by-case determination must currently be made regarding what repairs meet that standard. This has deterred some companies from conducting certain repairs because they are not sure whether they would need to go through NSR. EPA is proposing guidelines for particular industries to more clearly establish what activities meet this standard.

C Debottlenecking EPA is proposing a rule to specify how NSR will apply when a company modifies one part of a facility in such a way that throughput in other parts of the facility increases (i.e., implements a "debottlenecking" project). Under the current rules, determining whether NSR applies to such complex projects is difficult and can be time consuming.

C Aggregation Currently, when multiple projects are implemented in a short period of time, a detailed analysis must be performed to determine whether the projects should be treated separately or together (i.e., "aggregated") under NSR. EPA's proposal will establish two criteria that will guide this determination.

It is important to note that we are undertaking changes in the NSR program at the same time as we are moving forward on the President's historic Clear Skies Initiative. The Clear Skies Initiative is the most important new clean air initiative in a generation, and will cut power plant emissions of three of the worst air pollutants—nitrogen oxides, sulfur dioxide, and mercury—by 70 percent. The initiative will improve air quality and public health, protect wildlife, habitats and ecosystems. By using a proven, market-based approach, Clear Skies will make these reductions further, faster, cheaper, and with more certainty than the current Clear Air Act. In the next decade alone, Clear Skies will remove 35 million more tons of air pollution than the current Clean Air Act.

In summary, the NSR reforms will remove the obstacles to environmentally beneficial projects, simplify NSR requirements, encourage emissions reductions, promote pollution prevention, provide incentives for energy efficient improvements, and help assure worker and plant safety. Overall, our reforms will improve the program so that industry will be able to make improvements to their plants that will result in greater environmental protection without needing to go through a lengthy permitting process. Our actions are completely consistent with key provisions of the Clean Air Act designed to protect human health and the environment from the harmful effects of air pollution.

POSTSCRIPT

Should Existing Power Plants Be Required to Install State-of-the-Art Pollution Controls?

The EPA's revised NSR rules were announced in November 2002 and were to go into effect in March 2003 (with a comment period extended to May). It broadened the definition of "routine maintenance, repair, and replacement" activities that would not trigger NSR, made costs—not emissions—a trigger for NSR review, and redefined emissions levels to cover whole power plant sites rather than portions of those sites. The attorneys general of nine Northeastern states—including Spitzer—sued the EPA over the new rules in December. New York, Connecticut, Maine, Maryland, Massachusetts, New Hampshire, New Jersey, Rhode Island, and Vermont were soon joined by California, Delaware, Illinois, Pennsylvania, and Wisconsin. A month later, the attorneys general of Virginia, Indiana, Kansas, Nebraska, North Dakota, South Carolina, South Dakota, and Utah (many of which include aging power plants and other industrial sites that will benefit from the new rules) chimed in on the other side, warning that if the Spitzer et al. suit succeeds, it will increase the cost and difficulty of enforcing clean air rules and achieving pollution control goals. The case is reviewed by Scott Richards and Yvette Hurt, in "Federalism and the Environment," *State Government News* (February 2004).

At the end of February 2003, the environmental group Earthjustice filed another suit on behalf of a coalition that includes the American Lung Association, Communities for a Better Environment, Environmental Defense, the Natural Resources Defense Council (NRDC), and the Sierra Club. Their argument gained strength when it was reported that airborne particulate pollution causes DNA mutations. See Christopher M. Somers, Brian E. McCarry, Farideh Malek, and James S. Quinn, "Reduction of Particulate Air Pollution Lowers the Risk of Heritable Mutations in Mice," *Science* (May 14, 2004). Larry Morandi and Molly Stauffer, in "Winds of Change," *State Legislatures* (May 2003), summarize the debate over New Source Review and note that some states are taking their own approaches. Both New Hampshire and North Carolina, for instance, require emissions reductions but leave the method up to the companies affected; methods may include emissions trading.

Stephen L. Kass and Jean M. McCarroll, in "New Source Review Under the Clean Air Act," *New York Law Journal*, Part 1 (December 27, 2002) and Part 2 (February 28, 2003), conclude, "The EPA is now proposing a rule that would make costs, not emissions, a trigger for NSR review. That, in our view, is not consistent with the purpose of ... the Clean Air Act ... and is likely to retard, rather than advance, the nation's compliance with its air quality goals and its search for more efficient use of energy."

In April 2003, the National Academy of Public Administration issued the report "A Breath of Fresh Air: Reviving the New Source Review Program," (http://www.napawash.org/Pubs/Fresh%20Air%20Full%20Report.pdf), which concluded that NSR "is critical for protecting public health" and has worked well for controlling pollution from newly built power plants and industrial facilities but has failed when applied to older ones. The report recommended a performance-based system that would force the closure of older facilities that did not install best-available pollution controls and bring their emissions within legal limits within 10 years. After delays, the EPA announced at the end of August 2003 that the new NSR rules would go into effect in December 2003. Up to 20 percent of the cost of replacing a power plant's equipment would be considered "routine maintenance" and hence exempt from NSR. Lawsuits continued, and in December the U.S. Circuit Court of Appeals for the District of Columbia blocked the changes, at least until the lawsuits can be heard in court in late 2004 or 2005. The court said that the challengers had successfully "demonstrated the irreparable harm and likelihood of success" of their case (necessary criteria for keeping a rule from taking effect). It is, however, too early to say whether the federal government or its challengers will finally prevail.

By April 2004, there were signs of a turnaround in the EPA's pro-industry stance, for it was proposing measures to help states force older power plants to reduce haze-causing air pollution. But again, it is too early to say whether or not this indicates a genuine change in attitude.

ISSUE 12

Is It Time to Revive Nuclear Power?

YES: Stephen Ansolabehere et al., from "The Future of Nuclear Power," *An Interdisciplinary MIT Study* (MIT 2003)

NO: Karl Grossman, from "The Push to Revive Nuclear Power," *Synthesis/Regeneration* 28 (http://www.greens.org/s-r/28/28-21.html) (Spring 2002)

ISSUE SUMMARY

YES: Professor Stephen Ansolabehere, et al. argue that greatly expanded use of nuclear power should not be excluded as a way to meet future energy needs and reduce the carbon emissions that contribute to global warming.

NO: Professor of journalism Karl Grossman argues that to encourage the use of nuclear power is reckless. He concludes that it would be wiser to promote renewable energy and energy efficiency.

The technology of releasing for human use the energy that holds the atom together got off to an auspicious start. Its first significant application was military, and the deaths associated with the Hiroshima and Nagasaki explosions have ever since tainted the technology. It did not help that for the ensuing half century, millions of people grew up under the threat of nuclear Armageddon. But almost from the beginning, nuclear physicists and engineers wanted to put nuclear energy to more peaceful uses, largely in the form of power plants. Touted in the 1950s as an astoundingly cheap source of electricity, nuclear power soon proved to be more expensive than conventional sources, largely because safety concerns caused delays in the approval process and prompted elaborate built-in precautions. Many say that safety measures have worked well when needed—Three Mile Island, often cited as a horrific example of what can go wrong with nuclear power, released very little radioactive material to the environment. The Chernobyl disaster occurred when safety measures were ignored. In both cases, human error was more to blame than the technology itself. The related issue of nuclear waste (see Issue 19) has also raised fears and added expense to the technology.

194

It is clear that two factors—fear and expense—impede the wide adoption of nuclear power. If both could somehow be alleviated, it might become possible to gain the benefits of the technology. Among those benefits are that nuclear power does not burn oil, coal, nor any other fuel; does not emit air pollution and thus contribute to smog and haze; does not depend on foreign sources of fuel and thus weaken national independence; and does not emit carbon dioxide. The last may be the most important benefit at a time when society is concerned about global warming, and it is the one that prompted James Lovelock, creator of the Gaia Hypothesis and an inspiration to many environmentalists, to say, "If we had nuclear power we wouldn't be in this mess now, and whose fault was it? It was [the antinuclear environmentalists']." See his autobiography, *Homage to Gaia: The Life of an Independent Scientist* (Oxford University Press, 2001). The Organisation for Economic Co-operation and Development (OECD's) Nuclear Energy Agency, in "Nuclear Power and Climate Change," (Paris, France, 1998), available at http:// www.nea.fr/html/ndd/climate/climate.pdf, found that a greatly expanded deployment of nuclear power to combat global warming was both technically and economically feasible. In 2000 Robert C. Morris published *The Environmental Case for Nuclear Power: Economic, Medical, and Political Considerations* (Paragon House). In August 2000 *USA Today Magazine* published "A Nuclear Solution to Global Warming?" "The time seems right to reconsider the future of nuclear power," say James A. Lake, Ralph G. Bennett, and John F. Kotek, in "Next-Generation Nuclear Power," *Scientific American* (January 2002). See also I. Fells, "Clean and Secure Energy for the Twenty-First Century," *Proceedings of the Institution of Mechanical Engineers, Part A—Power & Energy* (August 1, 2002). This is not just a Western sentiment, for nuclear power and its antiwarming effects may have special meaning in developing countries. See R. Ramachandran, "The Case for Nuclear Power," *Frontline* (November 26–December 3, 2002), available at http:// www.flonnet.com/fl1924/stories/20021206005911700.htm.

In the following selections, Stephen Ansolabehere, et al. argue that greatly expanded use of nuclear power should not be excluded as a way to meet future energy needs and reduce the carbon emissions that contribute to global warming, although due attention must be paid to reducing costs and risks. Karl Grossman argues that to encourage the use of nuclear power is reckless. "Instead of promoting dangerous and dirty forms of energy, the United States should be a world leader in promoting renewable energy and energy efficiency," he concludes.

Stephen Ansolabehere et al.  **YES**

The Future of Nuclear Power— Overview and Conclusions

The generation of electricity from fossil fuels, notably natural gas and coal, is a major and growing contributor to the emission of carbon dioxide—a greenhouse gas that contributes significantly to global warming. We share the scientific consensus that these emissions must be reduced and believe that the U.S. will eventually join with other nations in the effort to do so.

At least for the next few decades, there are only a few realistic options for reducing carbon dioxide emissions from electricity generation:

- increase efficiency in electricity generation and use;
- expand use of renewable energy sources such as wind, solar, biomass, and geothermal;
- capture carbon dioxide emissions at fossil-fueled (especially coal) electric generating plants and permanently sequester the carbon; and
- increase use of nuclear power.

The goal of this interdisciplinary MIT study is not to predict which of these options will prevail or to argue for their comparative advantages. In *our view, it is likely that we shall need all of these options and accordingly it would be a mistake at this time to exclude any of these four options from an overall carbon emissions management strategy.* Rather we seek to explore and evaluate actions that could be taken to maintain nuclear power as one of the significant options for meeting future world energy needs at low cost and in an environmentally acceptable manner.

In 2002, nuclear power supplied 20% of United States and 17% of world electricity consumption. Experts project worldwide electricity consumption will increase substantially in the coming decades, especially in the developing world, accompanying economic growth and social progress. However, official forecasts call for a mere 5% increase in nuclear electricity generating capacity worldwide by 2020 (and even this is questionable), while electricity use could grow by as much as 75%. These projections entail little new nuclear plant construction and reflect both economic considerations and growing anti-nuclear

From An *Interdisciplinary MIT Study*, 2003. Copyright © 2003 by MIT Press. Reprinted by permission.

sentiment in key countries. The limited prospects for nuclear power today are attributable, ultimately, to four unresolved problems:

- *Costs: nuclear power has higher overall lifetime costs* compared to natural gas with combined cycle turbine technology (CCGT) and coal, at least in the absence of a carbon tax or an equivalent "cap and trade" mechanism for reducing carbon emissions;
- *Safety: nuclear power has perceived adverse safety, environmental, and health effects*, heightened by the 1979 Three Mile Island and 1986 Chernobyl reactor accidents, but also by accidents at fuel cycle facilities in the United States, Russia, and Japan. There is also growing concern about the safe and secure transportation of nuclear materials and the security of nuclear facilities from terrorist attack;
- *Proliferation: nuclear power entails potential security risks*, notably the possible misuse of commercial or associated nuclear facilities and operations to acquire technology or materials as a precursor to the acquisition of a nuclear weapons capability. Fuel cycles that involve the chemical reprocessing of spent fuel to separate weapons-usable plutonium and uranium enrichment technologies are of special concern, especially as nuclear power spreads around the world;
- *Waste: nuclear power has unresolved challenges in long-term management of radioactive wastes.* The United States and other countries have yet to implement final disposition of spent fuel or high level radioactive waste streams created at various stages of the nuclear fuel cycle. Since these radioactive wastes present some danger to present and future generations, the public and its elected representatives, as well as prospective investors in nuclear power plants, properly expect continuing and substantial progress towards solution to the waste disposal problem. Successful operation of the planned disposal facility at Yucca Mountain would ease, but not solve, the waste issue for the U.S. and other countries if nuclear power expands substantially.

Today, nuclear power is not an economically competitive choice. Moreover, unlike other energy technologies, nuclear power requires significant government involvement because of safety, proliferation, and waste concerns. If in the future carbon dioxide emissions carry a significant "price," however, nuclear energy could be an important—indeed vital—option for generating electricity. We do not know whether this will occur. But *we believe the nuclear option should be retained, precisely because it is an important carbon-free source of power that can potentially make a significant contribution to future electricity supply.*

To preserve the nuclear option for the future requires overcoming the four challenges described above—costs, safety, proliferation, and wastes. These challenges will escalate if a significant number of new nuclear generating plants are built in a growing number of countries. The effort to overcome these challenges, however, is justified only if nuclear power can potentially contribute significantly to reducing global warming, which entails major expansion of nuclear power. In effect, preserving the nuclear option for the future means planning for growth, as well as for a future in which nuclear energy is a competitive, safer, and more secure source of power.

To explore these issues, our study postulates a *global growth scenario* that by mid-century would see 1000 to 1500 reactors of 1000 megawatt-electric (MWe) capacity each deployed worldwide, compared to a capacity equivalent to 366 such reactors now in service. Nuclear power expansion on this scale requires U.S. leadership, continued commitment by Japan, Korea, and Taiwan, a renewal of European activity, and wider deployment of nuclear power around the world....

This scenario would displace a significant amount of carbon-emitting fossil fuel generation. In 2002, carbon equivalent emission from human activity was about 6,500 million tonnes per year; these emissions will probably more than double by 2050. The 1000 GWe [gigawatt-electric] of nuclear power postulated here would avoid annually about 800 million tonnes of carbon equivalent if the electricity generation displaced was gas-fired and 1,800 million tonnes if the generation was coal-fired, assuming no capture and sequestration of carbon dioxide from combustion sources.

Fuel Cycle Choices

A critical factor for the future of an expanded nuclear power industry is the choice of the fuel cycle—what type of fuel is used, what types of reactors "burn" the fuel, and the method of disposal of the spent fuel. This choice affects all four key problems that confront nuclear power—costs, safety, proliferation risk, and waste disposal. For this study, we examined three representative nuclear fuel cycle deployments:

- *conventional thermal reactors operating in a "once through" mode*, in which discharged spent fuel is sent directly to disposal;
- *thermal reactors with reprocessing in a "closed" fuel cycle*, which means that waste products are separated from unused fissionable material that is re-cycled as fuel into reactors. This includes the fuel cycle currently used in some countries in which plutonium is separated from spent fuel, fabricated into a mixed plutonium and uranium oxide fuel, and recycled to reactors for one pass;
- *fast reactors with reprocessing in a balanced "closed" fuel cycle*, which means thermal reactors operated world-wide in "once-through" mode and a balanced number of fast reactors that destroy the actinides separated from thermal reactor spent fuel. The fast reactors, reprocessing, and fuel fabrication facilities would be co-located in secure nuclear energy "parks" in industrial countries.

Closed fuel cycles extend fuel supplies. The viability of the once-through alternative in a global growth scenario depends upon the amount of uranium resource that is available at economically attractive prices. *We believe that the world-wide supply of uranium ore is sufficient to fuel the deployment of 1000 reactors over the next half century* and to maintain this level of deployment over a 40 year lifetime of this fleet. This is an important foundation of our study, based upon currently available information and the history of natural resource supply....

Our analysis leads to a significant conclusion: *The once-through fuel cycle best meets the criteria of low costs and proliferation resistance.* Closed fuel cycles may have an advantage from the point of view of long-term waste disposal and, if it ever becomes relevant, resource extension. But closed fuel cycles will be more expensive than once-through cycles, until ore resources become very scarce. This is unlikely to happen, even with significant growth in nuclear power, until at least the second half of this century, and probably considerably later still. Thus our most important recommendation is:

> For the next decades, government and industry in the U.S. and elsewhere should give priority to the deployment of the once-through fuel cycle, rather than the development of more expensive closed fuel cycle technology involving reprocessing and new advanced thermal or fast reactor technologies.

This recommendation implies a major re-ordering of priorities of the U.S. Department of Energy [DOE] nuclear R&D [research and development] programs.

Public Attitudes Toward Nuclear Power

Expanded deployment of nuclear power requires public acceptance of this energy source. Our review of survey results shows that a majority of Americans and Europeans oppose building new nuclear power plants to meet future energy needs. To understand why, we surveyed 1350 adults in the US about their attitudes toward energy in general and nuclear power in particular. Three important and unexpected results emerged from that survey:

- The U.S. public's attitudes are informed almost entirely by their perceptions of the technology, rather than by politics or by demographics such as income, education, and gender.
- The U.S. public's views on nuclear waste, safety, and costs are critical to their judgments about the future deployment of this technology. Technological improvements that lower costs and improve safety and waste problems can increase public support substantially.
- In the United States, people do not connect concern about global warming with carbon-free nuclear power. There is no difference in support for building more nuclear power plants between those who are very concerned about global warming and those who are not. Public education may help improve understanding about the link between global warming, fossil fuel usage, and the need for low-carbon energy sources.

There are two implications of these findings for our study: first, the U.S. public is unlikely to support nuclear power expansion without substantial improvements in costs and technology. Second, the carbon-free character of nuclear power, the major motivation for our study, does not appear to motivate the U.S. general public to prefer expansion of the nuclear option.

Economics

Nuclear power will succeed in the long run only if it has a lower cost than competing technologies. This is especially true as electricity markets become progressively less subject to economic regulation in many parts of the world. We constructed a model to evaluate the real cost of electricity from nuclear power versus pulverized coal plants and natural gas combined cycle plants (at various projected levels of real lifetime prices for natural gas), over their economic lives. These technologies are most widely used today and, absent a carbon tax or its equivalent, are less expensive than many renewable technologies....

[C]ost improvements for nuclear power [are] plausible, but not proven. The model results make clear why electricity produced from new nuclear power plants today is not competitive with electricity produced from coal or natural gas-fueled CCGT plants with low or moderate gas prices, unless *all* cost improvements for nuclear power are realized. The cost comparison becomes worse for nuclear if the capacity factor falls. It is also important to emphasize that the nuclear cost structure is driven by high up-front capital costs, while the natural gas cost driver is the fuel cost; coal lies in between nuclear and natural gas with respect to both fuel and capital costs.

Nuclear does become more competitive by comparison if the social cost of carbon emissions is internalized, for example through a carbon tax or an equivalent "cap and trade" system.... The ultimate cost will depend on both societal choices (such as how much carbon dioxide emission to permit) and technology developments, such as the cost and feasibility of large-scale carbon capture and long-term sequestration.... [C]osts in the range of $100 to $200/tonne C would significantly affect the relative cost competitiveness of coal, natural gas, and nuclear electricity generation.

The carbon-free nature of nuclear power argues for government action to encourage maintenance of the nuclear option, particularly in light of the regulatory uncertainties facing the use of nuclear power and the unwillingness of investors to bear the risk of introducing a new generation of nuclear facilities with their high capital costs.

We recommend three actions to improve the economic viability of nuclear power:

> The government should cost share for site banking for a number of plants, certification of new plant designs by the Nuclear Regulatory Commission, and combined construction and operating licenses for plants built immediately or in the future; we support U.S. Department of Energy initiatives on these subjects.
>
> The government should recognize nuclear as carbon-free and include new nuclear plants as an eligible option in any federal or state mandatory renewable energy portfolio (i.e., a "carbon-free" portfolio) standard.
>
> The government should provide a modest subsidy for a small set of "first mover" commercial nuclear plants to demonstrate cost and regulatory feasibility in the form of a production tax credit.

We propose a production tax credit of up to $200 per kWe of the construction cost of up to 10 "first mover" plants. This benefit might be paid out at about 1.7 cents per kWe-hr, over a year and a half of full-power plant operation. We prefer the production tax credit mechanism because it offers the greatest incentive for projects to be completed and because it can be extended to other carbon free electricity technologies, for example renewables, (wind currently enjoys a 1.7 cents per kWe-hr tax credit for ten years) and coal with carbon capture and sequestration. The credit of 1.7 cents per kWe-hr is equivalent to a credit of $70 per avoided metric ton of carbon if the electricity were to have come from coal plants (or $160 from natural gas plants). Of course, the carbon emission reduction would then continue without public assistance for the plant life (perhaps 60 years for nuclear). If no new nuclear plant is built, the government will not pay a subsidy.

These actions will be effective in stimulating additional investment in nuclear generating capacity if, and only if, the industry can live up to its own expectations of being able to reduce considerably capital costs for new plants.

Advanced fuel cycles add considerably to the cost of nuclear electricity.We considered reprocessing and one-pass fuel recycle with current technology, and found the fuel cost, including waste storage and disposal charges, to be about 4.5 times the fuel cost of the once-through cycle. Thus use of advanced fuel cycles imposes a significant economic penalty on nuclear power.

Safety

We believe the safety standard for the global growth scenario should maintain today's standard of less than one serious release of radioactivity accident for 50 years from all fuel cycle activity. This standard implies a ten-fold reduction in the expected frequency of serious reactor core accidents, from 10^{-4}/reactor year to 10^{-5}/reactor year. This reactor safety standard should be possible to achieve in new light water reactor plants that make use of advanced safety designs. International adherence to such a standard is important, because an accident in any country will influence public attitudes everywhere. The extent to which nuclear facilities should be hardened to possible terrorist attack has yet to be resolved.

We do not believe there is a nuclear plant design that is totally risk free. In part, this is due to technical possibilities; in part due to workforce issues. Safe operation requires effective regulation, a management committed to safety, and a skilled work force.

The high temperature gas-cooled reactor is an interesting candidate for reactor research and development because there is already some experience with this system, although not all of it is favorable. This reactor design offers safety advantages because the high heat capacity of the core and fuel offers longer response times and precludes excessive temperatures that might lead to release of fission products; it also has an advantage compared to light water reactors in terms of proliferation resistance.

Because of the accidents at Three Mile Island in 1979 and Chernobyl in 1986, a great deal of attention has focused on reactor safety. However, the safety record of reprocessing plants is not good, and there has been little safety analysis of fuel cycle facilities using, for example, the probabilistic risk assessment method. More work is needed here.

Our principal recommendation on safety is:

> The government should, as part of its near-term R&D program, develop more
> fully the capabilities to analyze life-cycle health and safety impacts of fuel

cycle facilities and focus reactor development on options that can achieve enhanced safety standards and are deployable within a couple of decades.

Waste Management

The management and disposal of high-level radioactive spent fuel from the nuclear fuel cycle is one of the most intractable problems facing the nuclear power industry throughout the world. No country has yet successfully implemented a system for disposing of this waste. We concur with the many independent expert reviews that have concluded that geologic repositories will be capable of safely isolating the waste from the biosphere. However, implementation of this method is a highly demanding task that will place great stress on operating, regulatory, and political institutions.

For fifteen years the U.S. high-level waste management program has focused almost exclusively on the proposed repository site at Yucca Mountain in Nevada. Although the successful commissioning of the Yucca Mountain repository would be a significant step towards the secure disposal of nuclear waste, we believe that a broader, strategically balanced nuclear waste program is needed to prepare the way for a possible major expansion of the nuclear power sector in the U.S. and overseas.

The global growth scenario, based on the once-through fuel cycle, would require multiple disposal facilities by the year 2050. To dispose of the spent fuel from a steady state deployment of one thousand 1 GWe reactors of the light water type, new repository capacity equal to the nominal storage capacity of Yucca Mountain would have to be created somewhere in the world every three to four years. This requirement, along with the desire to reduce long-term risks from the waste, prompts interest in advanced, closed fuel cycles. These schemes would separate or partition plutonium and other actinides— and possibly certain fission products—from the spent fuel and transmute them into shorter-lived and more benign species. The goals would be to reduce the thermal load from radioactive decay of the waste on the repository, thereby increasing its storage capacity, and to shorten the time for which the waste must be isolated from the biosphere.

We have analyzed the waste management implications of both once-through and closed fuel cycles, taking into account each stage of the fuel cycle and the risks of radiation exposure in both the short and long-term. *We do not believe that a convincing case can be made on the basis of waste management considerations alone that the benefits of partitioning and transmutation will outweigh the attendant safety, environmental, and security risks and economic costs.* Future technology developments could change the balance of expected costs, risks, and benefits. For our fundamental conclusion to change, however, not only would the expected long term risks from geologic repositories have to be significantly higher than those indicated in current assessments, but the incremental costs and short-term safety and environmental risks would have to be greatly reduced relative to current expectations and experience.

We further conclude that waste management strategies in the once-through fuel cycle are potentially available that could yield long-term risk reductions at least as great as those claimed for waste partitioning and transmutation, with fewer short-term risks and lower development and deployment costs. These include both incremental improvements to the current mainstream mined repositories approach and more far-reaching innovations such as deep borehole disposal. Finally, replacing the current ad hoc approach to spent fuel storage at reactor sites with an explicit strategy to store spent fuel for a period of several decades will create additional flexibility in the waste management system.

Our principal recommendations on waste management are:

The DOE should augment its current focus on Yucca Mountain with a balanced long-term waste management R&D program.

A research program should be launched to determine the viability of geologic disposal in deep boreholes within a decade.

A network of centralized facilities for storing spent fuel for several decades should be established in the U.S. and internationally.

Nonproliferation

Nuclear power should not expand unless the risk of proliferation from operation of the commercial nuclear fuel cycle is made acceptably small. We believe that nuclear power can expand as envisioned in our global growth scenario with acceptable incremental proliferation risk, provided that reasonable safeguards are adopted and that deployment of reprocessing and enrichment are restricted. The international community must prevent the acquisition of weapons-usable material, either by diversion (in the case of plutonium) or by misuse of fuel cycle facilities (including related facilities, such as research reactors or hot cells). Responsible governments must control, to the extent possible, the know-how relevant to produce and process either highly enriched uranium (enrichment technology) or plutonium.

Three issues are of particular concern: existing stocks of *separated* plutonium around the world that are directly usable for weapons; nuclear facilities, for example in Russia, with inadequate controls; and transfer of technology, especially enrichment and reprocessing technology, that brings nations closer to a nuclear weapons capability. The proliferation risk of the global growth scenario is underlined by the likelihood that use of nuclear power would be introduced and expanded in many countries in different security circumstances.

An international response is required to reduce the proliferation risk. The response should:

- re-appraise and strengthen the institutional underpinnings of the IAEA safeguards regime in the near term, including sanctions;
- guide nuclear fuel cycle development in ways that reinforce shared nonproliferation objectives.

Accordingly, we recommend:

The International Atomic Energy Agency (IAEA) should focus overwhelmingly on its safeguards function and should be given the authority to carry out inspections beyond declared facilities to suspected illicit facilities;

Greater attention must be given to the proliferation risks at the front end of the fuel cycle from enrichment technologies;

IAEA safeguards should move to an approach based on continuous materials protection, control and accounting using surveillance and containment systems, both in facilities and during transportation, and should implement safeguards in a risk-based framework keyed to fuel cycle activity;

Fuel cycle analysis, research, development, and demonstration efforts must include explicit analysis of proliferation risks and measures defined to minimize proliferation risks;

International spent fuel storage has significant nonproliferation benefits for the growth scenario and should be negotiated promptly and implemented over the next decade.

Analysis, Research, Development, and Demonstration Program

The U.S. Department of Energy (DOE) analysis, research, development, and demonstration (ARD&D) program should support the technology path leading to the global growth scenario and include diverse activities that balance risk and time scales, in pursuit of the strategic objective of preserving the nuclear option. For technical, economic, safety, and public acceptance reasons, the highest priority in fuel cycle ARD&D, deserving first call on available funds, lies with efforts that enable robust *deployment of the once-through fuel cycle*. The current DOE program does not have this focus....

Accordingly, we recommend:

The U.S. Department of Energy should focus its R&D program on the once-through fuel cycle;

The U.S. Department of Energy should establish a Nuclear System Modeling project to carryout the analysis, research, simulation, and collection of engineering data needed to evaluate all fuel cycles from the viewpoint of cost, safety, waste management, and proliferation resistance;

The U.S. Department of Energy should undertake an international uranium resource evaluation program;

The U.S. Department of Energy should broaden its waste management R&D program;

The U.S. Department of Energy should support R&D that reduces Light Water Reactor (LWR) costs and for development of the HTGR [High Temperature Gas Reactor] for electricity application.

We believe that the ARD&D program proposed here is aligned with the strategic objective of enabling a credible growth scenario over the next several decades. Such a ARD&D program requires incremental budgets of almost $400 million per year over the next 5 years, and at least $460 million per year for the 5-10 year period.

NO

Karl Grossman

The Push to Revive Nuclear Power

The [George W.] Bush administration and the nuclear industry are making an intense push to "revive" nuclear power in the United States. Diane D'Arrigo of the Nuclear Information and Resource Service (NIRS) says "relapse" is the better term: "It's the push to relapse," she says.

As Bob Alvarez, executive director of the group Standing for Truth About Radiation says, "It's like reviving Frankenstein—this is the sequel." For years—ever since the accidents at Three Mile Island and Chernobyl shattered public trust in atomic power—nuclear power advocates in government and industry have been laying the groundwork for a comeback in the US.

As Bush's Secretary of Treasury Paul O'Neill told *The Wall Street Journal*: "If you set aside Three Mile Island and Chernobyl, the safety record of nuclear is really good." (Yes, Mrs. Lincoln, apart from that, how did you enjoy the show?) The Bush administration struck a close working relationship with the nuclear industry well before taking office. The energy "transition" advisors included Joseph Colvin, president of the Nuclear Energy Institute, the self-described "policy organization of the nuclear energy and technologies industry," and other nuclear industry biggies. There was no one representing renewable energy or environmental organizations.

Two weeks after being sworn in, Bush set up a "National Energy Policy Development Group" and appointed Vice President [Dick] Cheney as its chairman. The group included O'Neill. Behind closed doors, it huddled with fat energy industry cats—indeed, the General Accounting Office is now in the process of pursuing an unprecedented lawsuit because of Cheney's refusal to disclose who this government panel met with before setting policy.

The panel, 10 weeks after being organized, issued its report declaring how it "supports the expansion of nuclear energy in the United States." The National Energy Policy plan would substantially increase the use of nuclear power in the US both by building new nuclear power plants—many to be constructed on existing nuclear plant sites—and extending the 40-year licenses of currently operating plants by another 20 years.

The National Energy Policy says: "Many US nuclear plants sites were designed to host 4 to 6 reactors, and most operate only 2 or 3; many sites across the country could host additional plants." Further, "Building new generators on existing sites avoids many complex issues associated with building plants on new sites." It would also magnify the impacts of an accident—if one nuclear plant in a cluster of plants undergoes a catastrophic accident resulting in a site evacuation and abandonment of control rooms, there is then the potential for a "cascading loss" involving additional plants, stresses Paul Gunter, who heads NIRS' Reactor Watchdog Project.

"No one foresaw" nuclear plants "running for more than 40 years," says Alvarez, who was senior policy advisor to the DOE [U.S. Department of Energy] secretary from 1993 to 1999. "These reactors are just like old machines but they are ultra-hazardous," he says, and by pushing their operating span to 60 years, "disaster is being invited." The Bush-Cheney administration National Energy Policy supports purportedly "new and improved" nuclear plants, "advanced" nukes. It says, "Advanced reactor technology promises to improve nuclear safety." The administration is especially bullish on the gas-cooled, pebble bed reactor, which it claims has inherent safety features. In fact, says Gunter, the pebble bed reactor is not new, it's just "old wine in a new bottle." It's a "hybrid" of the gas-cooled high-temperature design that "has appeared and been rejected in England, Germany and the United States." And far from being "inherently safe," a reactor of similar design, a THTR300 in the Ruhr Valley in Germany, spewed out substantial amounts of radioactivity in a 1986 accident leading to its permanent closure.

The new nuclear push would be pursued through what's called "one-step" licensing. This was part of an Energy Policy Act bill approved by a Democratic-controlled Congress in 1992—381 to 37—and signed into law by the first President George Bush.

"One-step" licensing allows the Nuclear Regulatory Commission (NRC) to hold a single hearing for a "combined construction and operating license." No longer can nuclear plant projects be slowed down or stopped at a separate operating license proceeding at which evidence of construction defects is revealed. As *The New York Times* described passage of the Energy Policy Act in a back-of-the paper story in 1992, "Nuclear power lobbyists called the bill their biggest victory in Congress since the Three Mile Island accident." As NIRS reported in its *Nuclear Monitor* in 1992: "As the bill wound its way through the Senate and House, the nuclear industry won nearly every vote that mattered, proving that Congress remains captive to industry lobbying and political contributions over public opinion." That has not changed.

Public Citizen's Critical Mass Energy and Environment Program has documented how the nuclear industry regularly showers Congress—and this includes members of both major parties—with political contributions. Likewise, nuclear industry money pours into presidential campaigns.

The website of the Nuclear Energy Institute—www.nei.org—includes a page of "Endorsements of Nuclear Energy" and among those quoted are Al Gore: "Nuclear power, designed well, regulated properly, cared for meticulously, has a place in the world's energy supply."

Gore's running mate, Senator Joseph Lieberman, is quoted as saying at a Senate hearing in 1998: "I am a supporter of nuclear energy." To make sure the public hardly participates even in the "one-step" process, the NRC is now involved in a "rulemaking" to undo what it through the years interpreted as the public's right to formal trial-type hearings on nuclear plant licensing. It seeks to "deformalize" the hearings, eliminating due process procedures. Documents would be restricted to what the NRC staff and company deem relevant, and instead of cross-examining witnesses, people will have to submit written questions as suggestions to the NRC presiding officer for he or she to ask—at their discretion—at a hearing.

Also to help in a nuclear power comeback is the effort to alter the standards for radiation exposure. As more and more has been learned about radioactivity, the realization came that any amount can kill, that there is no "safe" level. This is called the "linear no-threshold theory." Now nuclear advocates in government and industry want to alter the standards premised on a contention that low doses of radiation are not so bad after all. There is even interest in a long-rejected notion called "hormesis"—that a little radiation is good for people, that it helps exercise the immune system. The instrument for making the changes is a new Biological Effects of Ionizing Radiation (BEIR) panel of the National Academy of Sciences that is to make recommendations to the federal government. It is "stacked," notes Diane D'Arrigo of NIRS, with "radiation advocates."

Nuclear waste is another obstacle the nuclear proponents in government and industry are seeking to get around. The Bush administration is now moving to open Yucca Mountain in Nevada as a repository and also use Utah's Skull Valley Goshute Reservation and possibly other Native American reservations. For what is considered "low-level" waste, the strategy is to "recycle" it—to smelt metals down and incorporate irradiated material into consumer items.

The huge problem with using Yucca Mountain, which the government began exploring as a high-level nuclear waste repository in the 1980s, is that it is on or near 32 earthquake faults and, notes D'Arrigo, has a "history and prospects of volcanoes and a likelihood of flooding and leakage." In 1997, tribal leaders of the Goshute Reservation, as the Goshute's website notes, "leased land to a private group of electrical utilities for the temporary storage of 40,000 metric tons of spent nuclear fuel." Some members of the tribe are fighting the deal in court demanding to know who got what for what. To nuclear advocates in government and industry, collaborating with Indian reservations as sovereign nations is a way to unload atomic garbage. Critics describe it as a new form of environmental racism"—"nuclear racism"—seeking to take advantage of the poverty of Native Americans.

The drive to "recycle" nuclear waste has been percolating for years. In 1980, the NRC first proposed that irradiated "metal scrap could be converted," that "radioactive waste burial costs could be avoided [and] the smelted scrap could be made into any number of consumer or capital equipment products such as automobiles, appliances, furniture, utensils, personal items and coins." Some thought that the push for radioactive quarters and hot Pontiacs was too crazy to be true.

Meanwhile, those behind the nuclear push have moved to extend a key piece of US law that facilitated the nuclear power industry in the first place: the Price-Anderson Act, the law that drastically limits the amount of money people can collect as a result of a nuclear power plant disaster. It was enacted in 1957 as a temporary measure to give a boost to setting up a nuclear power industry. It originally limited in the event of a nuclear plant accident to $560 million with the federal government paying the first $500 million. Price-Anderson has been extended and extended, and now it's being extended once more—to provide a financial umbrella for the push to revive nuclear power. As Michael Mariotte has pointed out: "The renewal of Price-Anderson is only to build new reactors. That's the issue. Existing nuclear plants are covered by the present law." The new Price-Anderson liability limit would be $8.6 billion, a fraction of what the NRC itself has concluded would be the financial consequences of a nuclear plant accident. Those figures are contained in a 1982 report done for the NRC by the DOE's Sandia National Laboratories and titled "Calculation of Reactor Accident Consequences for US Nuclear Power Plants." It calculates—in 1980 dollars—costs as a result of a nuclear plant disaster as high as $274 billion for Indian Point 2 and $314 billion at the Indian Point 3 nuclear plants. The number of "early fatalities"—46,000 as a result of Indian Point 2 undergoing a meltdown with breach of containment, 50,000 for Indian Point 3.

And what are the chances of such a disaster occurring? In 1985, the NRC was asked by a House oversight committee to determine the "probability" of a "severe core melt accident" in the "next 20 years for those reactors now operating and those expected to operate during that time." The NRC concluded: "The crude cumulative probability of such an accident would be 45%." That disaster has not come...yet. "Luck" is the only reason it hasn't, says David Lochbaum of the Union of Concerned Scientists. But the drive to revive nuclear power, the push to relapse, will, if it succeeds, help make inevitable that catastrophe—along with extending the damage of every aspect of the nuclear power chain, from mining to milling to transportation to fuel enrichment and fabrication to reactor operation and the "routine" emissions of radioactivity and then atomic waste management in perpetuity.

And, new—but not really new—is the specter of nuclear plants as terrorist targets. In 1980, a landmark book by Bennett Ramberg was published, *Nuclear Power Plants as Weapons for the Enemy: An Unrecognized Military Peril*. Despite the "multiplication of nuclear power plants," it begins, "little public consideration has been given to their vulnerability in time of war."

Dr. Ramberg, now research director of the Los Angeles-based Committee to Bridge the Gap, said in a post-9/11 presentation at the National Press Club: "I presented my findings to the Nuclear Regulatory Commission raising questions about the vulnerability of American reactors to terrorist action. The commission dismissed my concerns." Indeed, in a rule-making in 1982, an Atomic Safety and Licensing Board of the NRC, in considering an operating license for the Shearon Harris nuclear power plant in North Carolina, dismissed a contention by an intervenor, Wells Eddleman, that the plant's safety analysis was deficient because it did not consider the "consequences of terrorists commandeering a very large airplane...and diving it into the containment." The NRC board declared: "Reactors could not be effectively protected against such attacks without turning them into virtually impregnable fortresses at much higher cost... The applicants are not required to design against such things as artillery bombardments, missiles with nuclear warheads, or kamikaze dives by large airplanes."

Meanwhile, new since 1982 is the full arrival of safe, clean, renewable energy technologies. The need is for broad-scale implementation. Wind power, solar energy, hydrogen fuel technologies including fuel cells, among other renewable energy technologies, are more than ready after years of dramatic advances. Coupled with energy efficiency, they can be tapped and widely utilized—and render nuclear power completely unnecessary.

As NIRS, Public Citizen's Critical Mass Energy and Environment Program, Greenpeace USA, Safe Energy Communication Council, and the Global Resource Action Center for the Environment said of the National Energy Policy: "The Bush/Cheney administration is recklessly promoting the building of new nuclear plants to address an energy crisis that in large part is being manufactured by the energy corporations that will benefit from building new power plants... We believe that instead of promoting dangerous and dirty forms of energy, the United States should be a world leader in promoting renewable energy and energy efficiency. Let us not sell our children's future." Amen.

POSTSCRIPT

Is It Time to Revive Nuclear Power?

In 2000 there were 100 nuclear reactors in use in the United States and 352 in the developed world as a whole. There were only 15 in developing nations. Ansolabehere et al. recognize in their report that even a very large increase in the number of nuclear power plants—to 1,000–1500—will not stop all releases of carbon dioxide. In fact, if carbon emissions double by 2050 as they expect, from 6,500 to 13,000 million metric tons per year, the 1,800 million metric tons of carbon not emitted because of nuclear power will seem relatively insignificant. Christine Laurent, in "Beating Global Warming With Nuclear Power?" *UNESCO Courier* (February 2001), notes, "For several years, the nuclear energy industry has attempted to cloak itself in different ecological robes. Its credo: nuclear energy is a formidable asset in battle against global warming because it emits very small amounts of greenhouse gases. This stance, first presented in the late 1980s when the extent of the phenomenon was still the subject of controversy, is now at the heart of policy debates over how to avoid droughts, downpours and floods." Laurent adds that it makes more sense to focus on reducing carbon emissions by reducing energy consumption.

The MIT study is clearly part of the continuing debate. Upon its release, it received a good deal of press, which helped bring the issue to public attention. Most press reports appropriately stressed that the study does not say that nuclear power is *the* answer to energy supply and global warming problems. As Richard A. Meserve said in an editorial (in "Global Warming and Nuclear Power," *Science* (January 23, 2004), "For those who are serious about confronting global warming, nuclear power should be seen as part of the solution. Although it is unlikely that many environmental groups will become enthusiastic proponents of nuclear power, the harsh reality is that any serious program to address global warming cannot afford to jettison any technology prematurely.... The stakes are large, and the scientific and educational community should seek to ensure that the public understands the critical link between nuclear power and climate change."

Alvin M. Weinberg, former director of the Oak Ridge National Laboratory, notes in "New Life for Nuclear Power," *Issues in Science and Technology* (Summer 2003) that to make a serious dent in carbon emissions would require perhaps four times as many reactors as suggested in the MIT study. The accompanying safety and security problems would be challenging. If the challenges can be met, says John J. Taylor, retired vice president for nuclear power at the Electric Power Research Institute, in "The Nuclear Power Bargain," *Issues in Science and Technology* (Spring 2004), there are a great many potential benefits.

Grossman is by no means alone in his resistance to any expansion of the role of nuclear power. His essay is one among many that see chiefly danger in nuclear power. See "Fact and Fission," *The Economist* (July 19, 2003). Environmental groups such as Friends of the Earth are adamantly opposed to nuclear power, saying, "Those who back nuclear over renewables and increased energy efficiency completely fail to acknowledge the deadly radioactive legacy nuclear power has created and continues to create." (See the press release Nuclear Power Revival Plan Slammed," (April 18, 2004), available at `http://www.foe-scotland.org.uk/press/pr20040408.html`.)

On the Internet . . .

The Population Council

Established in 1952, The Population Council is "an international, nonprofit institution that conducts research on three fronts: biomedical, social science, and public health. This research—and the information it produces—helps change the way people think about problems related to reproductive health and population growth." Many of the council's publications are available online.

http://www.popcouncil.org/

United Nations Population Division

The United Nations Population Division is responsible for monitoring and appraising a broad range of areas in the field of population. This site offers a wealth of recent data and links.

http://www.un.org/esa/population/unpop.htm

The Agriculture Network Information Center

The Agriculture Network Information Center is a guide to quality agricultural (including biotechnology) information on the Internet as selected by the National Agricultural Library, Land-Grant Universities, and other institutions.

http://www.agnic.org/

Agriculture: Genetic Resources and GMOs

The Agriculture portion of the European Union's portal Web site (EUROPA) provides information on different subjects associated with the genetic base for agricultural activities.

http://www.europa.eu.int/comm/agriculture/res/
index_en.htm

Earth Policy Institute

The purpose of the Earth Policy Institute is to provide a vision of what an environmentally sustainable economy will look like; a roadmap of how to get from here to there; and an ongoing assessment of this effort, to identify where progress is being made and where it is not.

http://www.earth-policy.org/

SeaWeb

SeaWeb is a project designed to raise awareness of the world's oceans and the life within them. According to SeaWeb, as more people understand the ocean's critical role in everyday life and the future of the planet, they will take actions to conserve the ocean and the web of life it supports. This site provides access to a great deal of information related to fisheries.

http://www.seaweb.org

PART 4

Food and Population

*T*o many "sustainability" means enabling the natural world—plants and animals, forests and coral reefs, fresh water and landscapes—to continue to exist more or less (mostly less) as it did before human beings multiplied, developed technology, and began to cause extinctions, air and water pollution, soil erosion, desertification, climate change, and so on. To others "sustainability" means enabling humankind to continue to survive and thrive, even keeping up its history of growth, technological development, and energy use—as if the environment and its resources were infinite.

Because the two visions of "sustainability" are logically incompatible, mankind must struggle to find some sort of middle ground. Must the numbers of people on the planet be reduced? Must the use of technology or mankind's standard of living be curtailed? If not, how can we continue to feed everyone? And if people wish to eat wild food such as fish, how can their appetite be kept from destroying the source?

- Is Limiting Population Growth a Key Factor in Protecting the Global Environment?

- Is Genetic Engineering an Environmentally Sound Way to Increase Food Production?

- Are Marine Reserves Needed to Protect Global Fisheries?

ISSUE 13

Is Limiting Population Growth a Key Factor in Protecting the Global Environment?

YES: Lester R. Brown, from "Rescuing a Planet Under Stress," *The Humanist* (November/December 2003)

NO: Stephen Moore, from "Body Count," *National Review* (October 25, 1999)

ISSUE SUMMARY

YES: Lester R. Brown, president of the Earth Policy Institute, argues that stabilizing world population is central to preventing overconsumption of environmental resources.

NO: Stephen Moore, director of the Cato Institute, argues that human numbers pose no threat to human survival or the environment but that efforts to control population do threaten human freedom and worth.

In 1798 the British economist Thomas Malthus published his *Essay on the Principle of Population*. In it, he pointed with alarm at the way the human population grew geometrically (a hockey-stick curve of increase) and at how agricultural productivity grew only arithmetically (a straight-line increase). It was obvious, he said, that the population must inevitably outstrip its food supply and experience famine. Contrary to the conventional wisdom of the time, population growth was not necessarily a good thing. Indeed, it led inexorably to catastrophe. For many years, Malthus was something of a laughingstock. The doom he forecast kept receding into the future as new lands were opened to agriculture, new agricultural technologies appeared, new ways of preserving food limited the waste of spoilage, and the developed nations underwent a "demographic transition" from high birth rates and high death rates to low birth rates and low death rates.

Demographers initially attributed the demographic transition to increasing prosperity and predicted that as prosperity increased in countries whose populations were rapidly growing, birth rates would surely fall. Later, some scholars analyzed the historical data and concluded that the transition had actually preceded prosperity. The two views have contrasting implications for public policy designed to slow population growth—economic aid or family planning aid—but neither has worked very well. In 1994 the UN Conference on Population and Development, which was held in Cairo, Egypt, concluded that better results would follow from improving women's access to education and health care.

Should we be trying to slow or reverse population growth? In the 1968 book *The Population Bomb* (Ballantine Books), Paul R. Ehrlich warned that unrestricted population growth would lead to both human and environmental disaster. But some religious leaders oppose population control because family planning is against God's will. Furthermore, minority groups and developing nations contend that they are unfairly targeted by family planning programs.

The world's human population has grown tremendously. In Malthus's time, there were about 1 billion human beings on earth. By 1950 there were a little over 2.5 billion. In 1999 the tally passed 6 billion. By 2025 it will be over 8 billion. Statistics like these are positively frightening. By 2050 the UN expects the world population to be about 9 billion and to still be rising. While global agricultural production has also increased, it has not kept up with rising demand, and—because of the loss of topsoil to erosion, the exhaustion of aquifers for irrigation water, and the high price of energy for making fertilizer (among other things)—the prospect of improvement seems exceedingly slim to many observers. Paul R. Ehrlich and Anne H. Ehrlich argue in "The Population Explosion: Why We Should Care and What We Should Do About It," *Environmental Law* (Winter 1997) that "population growth may be the paramount force moving humanity inexorably towards disaster." They therefore maintain that it is essential to reduce the impact of population in terms of both numbers and resource consumption.

What will happen to the environment? The earth already faces global climate change, air pollution, resource depletion, loss of species, and more, and the more people there are on the planet, the greater the effect they must have. Some say that current trends suggest that population will level off and even decline before catastrophe strikes, but most observers are not that optimistic. However, there is hope. For example, the United Nations Development Programme, the United Nations Environment Programme, the World Bank, and the World Resources Institute analyzed world ecosystems and concluded that although there are many signs of trouble, once overuse is controlled, many ecosystems can recover. See *World Resources 2000-2001: People and Ecosystems: The Fraying Web of Life* (World Resources Institute, 2000).

In the following selections, Lester R. Brown argues that humanity is consuming the resource base on which it depends for food and that stabilizing the human population is central to preventing disaster—economic, human, and environmental. Stephen Moore, on the other hand, argues that human numbers pose no threat to human survival or the environment but that efforts to control population do threaten human freedom and worth.

215

Lester R. Brown

 **YES**

Rescuing a Planet Under Stress

Understanding the Problem

As world population has doubled and as the global economy has expanded sevenfold over the last half-century, our claims on the Earth have become excessive. We are asking more of the Earth than it can give on an ongoing basis.

We are harvesting trees faster than they can regenerate, overgrazing rangelands and converting them into deserts, overpumping aquifers, and draining rivers dry. On our cropland, soil erosion exceeds new soil formation, slowly depriving the soil of its inherent fertility. We are taking fish from the ocean faster than they can reproduce.

We are releasing carbon dioxide into the atmosphere faster than nature can absorb it, creating a greenhouse effect. As atmospheric carbon dioxide levels rise, so does the earth's temperature. Habitat destruction and climate change are destroying plant and animal species far faster than new species can evolve, launching the first mass extinction since the one that eradicated the dinosaurs sixty-five million years ago.

Throughout history, humans have lived on the Earth's sustainable yield—the interest from its natural endowment. But now we are consuming the endowment itself. In ecology, as in economics, we can consume principal along with interest in the short run but in the long run it leads to bankruptcy.

In 2002 a team of scientists led by Mathis Wackernagel, an analyst at Redefining Progress, concluded that humanity's collective demands first surpassed the Earth's regenerative capacity around 1980. Their study, published by the U.S. National Academy of Sciences, estimated that our demands in 1999 exceeded that capacity by 20 percent. We are satisfying our excessive demands by consuming the Earth's natural assets, in effect creating a global bubble economy.

Bubble economies aren't new. U.S. investors got an up-close view of this when the bubble in high-tech stocks burst in 2000 and the NASDAQ, an indicator of the value of these stocks, declined by some 75 percent. According to the *Washington Post*, Japan had a similar experience in 1989 when the real estate bubble burst, depreciating stock and real estate assets by 60 percent. The bad-debt fallout and other effects of this collapse have left the once-dynamic Japanese economy dead in the water ever since.

The bursting of these two bubbles affected primarily people living in the United States and Japan but the global bubble economy that is based on the overconsumption of the Earth's natural capital assets will affect the entire world. When the food bubble economy, inflated by the overpumping of aquifers, bursts, it will raise food prices worldwide. The challenge for our generation is to deflate the economic bubble before it bursts.

Unfortunately, since September 11, 2001, political leaders, diplomats, and the media worldwide have been preoccupied with terrorism and, more recently, the occupation of Iraq. Terrorism is certainly a matter of concern, but if it diverts us from the environmental trends that are undermining our future until it is too late to reverse them, Osama bin Laden and his followers will have achieved their goal in a way they couldn't have imagined.

In February 2003, United Nations demographers made an announcement that was in some ways more shocking than the 9/11 attack: the worldwide rise in life expectancy has been dramatically reversed for a large segment of humanity—the seven hundred million people living in sub-Saharan Africa. The HIV epidemic has reduced life expectancy among this region's people from sixty-two to forty-seven years. The epidemic may soon claim more lives than all the wars of the twentieth century. If this teaches us anything, it is the high cost of neglecting newly emerging threats.

The HIV epidemic isn't the only emerging mega-threat. Numerous nations are feeding their growing populations by overpumping their aquifers—a measure that virtually guarantees a future drop in food production when the aquifers are depleted. In effect, these nations are creating a food bubble economy—one where food production is artificially inflated by the unsustainable use of groundwater.

Another mega-threat—climate change—isn't getting the attention it deserves from most governments, particularly that of the United States, the nation responsible for one-fourth of all carbon emissions. Washington, D.C., wants to wait until all the evidence on climate change is in, by which time it will be too late to prevent a wholesale warming of the planet. Just as governments in Africa watched HIV infection rates rise and did little about it, the United States is watching atmospheric carbon dioxide levels rise and doing little to check the increase.

Other mega-threats being neglected include eroding soils and expanding deserts, which jeopardize the livelihood and food supply of hundreds of millions of the world's people. These issues don't even appear on the radar screen of many national governments.

Thus far, most of the environmental damage has been local: the death of the Aral Sea, the burning rainforests of Indonesia, the collapse of the Canadian cod fishery, the melting of the glaciers that supply Andean cities with water, the dust bowl forming in northwestern China, and the depletion of the U.S. great plains aquifer. But as these local environmental events expand and multiply, they will progressively weaken the global economy, bringing closer the day when the economic bubble will burst.

Humanity's demands on the Earth have multiplied over the last half-century as our numbers have increased and our incomes have risen. World population grew from 2.5 billion in 1950 to 6.1 billion in 2000. The growth during those fifty years exceeded that during the four million years since our ancestors first emerged from Africa.

Incomes have risen even faster than population. According to Erik Assadourian's *Vital Signs* 2003 article, "Economic Growth Inches Up," income per person worldwide nearly tripled from 1950 to 2000. Growth in population and the rise in incomes together expanded global economic output from just under $7 trillion (in 2001 dollars) of goods and services in 1950 to $46 trillion in 2000—a gain of nearly sevenfold.

Population growth and rising incomes together have tripled world grain demand over the last half-century, pushing it from 640 million tons in 1950 to 1,855 million tons in 2000, according to the U.S. Department of Agriculture (USDA). To satisfy this swelling demand, farmers have plowed land that was highly erodible—land that was too dry or too steeply sloping to sustain cultivation. Each year billions of tons of topsoil are being blown away in dust storms or washed away in rainstorms, leaving farmers to try to feed some seventy million additional people but with less topsoil than the year before.

Demand for water also tripled as agricultural, industrial, and residential uses increased, outstripping the sustainable supply in many nations. As a result, water tables are falling and wells are going dry. Rivers are also being drained dry, to the detriment of wildlife and ecosystems.

Fossil fuel use quadrupled, setting in motion a rise in carbon emissions that is overwhelming nature's capacity to fix carbon dioxide. As a result of this carbon-fixing deficit, atmospheric carbon dioxide concentrations climbed from 316 parts per million (ppm) in 1959, when official measurement began, to 369 ppm in 2000, according to a report issued by the Scripps Institution of Oceanography at the University of California.

The sector of the economy that seems likely to unravel first is food. Eroding soils, deteriorating rangelands, collapsing fisheries, falling water tables, and rising temperatures are converging to make it more difficult to expand food production fast enough to keep up with demand. According to the USDA, in 2002 the world grain harvest of 1,807 million tons fell short of world grain consumption by 100 million tons, or 5 percent. This shortfall, the largest on record, marked the third consecutive year of grain deficits, dropping stocks to the lowest level in a generation.

Now the question is: can the world's farmers bounce back and expand production enough to fill the hundred-million-ton shortfall, provide for the more than seventy million people added each year, and rebuild stocks to a more secure level? In the past, farmers responded to short supplies and higher grain prices by planting more land and using more irrigation water and fertilizer. Now it is doubtful that farmers can fill this gap without further depleting aquifers and jeopardizing future harvests.

At the 1996 World Food Summit in Rome, Italy, hosted by the UN Food and Agriculture Organization (FAO), 185 nations plus the European community agreed to reduce hunger by half by 2015. Using 1990–1992 as a base, governments set the goal of cutting the number of people who were hungry—860 million—by roughly 20 million per year. It was an exciting and worthy goal, one that later became one of the UN Millennium Development Goals.

But in its late 2002 review of food security, the UN issued a discouraging report:

> This year we must report that progress has virtually ground to a halt. Our latest estimates, based on data from the years 1998–2000, put the number of undernourished people in the world at 840 million....a decrease of barely 2.5 million per year over the eight years since 1990–92.

Since 1998–2000, world grain production per person has fallen 5 percent, suggesting that the ranks of the hungry are now expanding. As noted earlier, life expectancy is plummeting in sub-Saharan Africa. If the number of hungry people worldwide is also increasing, then two key social indicators are showing widespread deterioration in the human condition.

The ecological deficits just described are converging on the farm sector, making it more difficult to sustain rapid growth in world food output. No one knows when the growth in food production will fall behind that of demand, driving up prices, but it may be much sooner than we think. The triggering events that will precipitate future food shortages are likely to be spreading water shortages interacting with crop-withering heat waves in key food-producing regions. The economic indicator most likely to signal serious trouble in the deteriorating relationship between the global economy and the Earth's ecosystem is grain prices.

Food is fast becoming a national security issue as growth in the world harvest slows and as falling water tables and rising temperatures hint at future shortages. According to the USDA more than one hundred nations import part of the wheat they consume. Some forty import rice. While some nations are only marginally dependent on imports, others couldn't survive without them. Egypt and Iran, for example, rely on imports for 40 percent of their grain supply. For Algeria, Japan, South Korea, and Taiwan, among others, it is 70 percent or more. For Israel and Yemen, over 90 percent. Just six nations—Argentina, Australia, Canada, France, Thailand, and the United States—supply 90 percent of grain exports. The United States alone controls close to half of world grain exports, a larger share than Saudi Arabia does of oil.

Thus far the nations that import heavily are small and middle-sized ones. But now China, the world's most populous nation, is soon likely to turn to world markets in a major way. As reported by the International Monetary Fund, when the former Soviet Union unexpectedly turned to the world market in 1972 for roughly a tenth of its grain supply following a weather-reduced harvest, world wheat prices climbed from $1.90 to $4.89 a bushel. Bread prices soon rose, too.

If China depletes its grain reserves and turns to the world grain market to cover its shortfall—now forty million tons per year—it could destabilize world grain markets overnight. Turning to the world market means turning to the United States, presenting a potentially delicate geopolitical situation in which 1.3 billion Chinese consumers with a $100-billion trade surplus with the United States will be competing with U.S. consumers for U.S. grain. If this leads to rising food prices in the United States, how will the government respond? In times past, it could have restricted exports, even imposing an export embargo, as it did with soybeans to Japan in 1974. But today the United States has a stake in a politically stable China. With an economy growing at 7 to 8 percent a year, China is the engine that is powering not only the Asian economy but, to some degree, the world economy.

For China, becoming dependent on other nations for food would end its history of food self-sufficiency, leaving it vulnerable to world market uncertainties. For Americans, rising food prices would be the first indication that the world has changed fundamentally and that they are being directly affected by the growing grain deficit in China. If it seems likely that rising food prices are being driven in part by crop-withering temperature rises, pressure will mount for the United States to reduce oil and coal use.

For the world's poor—the millions living in cities on $1 per day or less and already spending 70 percent of their income on food—rising grain prices would be life threatening. A doubling of world grain prices today could impoverish more people in a shorter period of time than any event in history. With desperate people holding their governments responsible, such a price rise could also destabilize governments of low-income, grain-importing nations.

Food security has changed in other ways. Traditionally it was largely an agricultural matter. But now it is something that our entire society is responsible for. National population and energy policies may have a greater effect on food security than agricultural policies do. With most of the three billion people to be added to world population by 2050 (as estimated by the UN) being born in nations already facing water shortages, child-bearing decisions may have a greater effect on food security than crop planting decisions. Achieving an acceptable balance between food and people today depends on family planners and farmers working together.

Climate change is the wild card in the food security deck. The effect of population and energy policies on food security differ from climate in one important respect: population stability can be achieved by a nation acting unilaterally. Climate stability cannot.

Instituting the Solution

Business as usual—Plan A—clearly isn't working. The stakes are high, and time isn't on our side. The good news is that there are solutions to the problems we are facing. The bad news is that if we continue to rely on timid, incremental responses our bubble economy will continue to grow until eventually it bursts. A new approach is necessary—a Plan B—an urgent reordering of priorities and a restructuring of the global economy in order to prevent that from happening.

Plan B is a massive mobilization to deflate the global economic bubble before it reaches the bursting point. Keeping the bubble from bursting will require an unprecedented degree of international cooperation to stabilize population, climate, water tables, and soils—and at wartime speed. Indeed, in both scale and urgency the effort required is comparable to the U.S. mobilization during World War II.

Our only hope now is rapid systemic change—change based on market signals that tell the ecological truth. This means restructuring the tax system by lowering income taxes and raising taxes on environmentally destructive activities, such as fossil fuel burning, to incorporate the ecological costs. Unless we can get the market to send signals that reflect reality, we will continue making faulty decisions as consumers, corporate planners, and government policymakers. Ill-informed economic decisions and the economic distortions they create can lead to economic decline.

Stabilizing the world population at 7.5 billion or so is central to avoiding economic breakdown in nations with large projected population increases that are already overconsuming their natural capital assets. According to the Population Reference Bureau, some thirty-six nations, all in Europe except Japan, have essentially stabilized their populations. The challenge now is to create the economic and social conditions and to adopt the priorities that will lead to population stability in all remaining nations. The keys here are extending primary education to all children, providing vaccinations and basic health care, and offering reproductive health care and family planning services in all nations.

Shifting from a carbon-based to a hydrogen-based energy economy to stabilize climate is now technologically possible. Advances in wind turbine design and in solar cell manufacturing, the availability of hydrogen generators, and the evolution of fuel cells provide the technologies needed to build a climate-benign hydrogen economy. Moving quickly from a carbon-based to a hydrogen-based energy economy depends on getting the price right, on incorporating the indirect costs of burning fossil fuels into the market price.

On the energy front, Iceland is the first nation to adopt a national plan to convert its carbon-based energy economy to one based on hydrogen. Denmark and Germany are leading the world into the age of wind. Japan has emerged as the world's leading manufacturer and user of solar cells. With its commercialization of a solar roofing material, it leads the world in electricity generation from solar cells and is well positioned to assist in the electrification of villages in the developing world. The Netherlands leads the industrial world in exploiting the bicycle as an alternative to the automobile. And the Canadian province of Ontario is emerging as a leader in phasing out coal. It plans to replace its five coal-fired power plants with gas-fired plants, wind farms, and efficiency gains.

Stabilizing water tables is particularly difficult because the forces triggering the fall have their own momentum, which must be reversed. Arresting the fall depends on quickly raising water productivity. In pioneering drip irrigation technology, Israel has become the world leader in the efficient use of agricultural water. This unusually labor-intensive irrigation practice, now being used to produce high-value crops in many nations, is ideally suited where water is scarce and labor is abundant.

In stabilizing soils, South Korea and the United States stand out. South Korea, with once denuded mountainsides and hills now covered with trees, has achieved a level of flood control, water storage, and hydrological stability that is a model for other nations. Beginning in the late 1980s, U.S. farmers systematically retired roughly 10 percent of the most erodible cropland, planting the bulk of it to grass, according to the USDA. In addition, they lead the world in adopting minimum-till, no-till, and other soil-conserving practices. With this combination of programs and practices, the United States has reduced soil erosion by nearly 40 percent in less than two decades.

Thus all the things we need to do to keep the bubble from bursting are now being done in at least a few nations. If these highly successful initiatives are adopted worldwide, and quickly, we can deflate the bubble before it bursts.

Yet adopting Plan B is unlikely unless the United States assumes a leadership position, much as it belatedly did in World War II. The nation responded to the aggression of Germany and Japan only after it was directly attacked at Pearl Harbor on December 7, 1941. But respond it did. After an all-out mobilization, the U.S. engagement helped turn the tide, leading the Allied Forces to victory within three and a half years.

This mobilization of resources within a matter of months demonstrates that a nation and, indeed, the world can restructure its economy quickly if it is convinced of the need to do so. Many people—although not yet the majority—are already convinced of the need for a wholesale restructuring of the economy. The issue isn't whether most people will eventually be won over but whether they will be convinced before the bubble economy collapses.

History judges political leaders by whether they respond to the great issues of their time. For today's leaders, that issue is how to deflate the world's bubble economy before it bursts. This bubble threatens the future of everyone, rich and poor alike. It challenges us to restructure the global economy, to build an eco-economy.

We now have some idea of what needs to be done and how to do it. The UN has set social goals for education, health, and the reduction of hunger and poverty in its Millennium Development Goals. My latest book, *Plan B*, offers a sketch for the restructuring of the energy economy to stabilize atmospheric carbon dioxide levels, a plan to stabilize population, a strategy for raising land productivity and restoring the earth's vegetation, and a plan to raise water productivity worldwide. The goals are essential and the technologies are available.

We have the wealth to achieve these goals. What we don't yet have is the leadership. And if the past is any guide to the future, that leadership can only come from the United States. By far the wealthiest society that has ever existed, the United States has the resources to lead this effort.

Yet the additional external funding needed to achieve universal primary education in the eighty-eight developing nations that require help is conservatively estimated by the World Bank at $15 billion per year. Funding for an adult literacy program based largely on volunteers is estimated at $4 billion. Providing for the most basic health care is estimated at $21 billion by the World Health Organization. The additional funding needed to provide reproductive health and family planning services to all women in developing nations is $10 billion a year.

Closing the condom gap and providing the additional nine billion condoms needed to control the spread of HIV in the developing world and Eastern Europe requires $2.2 billion—$270 million for condoms and $1.9 billion for AIDS prevention education and condom distribution. The cost per year of extending school lunch programs to the forty-four poorest nations is $6 billion per year. An additional $4 billion per year would cover the cost of assistance to preschool children and pregnant women in these nations.

In total, this comes to $62 billion. If the United States offered to cover one-third of this additional funding, the other industrial nations would almost certainly be willing to provide the remainder, and the worldwide effort to eradicate hunger, illiteracy, disease, and poverty would be under way.

The challenge isn't just to alleviate poverty, but in doing so to build an economy that is compatible with the Earth's natural systems—an eco-economy, an economy that can sustain progress. This means a fundamental restructuring of the energy economy and a substantial modification of the food economy. It also means raising the productivity of energy and shifting from fossil fuels to renewables. It means raising water productivity over the next half-century, much as we did land productivity over the last one.

It is easy to spend hundreds of billions in response to terrorist threats but the reality is that the resources needed to disrupt a modern economy are small, and a Department of Homeland Security, however heavily funded, provides only minimal protection from suicidal terrorists. The challenge isn't just to provide a high-tech military response to terrorism but to build a global society that is environmentally sustainable, socially equitable, and democratically based—one where there is hope for everyone. Such an effort would more effectively undermine the spread of terrorism than a doubling of military expenditures.

We can build an economy that doesn't destroy its natural support systems, a global community where the basic needs of all the Earth's people are satisfied, and a world that will allow us to think of ourselves as civilized. This is entirely doable. To paraphrase former President Franklin Roosevelt at another of those hinge points in history, let no one say it cannot be done.

The choice is ours—yours and mine. We can stay with business as usual and preside over a global bubble economy that keeps expanding until it bursts, leading to economic decline. Or we can adopt Plan B and be the generation that stabilizes population, eradicates poverty, and stabilizes climate. Historians will record the choice—but it is ours to make.

NO

Stephen Moore

Body Count

$\mathbf{A}$t a Washington reception, the conversation turned to the merits of small families. One woman volunteered that she had just read Bill McKibben's environmental tome, *Maybe One*, on the benefits of single-child families. She claimed to have found it "ethically compelling." I chimed in: "Even one child may put too much stress on our fragile ecosystem. McKibben says 'maybe one.' I say, why not none?" The response was solemn nods of agreement, and even some guilt-ridden whispers between husbands and wives.

McKibben's acclaimed book is a tribute to the theories of British economist Thomas Malthus. Exactly 200 years ago, Malthus—the original dismal scientist—wrote that "the power of population is ... greater than the power in the earth to produce subsistence for man." McKibben's application of this idea was to rush out and have a vasectomy. He urges his fellow greens to do the same—to make single-child families the "cultural norm" in America.

Now, with the United Nations proclaiming that this month we will surpass the demographic milestone of 6 billion people, the environmental movement and the media can be expected to ask: Do we really need so many people? A recent AP headline lamented: "Century's growth leaves Earth crowded—and noisy." Seemingly, Malthus has never had so many apostles.

In a rational world, Malthusianism would not be in a state of intellectual revival, but thorough disrepute. After all, virtually every objective trend is running in precisely the opposite direction of what the widely acclaimed Malthusians of the 1960s—from Lester Brown to Paul Ehrlich to the Club of Rome—predicted. Birth rates around the world are lower today than at any time in recorded history. Global per capita food production is much higher than ever before. The "energy crisis" is now such a distant memory that oil is virtually the cheapest liquid on earth. These facts, collectively, have wrecked the credibility of the population-bomb propagandists.

Yet the population-control movement is gaining steam. It has won the hearts and wallets of some of the most influential leaders inside and outside government today. Malthusianism has evolved into a multi-billion-dollar industry and a political juggernaut.

Today, through the U.S. Agency for International Development (AID), the State Department, and the World Bank, the federal government pumps some 350 million tax dollars a year into population-containment activities. The Clinton administration would be spending at least twice that amount if not for the efforts of two Republican congressmen, Chris Smith of New Jersey and Todd Tiahrt of Kansas, who have managed to cut off funding for the most coercive birth-reduction initiatives.

Defenders of the U.N. Population Fund (UNFPA) and other such agencies insist that these programs "protect women's reproductive freedom," "promote the health of mothers," and "reduce infant mortality." Opponents of international "family planning," particularly Catholic organizations, are tarred as anti-abortion fanatics who want to deprive poor women of safe and cheap contraception. A 1998 newspaper ad by Planned Parenthood, entitled "The Right Wing Coup in Family Planning," urged continued USAID funding by proclaiming: "The very survival of women and children is at stake in this battle." Such rhetoric is truly Orwellian, given that the entire objective of government sponsored birth-control programs has been to invade couples' "reproductive rights" in order to limit family size. The crusaders have believed, from the very outset, that coercion is necessary in order to restrain fertility and avert global eco-collapse.

The consequences of this crusade are morally atrocious. Consider the one-child policy in China. Some 10 million to 20 million Chinese girls are demographically "missing" today because of "sex-selective abortion of female fetuses, female infant mortality (through infanticide or abandonment), and selective neglect of girls ages 1 to 4," according to a 1996 U.S. Census Bureau report. Girls account for over 90 percent of the inmates of Chinese orphanages—where children are left to die from neglect.

Last year, Congress heard testimony from Gao Xiao Duan, a former Chinese administrator of the one-couple, one-child policy. Gao testified that if a woman in rural China is discovered to be pregnant without a state-issued "birth-allowed certificate," she typically must undergo an abortion—no matter how many months pregnant she is. Gao recalled, "Once I found a woman who was nine months' pregnant but did not have a birth-allowed certificate. According to the policy, she was forced to undergo an abortion surgery. In the operating room, I saw how the aborted child's lips were sucking, how its limbs were stretching. A physician injected poison into its skull, and the child died and was thrown into the trash can."

The pro-choice movement is notably silent about this invasion of women's "reproductive rights." In 1989, Molly Yard, of the National Organization for Women, actually praised China's program as "among the most intelligent in the world." Stanford biologist Paul Ehrlich, the godfather of today's neo-Malthusian movement, once trumpeted China's population control as "remarkably vigorous and effective." He has congratulated Chinese rulers for their "grand experiment in the management of population."

Last summer, Lisa McRee of *Good Morning America* started an interview with Bill McKibben by asking, in all seriousness, "Is China's one-child policy a good idea for every country?" She might as well have asked whether every country should have gulags.

Gregg Easterbrook, writing in the Nov. 23, 1998 *New Republic*, correctly lambasted China for its "horrifying record on forced abortion and sterilization." But even the usually sensible Easterbrook offered up a limp apology for the one-child policy, writing that "China, which is almost out of arable land, had little choice but to attempt some degree of fertility constraint." Hong Kong has virtually no arable land, and 75 times the population density of mainland China, but has one of the best-fed populations in the world.

These coercive practices are spreading to other countries. Brian Clowes writes in the *Yale Journal of Ethics* that coercion has been used to promote family planning in at least 35 developing countries. Peru has started to use sterilization as a means of family planning, and doctors have to meet sterilization quotas or risk losing their jobs. The same is true in Mexico.

In disease-ridden African countries such as Nigeria and Kenya, hospitals often lack even the most rudimentary medical care, but are stocked to the rafters with boxes of contraceptives stamped "UNFPA" and "USAID." UNFPA boasts that, thanks to its shipments, more than 80 percent of the women in Haiti have access to contraceptives; this is apparently a higher priority than providing access to clean water, which is still unavailable to more than half of the Haitian population.

Population-control groups like Zero Population Growth and International Planned Parenthood have teamed up with pro-choice women in Congress—led by Carolyn Maloney of New York, Cynthia McKinney of Georgia, and Connie Morella of Maryland—to try to secure $60 million in U.S. funding for UNFPA over the next two years. Maloney pledges, "I'm going to do whatever it takes to restore funding for [UNFPA]" this year.

Support for this initiative is based on two misconceptions. The first is the excessively optimistic view that (in the words of a *Chicago Tribune* report) "one child zealotry in China is fading." The Population Research Institute's Steve Mosher, an authority on Chinese population activities, retorts, "This fantasy that things are getting better in China has been the constant refrain of the one-child apologists for at least the past twenty years." In fact, after UNFPA announced in 1997 that it was going back into China, state councillor Peng Peiyun defiantly announced, "China will not slacken our family-planning policy in the next century."

The second myth is that UNFPA has always been part of the solution, and has tried to end China's one-child policy. We are told that it is pushing Beijing toward more "female friendly" family planning. This, too, is false. UNFPA has actually given an award to China for its effectiveness in population-control activities—activities far from female-friendly. Worse, UNFPA's executive director, Nafis Sadik, is, like her predecessors, a longtime apologist for the China program and even denies that it is coercive. She is on record as saying— falsely—that "the implementation of the policy is purely voluntary. There is no such thing as a license to have a birth."

Despite UNFPA's track record, don't be surprised if Congress winds up re-funding it. The past 20 years may have demonstrated the intellectual bankruptcy of the population controllers, but their coffers have never been more flush.

American billionaires, past and present, have devoted large parts of their fortunes to population control. The modern-day population-control movement dates to 1952, when John D. Rockefeller returned from a trip to Asia convinced that the teeming masses he saw there were the single greatest threat to the earth's survival. He proceeded to divert hundreds of millions of dollars from his foundation to the goal of population stabilization. He was followed by David Packard (co-founder of Hewlett-Packard), who created a $9 billion foundation whose top priority was reducing world population. Today, these foundations are joined by organizations ranging from Zero Population Growth (ZPG) to Negative Population Growth (which advocates an optimal U.S. population size of 150 million–120 million fewer than now) to Planned Parenthood to the Sierra Club. The combined budget of these groups approaches $1 billion.

These organizations tend to be extremist. Take ZPG. Its board of directors passed a resolution declaring that "parenthood is not an inherent right but a privilege" granted by the state, and that "every American family has a right to no more than two children."

"Population growth is analogous to a plague of locusts," says Ted Turner, a major source of population-movement funding. "What we have on this earth today is a plague of people. Nature did not intend for there to be as many people as there are." Turner has also penned "The Ted Commandments," which include "a promise to have no more than two children or no more than my nation suggests." He recently reconsidered his manifesto, and now believes that the voluntary limit should be even lower—just *one* child. In Turner's utopia, there are no brothers, sisters, aunts, or uncles.

Turner's $1 billion donation to the U.N. is a pittance compared with the fortunes that Warren Buffett (net worth $36 billion) and Bill Gates (net worth roughly $100 billion) may bestow on the cause of population control. Buffett has announced repeatedly that he views overpopulation as one of the greatest crises in the world today. Earlier this year, Gates and his wife contributed an estimated $7 billion to their foundation, of which the funding of population programs is one of five major initiatives.

This is a massive misallocation of funds, for the simple reason that the overpopulation crisis is a hoax. It is true that world population has tripled over the last century. But the explanation is both simple and benign: First, life expectancy—possibly the best overall numerical measure of human well-being—has almost doubled in the last 100 years, and the years we are tacking on to life are both more active and more productive. Second, people are wealthier—they can afford better health care, better diets, and a cleaner environment. As a result, infant-mortality rates have declined nearly tenfold in this century. As the late Julian Simon often explained, population growth is a sign of mankind's greatest triumph—our gains against death.

We are told that this good news is really bad news, because human numbers are soon going to bump up against the planet's "carrying capacity." Pessimists worry that man is procreating as uncontrollably as John B. Calhoun's famous Norwegian rats, which multiply until they die off from lack of sustenance. Bill McKibben warns that "we are adding another New York City every month, a Mexico every year, and almost another India every decade."

But a closer look shows that these fears are unfounded. Fact: If every one of the 6 billion of us resided in Texas, there would be room enough for every family of four to have a house and one-eighth of an acre of land—the rest of the globe would be vacant. (True, if population growth continued, some of these people would eventually spill over into Oklahoma.)

In short, the population bomb has been defused. The birth rate in developing countries has plummeted from just over 6 children per couple in 1950 to just over 3 today. The major explanation for smaller family sizes, even in China, has been economic growth. The Reaganites were right on the mark when, in 1984, they proclaimed this truth to a distraught U.N. delegation in Mexico City. (The policy they enunciated has been memorably expressed in the phrase "capitalism is by far the best contraceptive.") The fertility rate in the developed world has fallen from 3.3 per couple in 1950 to 1.6 today. These low fertility rates presage declining populations. If, for example, Japan's birth rate is not raised at some point, in 500 years there will be only about 15 Japanese left on the planet.

Other Malthusian worries are similarly wrongheaded. Global food prices have fallen by half since 1950, even as world population has doubled. The dean of agricultural economists, D. Gale Johnson of the University of Chicago, has documented "a dramatic decline in famines" in the last 50 years. Fewer than half as many people die of famine each year now than did a century ago—despite a near-quadrupling of the population. Enough food is now grown in the world to provide every resident of the planet with almost four pounds of food a day. In each of the past three years, global food production has reached new heights.

Overeating is fast becoming the globe's primary dietary malady. "It's amazing to say, but our problem is becoming overnutrition," Ho Zhiqiuan, a Chinese nutrition expert, recently told *National Geographic*. "Today in China obesity is becoming common."

Millions are still hungry, and famines continue to occur—but these are the result of government policies or political malice, not inadequate global food production. As the International Red Cross has reported, "the loss of access to food resources [during famines] is generally the result of intentional acts" by governments.

Even if the apocalyptic types are correct and population grows to 12 billion in the 21st century, so what? Assuming that human progress and scientific advancement continue as they have, and assuming that the global march toward capitalism is not reversed, those 12 billion people will undoubtedly be richer, healthier, and better fed than the 6 billion of us alive today. After all, we 6 billion are much richer, healthier, and better fed than the 1 billion who lived in 1800 or the 2 billion alive in 1920.

The greatest threat to the planet is not too many people, but too much statism. The Communists, after all, were the greatest polluters in history. Economist Mikhail Bernstam has discovered that market-based economies are about two to three times more energy-efficient than Communist, socialist, Maoist, or "Third Way" economies. Capitalist South Korea has three times the population density of socialist North Korea, but South Koreans are well fed while 250,000 North Koreans have starved to death in the last decade.

Government-funded population programs are actually counterproductive, because they legitimize command-and-control decision-making. As the great development economist Alan Rufus Waters puts it, "Foreign aid used for population activities gives enormous resources and control apparatus to the local administrative elite and thus sustains the authoritarian attitudes corrosive to the development process."

This approach usually ends up making poor people poorer, because it distracts developing nations from their most pressing task, which is market reform. When Mao's China established central planning and communal ownership of agriculture, tens of millions of Chinese peasants starved to death. In 1980, after private ownership was established, China's agricultural output doubled in just ten years. If Chinese leaders over the past 30 years had concentrated on rapid privatization and market reform, it's quite possible that economic development would have decreased birth rates every bit as rapidly as the one-child policy.

The problem with trying to win this debate with logic and an arsenal of facts is that modern Malthusianism is not a scientific theory at all. It's a religion, in which the assertion that mankind is overbreeding is accepted as an article of faith. I recently participated in a debate before an anti-population group called Carrying Capacity Network, at which one scholar informed me that man's presence on the earth is destructive because *Homo sapiens* is the only species without a natural predator. It's hard to argue with somebody who despairs because mankind is alone at the top of the food chain.

At its core, the population-control ethic is an assault on the principle that every human life has intrinsic value. Malthusian activists tend to view human beings neither as endowed with intrinsic value, nor even as resources, but primarily as consumers of resources. No wonder that at last year's ZPG conference, the Catholic Church was routinely disparaged as "our enemy" and "the evil empire."

The movement also poses a serious threat to freedom. Decisions on whether to have children—and how many—are among are the most private of all human choices. If governments are allowed to control human reproduction, virtually no rights of the individual will remain inviolable by the state. The consequence, as we have seen in China, is the debasement of human dignity on a grand scale.

Another (true) scene from a party: A radiant pregnant woman is asked whether this is her first child. She says, no, in fact, it is her sixth. Yuppies gasp, as if she has admitted that she has leprosy. To have three kids—to be above replacement level—is regarded by many as an act of eco-terrorism.

But the good news for this pregnant woman, and the millions of others who want to have lots of kids, is that the Malthusians are simply wrong. There is no moral, economic, or environmental case for small families. Period.

If some choose to subscribe to a voluntary one-child policy, so be it. But the rest of us—Americans, Chinese, and everybody else—don't need or want Ted Turner or the United Nations to tell us how many kids to have. Congress should not be expanding "international family planning" funding, but terminating it.

Congress may want to consider a little-known footnote of history. In time, Thomas Malthus realized that his dismal population theories were wrong. He awoke to the reality that human beings are not like Norwegian rats at all. Why? Because, he said, man is "impelled" by "reason" to solve problems, and not to "bring beings into the world for whom he cannot provide the means of support." Amazingly, 200 years later, his disciples have yet to grasp this lesson.

POSTSCRIPT

Is Limiting Population Growth a Key Factor in Protecting the Global Environment?

Some people believe that the answer to the human population crisis lies in reducing or stabilizing human numbers. Others believe that the answer lies in how people live—the technology they use, or their economic system, or social justice. If we do face a crisis, the answer is not so simple, as Joel E. Cohen discusses in *How Many People Can the Earth Support?* (W. W. Norton, 1996). The key to answering Cohen's titular question is human choice, and the choices are ones that must be made in the near future.

Some commentators still stress the role of population by itself; see Werner Fornos, "No Vacancy," *The Humanist* (July/August 1998). Malcolm Potts, in "The Unmet Need for Family Planning," *Scientific American* (January 2000), stresses the need for improved control of human fertility. Andrew R. B. Ferguson, in "Perceiving the Population Bomb," *World Watch* (July/August 2001), sets the maximum sustainable human population at 2.1 billion. Another important writer on the topic is Garrett Hardin, whose influential essay "The Tragedy of the Commons," *Science* (December 13, 1968) describes the consequences of using self-interest alone to guide the exploitation of publicly owned resources, such as air and water.

Cohen notes in "Human Population: The Next Half Century," *Science* (November 14, 2003) that human population growth is slowing, and by 2050 the world may hold as few as two billion more people than it does today or as many as four billion more. Norman E. Borlaug, in "Feeding a World of 10 Billion People: The Miracle Ahead," in Ronald Bailey, ed., *Global Warming and Other Eco-Myths: How the Environmental Movement Uses False Science to Scare Us to Death* (Prima, 2002), says that agricultural technology—including genetic modification of crops—is able, if widely and equitably distributed, to feed even the larger of Cohen's projected populations. In the same book, Nicholas Eberstadt, in "Population, Resources, and the Quest to 'Stabilize Human Population': Myths and Realities," says that "overpopulation" is not a problem of numbers but of poverty; as prosperity spreads, resource problems vanish because the greatest resource is human intelligence.

The United Nations Environmental Programme's *Global Environmental Outlook 3* (Earthscan, 2002), produced as a "global state of the environment report" in preparation for the World Summit on Sustainable Development in Johannesburg, South Africa, states, "Population size to a great extent governs demand for natural resources and material flows. Population growth enlarges the challenge of improving living standards and providing essential social services, including housing, transport, sanitation, health, education, jobs and security. It can also make it harder to deal with poverty."

ISSUE 14

Is Genetic Engineering an Environmentally Sound Way to Increase Food Production?

YES: Royal Society of London et al., from "Transgenic Plants and World Agriculture," A Report Prepared Under the Auspices of the Royal Society of London, the U.S. National Academy of Sciences, the Brazilian Academy of Sciences, the Chinese Academy of Sciences, the Indian National Science Academy, the Mexican Academy of Sciences, and the Third World Academy of Sciences (July 2000)

NO: Brian Halweil, from "The Emperor's New Crops," *World Watch* (July/August 1999)

ISSUE SUMMARY

YES: The national academies of science of the United Kingdom, the United States, Brazil, China, India, Mexico, and the Third World argue that genetically modified crops hold the potential to feed the world during the twenty-first century while also protecting the environment.

NO: Brian Halweil, a researcher at the Worldwatch Institute, argues that the genetic modification of crops threatens to produce pesticide-resistant insect pests and herbicide-resistant weeds, will victimize poor farmers, and is unlikely to feed the world.

In the early 1970s scientists first discovered that it was technically possible to move genes—the biological material that determines a living organism's physical traits—from one organism to another and thus (in principle) to give bacteria, plants, and animals new features. Most researchers in molecular genetics were excited by the potentialities that suddenly seemed within their reach. However, a few researchers—as well as many people outside the field— were disturbed by the idea; they thought that genetic mix-and-match games might spawn new diseases, weeds, and pests. Some people even argued that genetic engineering should be banned at the outset, before unforeseeable horrors were unleashed. Researchers in support of genetic experimentation responded by declaring a moratorium on their own work until suitable safeguards (in the form of government regulations) could be devised.

A 1987 National Academy of Sciences report said that genetic engineering posed no unique hazards. And, despite continuing controversy, by 1989 the technology had developed tremendously: researchers could obtain patents for mice with artificially added genes ("transgenic" mice); firefly genes had been added to tobacco plants to make them glow (faintly) in the dark; and growth hormone produced by genetically engineered bacteria was being used to grow low-fat pork and increase milk production in cows. The growing biotechnology industry promised more productive crops that made their own fertilizer and pesticide. Proponents argued that genetic engineering was in no significant way different from traditional selective breeding. Critics argued that genetic engineering was unnatural and violated the rights of both plants and animals to their "species integrity"; that expensive, high-tech, tinkered animals gave the competitive advantage to big agricultural corporations and drove small farmers out of business; and that putting human genes into animals, plants, or bacteria was downright offensive. See Betsy Hanson and Dorothy Nelkin, "Public Responses to Genetic Engineering," *Society* (November/December 1989).

In 1992 the U.S. Office of Science and Technology issued guidelines to bar regulations that are based on the assumption that genetically engineered crops pose greater risks than similar crops produced by traditional breeding methods. The result was the rapid commercial introduction of crops that were genetically engineered to make the bacterial insecticide Bt and to resist herbicides and disease, among other things. In 2003, some 70 engineered crop varieties were grown on over 68 million hectares in 18 countries. Sales of genetically engineered crop products are expected to reach $25 billion by 2010.

Skepticism about the benefits remains, but agricultural genetic engineering has proceeded at a breakneck pace, largely because, as Robert Shapiro, CEO of the Monsanto Corporation, said in June 1998, it "represents a potentially sustainable solution to the issue of feeding people." Many people are not reassured. They see potential problems in nutrition, toxicity, allergies, and ecology. Europe has paid more attention to the critics than the United States has, and the growing and marketing of genetically engineered crops has been either banned or severely restricted across the continent, although restrictions are now loosening.

The following selections illustrate the different current perspectives on the use of genetic engineering in agriculture. In the first selection, the national academies of science of the United Kingdom, the United States, Brazil, China, India, Mexico, and the Third World recognize that the use of genetically modified crops has some worrisome potentials that deserve further research, but they conclude that such crops hold the potential to feed the world during the twenty-first century while also protecting the environment. In the second selection, Brian Halweil argues that the genetic modification of crops threatens to produce pesticide-resistant insect pests and herbicide-resistant weeds, will victimize poor farmers, and is unlikely to feed the world.

233

 **YES**

Transgenic Plants and World Agriculture

During the 21st century, humankind will be confronted with an extraordinary set of challenges. By 2030, it is estimated that eight billion persons will populate the world—an increase of two billion people from today's population. Hunger and poverty around the globe must be addressed, while the life-support systems provided by the world's natural environment are maintained. Meeting these challenges will require new knowledge generated by continued scientific advances, the development of appropriate new technologies, and a broad dissemination of this knowledge and technology along with the capacity to use it throughout the world. It will also require that wise policies be implemented through informed decision-making on the part of national, state, and local governments in each nation.

Scientific advances require an open system of information exchange in which arguments are based on verifiable evidence. Although the primary goal of science is to increase our understanding of the world, knowledge created through science has had immense practical benefits. For example, through science, we have developed a more complete understanding of our natural environment, improved human health with new medicines, and discovered specific plant genes that control disease- or drought-resistance.

Biotechnology can be defined as the application of our knowledge and understanding of biology to meet practical needs. By this definition, biotechnology is as old as the growing of crops and the making of cheeses and wines. Today's biotechnology is largely identified with applications in medicine and agriculture based on our knowledge of the genetic code of life. Various terms have been used to describe this form of biotechnology including genetic engineering, genetic transformation, transgenic technology, recombinant DNA technology, and genetic modification technology. For the purposes of this report, which is focused on plants and products from plants, the term genetic modification technology, or GM technology is used.

GM technology was first developed in the 1970s. One of the most prominent developments, apart from the medical applications, has been the development of novel transgenic crop plant varieties. Many millions of hectares of commercially produced transgenic crops such as soybean, cotton, tobacco,

From Royal Society of London et al., "Transgenic Plants and World Agriculture," A Report Prepared Under the Auspices of the Royal Society of London, the U.S. National Academy of Sciences, the Brazilian Academy of Sciences, the Chinese Academy of Sciences, the Indian National Science Academy, the Mexican Academy of Sciences, and the World Academy of Sciences (July 2000). Copyright © 2000 by Royal Society of London et al. Reprinted with permission.

potato and maize have been grown annually in a number of countries including the USA (28.7 million hectares in 1999), Canada (4 million), China (0.3 million), and Argentina (6.7 million) (James 1999). However, there has been much debate about the potential benefits and risks that may result from the use of such crops.

The many crucial decisions to be made in the area of biotechnology in the next century by private corporations, governments, and individuals will affect the future of humanity and the planet's natural resources. These decisions must be based on the best scientific information in order to allow effective choices for policy options. It is for this reason that representatives of seven of the world's academies of science have come together to provide recommendations to the developers and overseers of GM technology and to offer scientific perspectives to the ongoing public debate on the potential role of GM technology in world agriculture....

The Need for GM Technology in World Agriculture

Today there are some 800 million people (18% of the population in the developing world) who do not have access to sufficient food to meet their needs (Pinstrup-Anderson and Pandya-Lorch 2000, Pinstrup-Anderson et al 1999), primarily because of poverty and unemployment. Malnutrition plays a significant role in half of the nearly 12 million deaths each year of children under five in developing countries (UNICEF 1998). In addition to lack of food, deficiencies in micro-nutrients (especially vitamin A, iodine and iron) are widespread. Furthermore, changes in the patterns of global climate and alterations in use of land will exacerbate the problems of regional production and demands for food. Dramatic advances are required in food production, distribution and access if we are going to address these needs. Some of these advances will occur from non-GM technologies, but others will come from the advantages offered by GM technologies.

Achieving the minimum necessary growth in total production of global staple crops—maize, rice, wheat, cassava, yams, sorghum, potatoes and sweet potatoes—without further increasing land under cultivation, will require substantial increases in yields per acre. Increases in production are also needed for other crops, such as legumes, millet, cotton, rape, bananas and plantains.

It is important to increase yield on land that is already intensively cultivated. However, increasing production is only one part of the equation. Income generation, particularly in low-income areas together with the more effective distribution of food stocks, are equally, if not more, important. GM technologies are relevant to both these elements of food security.

In developing countries, it is estimated that about 650 million of the poorest people live in rural areas where the local production of food is the main economic activity. Without successful agriculture, these people will have neither employment nor the resources they need for a better life. Farming the land, and in particular small-holder farming, is the engine of progress in the rural communities, particularly of less developed countries.

The domestication of plants for agricultural use was a long-term process with profound evolutionary consequences for many species. One of its most valuable results was the creation of a diversity of plants serving human needs. Using this stock of genetic variability through selection and breeding, the 'Green Revolution' produced many varieties that are used throughout the world. This work, carried out largely in publicly-supported research institutions, has resulted in our present high-yielding crop varieties. A good example of such selective breeding was the introduction of 'dwarf' genes into rice and wheat which, in conjunction with fertilizer applications, dramatically increased the yield of traditional food crops in the Indian sub-continent, China and elsewhere. Despite past successes, the rate of increase of food crop production has decreased recently (yield increase in the 1970s of 3% per annum has declined in the 1990s to approximately 1% per annum) (Conway *et al* 1999). There are still heavy losses of crops owing to biotic (e.g. pests and disease) and abiotic (e.g. salinity and drought) stresses. The genetic diversity of some crop plants has also decreased and there are species without wild relatives with which to cross-breed. There are fewer options available than previously to address current problems through traditional breeding techniques though it is recognised that these techniques will continue to be important in the future.

Increasing the amount of land available to cultivate crops, without having a serious impact on the environment and natural resources, is a limited option. Modern agriculture has increased production of food, but it has also introduced large-scale use of pesticides and fertilisers that are expensive and can potentially affect human health or damage the ecosystem. A major challenge faced by humankind today is how to increase world food production and people's access to food, which requires local and employment-intensive staples production, without further depleting non-renewable resources and causing environmental damage. In other words, how do we move towards sustainable agricultural practices that do not compromise the health and economic well-being of the current and future generations? In order to think in terms of sustainable agriculture, factors responsible for soil, water and environmental deterioration must be identified and corrective measures taken.

Research on transgenic crops, as with conventional plant breeding and selection by farmers, aims selectively to alter, add or remove a character of choice in a plant, bearing in mind the regional needs and opportunities. It offers the possibility of not only bringing in desirable characteristics from other varieties of the plant, but also of adding characteristics from other unrelated species. Thereafter the transgenic plant becomes a parent for use in traditional breeding. Modification of qualitative and quantitative characteristics such as the composition of protein, starch, fats or vitamins by modification of metabolic pathways has already been achieved in some species. Such modifications increase the nutritional status of the foods and may help to improve human health by addressing malnutrition and under-nutrition. GM technology has also shown its potential to address micro-nutrient deficiencies and thus reduce the national expenditure and resources required to implement the current supplementation programmes (Texas A&M University 1997). These nutritional improvements have rarely been achieved previously by traditional methods of plant breeding.

Transgenic plants with important traits such as pest and herbicide resistance are most necessary where no inherent resistance has been demonstrated within the local species. There is intense research on the development of resistance to viral, bacterial, and fungal diseases; modification of plant architecture (eg height) and development (eg early or late flowering or seed production); tolerance to abiotic stresses (eg salinity and drought); production of industrial chemicals (plant-based renewable resources); and the use of transgenic plant biomass for novel and sustainable sources of fuel. Other benefits from transgenic plants under study include increased flexibility in crop management, decreased dependency on chemical insecticides and soil disturbance, enhanced yields, easier harvesting and higher proportions of the crop available for trading. For the consumer this should lead to decreased cost of food and higher nutritive value.

A large proportion of developing world agriculture is in the hands of small-scale farmers whose interests must be taken into account. Concerns regarding GM technology range from its potential impact on human health and the environment to concerns about private sector monopolies of the technology. It is essential that such concerns are addressed if we are to reap the potential benefits of this new technology.

We conclude that steps must be taken to meet the urgent need for sustainable practices in world agriculture if the demands of an expanding world population are to be met without destroying the environment or natural resource base. In particular, GM technology, coupled with important developments in other areas, should be used to increase the production of main food staples, improve the efficiency of production, reduce the environmental impact of agriculture, and provide access to food for small-scale farmers.

Examples of GM Technology That Would Benefit World Agriculture

GM technology has been used to produce a variety of crop plants to date, primarily with 'market-led' traits, some of which have become commercially successful. Developments resulting in commercially produced varieties in countries such as the USA and Canada have centred on increasing shelf-life of fruits and vegetables, conferring resistance to insect pests or viruses, and producing tolerance to specific herbicides. While these traits have had benefits for farmers, it has been difficult for the consumers to see any benefit other than, in limited cases, a decreased price owing to reduced cost and increased ease of production (University of Illinois 1999; Falck-Zepeda et al 1999).

A possible exception is the development of GM technology that delays ripening of fruit and vegetables, thus allowing an increased length of storage. Farmers would benefit from this development by increased flexibility in production and harvest. Consumers would benefit by the availability of fruits and vegetables such as transgenic tomatoes modified to soften much more slowly than traditional varieties, resulting in improved shelf-life and decreased cost of production, higher quality and lower cost. It is possible that farmers in developing countries could benefit considerably from crops with delayed ripening or softening as this may allow them much greater flexibility in distribution than they have at present. In many cases small-scale farmers suffer heavy losses due to excessive or uncontrolled ripening or softening of fruit or vegetables.

The real potential of GM technology to help address some of the most serious concerns of world agriculture has only recently begun to be explored. The following examples show how GM technology can be applied to some of the specific problems of agriculture indicating the potential for benefits.

Pest Resistance

There is clearly a benefit to farmers if transgenic plants are developed that are resistant to a specific pest. For example, papaya-ringspot-virus-resistant Papaya has been commercialised and grown in Hawaii since 1996 (Gonsalves 1998). There may also be a benefit to the environment if the use of pesticides is reduced. Transgenic crops containing insect resistance genes from *Bacillus thuringiensis* have made it possible to reduce significantly the amount of insecticide applied on cotton in the USA. One analysis, for example, showed a reduction of five million acre-treatments (two-million-hectare-treatments) or about one million kilograms of chemicals insecticides in 1999 compared with 1998 (US National Research Council 2000). However, populations of pests and disease-causing organisms adapt readily and become resistant to pesticides, and there is no reason to suppose that this will not occur equally rapidly with transgenic plants. In addition, pest biotypes are different in various regions. For instance, insect resistant crops developed for use in the USA and Canada may be resistant to pests that are of no concern in developing countries, and this is true both for transgenic plants and those developed by conventional breeding techniques. Even where the same genes for insect or herbicide resistance are useful in different regions, typically these genes will need to be introduced into locally adapted cultivars. There is need, therefore, for more research on transgenic plants that have been made resistant to local pests to assess their sustainability in the face of increased selection pressures for ever more virulent pests.

Improved Yield

One of the major technologies that led to the 'Green Revolution' was the development of high-yielding semi-dwarf wheat varieties. The genes responsible for height reduction were the Japanese NORIN 10 genes introduced into Western wheats in the 1950s (Gibberellin-insensitive-dwarfing-genes). These genes had two benefits: they produced a shorter, stronger plant that could respond to more fertiliser without collapsing, and they increased yield directly by reducing cell elongation in the vegetative plant parts, thereby allowing the plant to invest more in the reproductive plant parts that are eaten. These genes have recently been isolated and demonstrated to act in exactly the same way when used to transform other crop plant species (Peng *et al* 1999, Worland et al 1999). This dwarfing technique can now potentially be used to increase productivity in any crop plant where the economic yield is in the reproductive rather than the vegetative parts.

Tolerance to Biotic and Abiotic Stresses

The development of crops that have an inbuilt resistance to biotic and abiotic stress would help to stabilise annual production. For example, Rice Yellow Mottle Virus (RYMV) devastates rice in Africa by destroying the majority of the crop directly, with a secondary effect on any surviving plants that makes them more susceptible to fungal infections. As a result this virus has seriously threatened rice production in Africa. Conventional approaches to the control of RYMV using traditional breeding methods have failed to introduce resistance from wild species to cultivated rice. Researchers have used a novel technique that mimics 'genetic immunisation' by creating transgenic rice plants that are resistant to RYMV (Pinto *et al* 1999). Resistant transgenic varieties are currently entering field trials to test the effectiveness of their resistance to RYMV. This could provide a solution to the threat of total crop failure in the sub-Saharan African rice growing regions.

Numerous other examples could be given to illustrate the range of current scientific research including transgenic plants modified to combat papaya ring spot virus (Souza *et al* 1999), blight resistant potatoes (Torres *et al* 1999) and rice bacterial leaf blight (Zhai *et al* 2000); or as an example of an abiotic stress, plants modified to overproduce citric acid in roots and provide better tolerance to aluminum in acid soils (de la Fuente *et al* 1997). These examples have clear commercial potential but it will be imperative to maintain publicly funded research in GM technology if their full benefits are to be realised. For example, while GM technology provides access to new gene pools for sources of resistance, it needs to be established that these sources of resistance will be more stable than the traditional intra-species sources.

Use of Marginalised Land

A vast landmass across the globe, both coastal as well as terrestrial has been marginalised because of excessive salinity and alkalinity. A salt tolerance gene from mangroves (*Avicennia marina*) has been identified, cloned and transferred to other plants. The transgenic plants were found to be tolerant to higher concentrations of salt. The gutD gene from *Escherichia coli* has also been used to generate salt-tolerant transgenic maize plants (Liu *et al* 1999). Such genes are a potential source for developing cropping systems for marginalised lands (MS Swaminathan, personal communication, 2000).

Nutritional Benefits

Vitamin A deficiency causes half a million children to become partially or totally blind each year (Conway and Toennissen 1999). Traditional breeding methods have been unsuccessful in producing crops containing a high vitamin A concentration and most national authorities rely on expensive and complicated supplementation programs to address the problem. Researchers have introduced three new genes into rice: two from daffodils and one from a microorganism. The transgenic rice exhibits an increased production of beta-carotene as a precursor to vitamin A and the seed is yellow in colour (Ye *et al* 2000). Such yellow, or golden, rice may be a useful tool to help treat the problem of vitamin A deficiency in young children living in the tropics.

Iron fortification is required because cereal grains are deficient in essential micro-nutrients such as iron. Iron deficiency causes anaemia in pregnant women and young children. About 400 million women of childbearing age suffer as a result and they are more prone to stillborn or underweight children and to mortality at childbirth. Anaemia has been identified as a contributing factor in over 20% of maternal deaths (after giving birth) in Asia and Africa (Conway 1999). Transgenic rice with elevated levels of iron has been produced using genes involved in the production of an iron-binding protein and in the production of an enzyme that facilitates iron availability in the human diet (Goto *et al* 1999, Lucca *et al* (it). (1999). These plants contain 2 to 4 times the levels of iron normally found in non-transgenic rice, but the bio-availability of this iron will need to be ascertained by further study.

Reduced Environmental Impact

Water availability and efficient usage have become global issues. Soils subjected to extensive tillage (ploughing) for controlling weeds and preparing seed beds are prone to erosion, and there is a serious loss of water content. Low tillage systems have been used for many years in traditional communities. There is a need to develop crops that thrive under such conditions, including the introduction of resistance to root diseases currently controlled by tillage and to herbicides that can be used as a substitute for tillage (Cook 2000). Applications in more developed countries show that GM technology offers a useful tool for the introduction of root disease resistance for conditions of reduced tillage. However, a careful cost-benefit analysis would be needed to ensure that maximum advantage is achieved. Regional differences in agricultural systems and the potential impact of substituting a traditional crop with a new transgenic one would also need to be carefully evaluated.

Other Benefits of Transgenic Plants

First generation transgenic varieties have benefited many farmers in the form of reduced production costs, higher yields, or both. In many cases, they have also benefited the environment because of reduced pesticide usage or by providing the means to grow crops with less tillage. Insects are responsible for huge losses to crops in the field and to harvested products in transit or storage, but health concerns for consumers and for environmental impact have limited the registration of many promising chemical pesticides. Genes for pest resistance carefully deployed in crops to avoid selecting for future pest resistance, provides alternative opportunities to reduce the use of chemical pesticides in many important crops. In addition, lowering the contamination of our food supply by pathogens that cause food safety problems (eg mycotoxins) would be beneficial to farmers and consumers alike.

Pharmaceuticals and Vaccines From Transgenic Plants

Vaccines are available for many of the diseases that cause widespread death or human discomfort in developing countries, but they are often expensive both to produce and use. The majority must be stored under conditions of refrigeration and administered by trained specialists, all of which adds to the expense. Even the cost of needles to administer vaccines is prohibitive in some countries. As a result, the vaccines often do not reach those in most need. Researchers are currently investigating the potential for GM technology to produce vaccines and pharmaceuticals in plants. This could allow easier access, cheaper production, and an alternative way to generate income. Vaccines against infectious diseases of the gastro-intestinal tract have been produced in plants such as potato and bananas (Thanavala *et al* 1995). Another appropriate target would be cereal grains. An anti-cancer antibody has recently been expressed in rice and wheat seeds that recognises cells of lung, breast and colon cancer and hence could be useful in both diagnosis and therapy in the future (Stoger *et al* 2000). Such technologies are at a very early stage in development and obvious concerns about human health and environmental safety during production must be investigated before such plants can be approved as speciality crops. Nevertheless, the development of transgenic plants to produce therapeutic agents has immense potential to help in solving problems of disease in developing countries.

About one third of medicines used today are derived from plants, one of the most famous examples being aspirin (the acetylated form of a natural plant product, salicylic acid). It is believed that less that 10% of medicinal plants have been identified and characterised, and the potential exists to use GM technology in a way that increases yields of these medicinal substances once identified. For example, the valuable anti-cancer agents vinblastine and vincristine are the only approved drugs for treatment of Hodgkin's lymphoma. Both products are derived from the Madagascar Periwinkle, which produces them in minute concentrations along with 80 to 100 very similar chemicals. The therapeutic compounds are therefore extremely expensive to produce. Currently, there is intensive research in progress to investigate the potential of GM technology to increase the yields of active compounds, or to allow their production in other plants that are easier to manage than the Periwinkle.

We recommend that transgenic crop research and development should focus on plants that will (i) improve production stability; (ii) give nutritional benefits to the consumer; (iii) reduce the environmental impacts of intensive and extensive agriculture; and (iv) increase the availability of pharmaceuticals and vaccines; while (v) developing protocols and regulations that ensure that transgenic crops designed for purposes other than food, such as pharmaceuticals, industrial chemicals, etc. do not spread or mix with either transgenic or non-transgenic food crops.

Transgenic Plants and Human Health and Safety

Through classical plant breeding techniques, present day cultivated crops have become significantly different from their wild counterparts. Many of these crops were originally less productive and at times unsuitable for human consumption. Over the years, traditional plant breeding and selection of these crops have resulted in plants that are more productive and nutritious. The advent of GM technology has allowed further development. To date, over 30 million hectares of transgenic crops have been grown and no human health problems associated specifically with the ingestion of transgenic crops or their products have been identified. However numerous potential concerns have been raised since the development of GM technology in the early 1970s. Such concerns have focused on the potential for allergic reactions to food products, the possible introduction or increase in production of toxic compounds as a result of the GM technology, and the use of antibiotic resistance as markers in the transformation process.

Every effort should be made to avoid the introduction of known allergens into food crops. Information concerning potential allergens and natural plant toxins should be made available to researchers, industry, regulators, and the general public. In order to facilitate this effort, public databases should be developed which facilitate access of all interested parties to data.

Traditional plant breeding methods include wide crosses with closely related wild species, and may involve a long process of crossing back to the commercial parent to remove undesirable genes. A feature of GM technology is that it involves the introduction of one or at most, a few, well-defined genes rather than the introduction of whole genomes or parts of chromosomes as in traditional plant breeding. This makes toxicity testing for transgenic plants more straightforward than for conventionally produced plants with new traits, because it is much clearer what the new features are in the modified plant. On the other hand, GM technology can introduce genes from diverse organisms, some of which have little history in the food supply.

Decisions regarding safety should be based on the nature of the product, rather than on the method by which it was modified. It is important to bear in mind that many of the crop plants we use contain natural toxins and allergens. The potential for human toxicity or allergenicity should be kept under scrutiny for any novel proteins produced in plants with the potential to become part of food or feed. Health hazards from food, and how to reduce them, are an issue in all countries, quite apart from any concerns about GM technology.

Since the advent of GM technology, researchers have used antibiotic resistance genes as selective markers for the process of genetic modification. The concern has been raised that the widespread use of such genes in plants could increase the antibiotic resistance of human pathogens. Kanamycin, one of the most commonly used resistance markers for plant transformation is still used for the treatment of the following human infections: bone, respiratory tract, skin, soft-tissue, and abdominal infections, complicated urinary tract infections, endocarditis, septicaemia, and enterococcal infections. Scientists now have the means to remove these marker genes before a crop plant is developed for commercial use (Zubko *et al* 2000). Developers should continue to move rapidly to remove all such markers from transgenic plants and to utilise alternative markers for the selection of new varieties. No definitive evidence exists that these antibiotic resistance genes cause harm to humans, but because of public concerns, all those involved in the development of transgenic plants should move quickly to eliminate these markers.

Ultimately, no credible evidence from scientists or regulatory institutions will influence popular public opinion unless there is public confidence in the institutions and mechanisms that regulate such products.

We recommend: (i) public health regulatory systems need to be put in place in every country to identify and monitor any potential adverse human health effects of transgenic plants, as for any other new variety. Such systems must remain fully adaptable to rapid advances in scientific knowledge. The possibility of long-term adverse effects should be kept in view when setting up such systems. This will require coordinated efforts between nations, the sharing of experience, and the standardisation of some types of risk assessments specifically related to human health; (ii) Information should be made available to the public concerning how their food supply is regulated and its safety ensured.

Transgenic Plants and the Environment

Modern agriculture is intrinsically destructive of the environment. It is particularly destructive of biological diversity, notably when practised in a very resource-inefficient way, or when it applies technologies that are not adapted to environmental features (soils, slopes, climatic regions) of a particular area. This is true of both small-scale and large-scale agriculture. The widespread application of conventional agricultural technologies such as herbicides, pesticides, fertilisers and tillage has resulted in severe environmental damage in many parts of the world. Thus the environmental risks of new GM technologies need to be considered in the light of the risks of continuing to use conventional technologies and other commonly used farming techniques.

Some agricultural practices in parts of the developing world maintain biological diversity. This is achieved by simultaneously cultivating several varieties of a crop and mixing them with other secondary crops, thus maintaining a highly diverse community of plants (Toledo *et al* 1995; Nations *et al* 1980; Whitmore *et al* 1992).

Most of the environmental concerns about GM technology in plants have derived from the possibility of gene flow to close relatives of the transgenic plant, the possible undesirable effects of the exotic genes or traits (eg insect resistance or herbicide tolerance), and the possible effect on non-target organisms.

As with the development of any new technology, a careful approach is warranted before development of a commercial product. It must be shown that the potential impact of a transgenic plant has been carefully analysed and that if it is not neutral or innocuous, it is preferable to the impact of the conventional agricultural technologies that it is designed to replace (Campbell *et al* 1997; May 1999; Toledo *et al* 1995).

Given the limited use of transgenic plants world-wide and the relatively constrained geographic and ecological conditions of their release, concrete information about their actual effects on the environment and on biological diversity is still very sparse. As a consequence there is no consensus as to the seriousness, or even the existence, of any potential environmental harm from GM technology. There is therefore a need for a thorough risk assessment of likely consequences at an early stage in the development of all transgenic plant varieties, as well as for a monitoring system to evaluate these risks in subsequent field tests and releases.

Risk assessments need base-line information including the biology of the species, its ecology and the identification of related species, the new traits resulting from GM technology, and relevant ecological data about the site(s) in which the transgenic plant is intended to be released. This information can be very difficult to obtain in highly diverse environments. Centres of origin or diversity of cultivated plants should receive careful consideration because there will be many wild relatives to which the new traits could be transferred (Ellstrand *et al* 1999; Mikkelsen *et al* 1996; Scheffler 1993; Van Raamsdonk *et al* 1997). For special environments, transgenic plants can be developed using technologies that minimise the possibilities of gene flow via pollen and its effects on wild relatives, through the use of male sterility methods or maternal inheritance resulting from chloroplast transformation (Daniell 1999; Daniell *et al* 1998; Scott & Wilkinson 1999).

Studies of gene transfer from conventional and transgenic plants to wild relatives and other plants in the ecosystem have so far concentrated on species of economic importance such as wheat, oilseed rape and barley. A virtual absence of data, particularly for species like maize, imposes the need to carefully and continuously monitor any possible effects of novel transgenic plants in the field (Hokanson *et al* 1997; Daniell *et al* 1998). In addition there is a continued need for research on the rates of gene transfer from traditional crops to indigenous species (Ellstrand *et al* 1999).

When monitoring a small-scale pilot release of a transgenic crop the following issues should be considered in addition to any concerns specific to a particular local environment:

(a) Does the existence of a transgenic plant with resistance for a particular pest or disease exacerbate the emergence of new resistant pests or diseases, and is this problem worse than that with the traditional alternative? (Riddick & Barbosa 1998; Hillbeck *et al* 1998; Birch *et al* 1999).

(b) If traits (eg salt tolerance, disease resistance, etc) are transferred to wild varieties, is there an expansion in the niche of these species that may result in the suppression of biological diversity in the surrounding areas?

(c) Would the widespread adoption of stress-tolerant plants promote a considerable increase in the use of land where formerly agriculture could not be practised in a way that destroys valuable natural ecosystems?

The risk assessments performed should be standardized for plants new to an environment. Most nations already have procedures for the approval and local release of new varieties of crop plants. Although these assessments are based primarily on the agronomic performance of the new variety compared with existing varieties, this approval process could serve as the beginning or model for a more formal risk assessment process to investigate the potential environmental impact of the new varieties, including those with transgenes.

Historically, both poverty and structural change in rural areas have resulted in severe environmental deterioration. The adoption of modern biotechnology should not accelerate this deterioration. It should instead be used in a way that reduces poverty and its deleterious effects on the environment.

We recommend that: (i) coordinated efforts be undertaken to investigate the potential environmental effects, both positive and negative, of transgenic plant technologies in their specific applications; (ii) all environmental effects should be assessed against the background of effects from conventional agricultural practices currently in use in places for which the transgenic crop has been developed or grown; and (iii) *in situ* and *ex situ* conservation of genetic resources for agriculture should be promoted that will guarantee the widespread availability of both conventional and transgenic varieties as germplasm for future plant breeding.

Membership of Working Group and Methodology

The following individuals represented the Councils of the Brazilian Academy of Sciences, the Chinese Academy of Sciences, the Indian National Science Academy, the Mexican Academy of Sciences, the Royal Society (UK), the Third World Academy of Sciences and the National Academy of Sciences of the USA during the preparation of this report. The text of the report was produced following meetings at the Royal Society (Chairman Professor Brian Heap FRS, Secretary Dr Rebecca Bowden) in London in July 1999 and February 2000 at which the issues covered in this report were discussed in detail.

The Brazilian Academy of Sciences
On working group:

- Dr Ernesto Paterniani
- Dr Fernando Perez
- Professor Fernando Reinach Professor
- Jose Galizia Tundisi

The Chinese Academy of Sciences
On working group:

- Professor Zhihong Xu
- Professor Rongxiang Fang
- Professor Qian Yingqian

The Indian National Science Academy
On working group:

- Professor R P Sharma
- Professor S K Sopory

Reviewers on behalf of Council:

- Professor P N Tandon, Chairman
- Dr H K Jain
- Dr Manju Sharma
- Dr R S Paroda
- Dr Anupam Varma
- Ms Suman Sahai
- Dr J Thomas
- Professor K Muralidhar

The Mexican Academy of Sciences
On working group:

- Mr Jorge Larson
- Dr Jorge Nieto Sotelo
- Professor Josè Sarukhàn

The Royal Society of London
On working group:

- Sir Aaron Klug OM PRS
- Professor Michael Gale FRS
- Professor Michael Lipton

Reviewed and approved by the Council of the Royal Society.

The Third World Academy of Sciences
On working group:

- Professor Muhammad Akhtar FRS

The National Academy of Sciences of the USA
(Staff Officer to NAS Delegation—Mr John Campbell)

On working group:

- Professor Bruce Alberts
- Professor F Sherwood Rowland
- Professor Luis Sequiera
- Professor R James Cook
- Professor Alex McCalla

References

Birch et al. (1999). *Molecular Breeding* **5** 75–83.

Campbell L H et al. (1997). The Joint Nature Conservation Committee, Report 227, UK.

Conway G. (1999). *Biotechnology, Food & Drought* in Proceedings of the World Commission on Water, Nov 1999.

Conway G. (1999). *The Doubly Green Revolution: Food for All in the 21st Century*. London: Penguin Books.

Conway G et al. (1999). *Nature* **402** C55–58.

Cook R J (2000) in *Agricultural Biotechnology and the Poor*, proceedings of an international conference, Washington DC, October 1999, 123–130 (G I Persley and M M Lantin, eds).

Daniell H. (1999). *Trends in Plant Science* **4(12)** 467–469.

Daniell H et al (1998). *Nature Biotechnology* **16** 345–348.

De la Fuente J M et al. (1997). *Science* **276 (5318)** 1566–1568.

Ellestrand N C et al. (1999). Annual Review of ecological Systems 30 539–563.

Falck-Zepeda B J et al. (1999). *International Service for the Acquisition of Agri-Biotech Applications* **14** 17.

Gonsalves D. (1998). *Annual Review of Phytopathology* **36** 415–437.

Goto F et al. (1999). *Nature Biotechnology* **17** 282–286.

Hillbeck A et al. (1998). *Environmental Entomology* **27** 480–487.

Hokanson et al. (1997). *Euphytica* **96** 397–403.

James C. (1999). *Global status of transgenic crops in* 1999. ISAAA: Ithaca, New York.

Liu Y et al. (1999). *Science in China (Series C)* **42** 90–95.

Lucca P et al (1999). In *Proceedings of General Meeting of the International Programme on Rice Biotechnology*, Sept 1999, Phuket, Thailand.

Mikkelsen T et al. (1996). *Nature* **380** 31.

Nations J et al. (1980). *Journal of Anthropological Research* **36** 1–30.

Peng J R et al. (1999). *Nature* **400** 256–261.

Pinstrup-Anderson P et al. (1999). *World food prospects: critical issues for the early 21st Century*. International Food Policy Research Institute: Washington D C, USA.

Pinstrup-Anderson P et al. (2000). *Meeting food needs in the 21st century: how many and who will be at risk?* Presented at AAAS Annual Meeting, Feb 2000, Washington D C, USA.

Pinto Y M et al. (1999) *Nature Biotechnology* **17** 702–707.

Riddick E et al. (1998). *Annals of the Entomology Society of America* **91** 303–307.

Scheffler J. (1993). *Transgenic Research* **2** 356–364.

Scott S E et al. (1999). *Nature Biotechnology* **17** 390–392.

Souza M T. (1999). *Analysis of the resistance in genetically engineered papaya against papaya ringspot potyvirus, partial characterisation of the PRSV. Brasil. Bahia isolate and development of transgenic papaya for Brazil.* PhD dissertation Cornell University, USA.

Stoger E et al. (2000). *Plant Molecular Biology* **42** 583–590.

Texas A & M University. (1997). *Report filed with USEPA for hearing on 21 May 1997.* Docket OPP-0478.

Thanavala Y et al. (1995). *Proceedings of the National Academy of Sciences of the USA* **92(8)** 3358–3361.

Toledo V M et al. (1995). *Interciencia* **20** 177–187.

Torres A C et al. (1999). Biotechnologia—*Ciencia & Disenvolvimento* **2 (7)** 74–77.

UNICEF. (1998). *The state of the world's children 1998: focus on nutrition.* United Nations: New York, USA.

U S National Research Council. (2000). *Genetically modified pest-protected plants: science and regulation.* National Academy Press: Washington, D C, USA.

University of Illinois. (1999). *The economics and politics of genetically modified organisms in agriculture: implications for WTO 2000.* USA: University of Illinois Bulletin 809, November 1999.

Van Raamsdonk L et al. (1997). *Acta Botanica Neerlandica* **48** 9–84.

Whitmore T M et al. (1992). *Annals of the Association of American Geographers* **82** 402–425.

Worland A J et al. (1999) *Nature* **400** 256–261.

Xudong Ye et al. (2000). *Science* **287** 303–306.

Zhai W et al. (2000). *Science in China (Series C)* **43** 361–368.

NO

Brian Halweil

The Emperor's New Crops

It's June 1998 and Robert Shapiro, CEO of Monsanto Corporation, is delivering a keynote speech at "BIO 98," the annual meeting of the Biotechnology Industry Organization. "Somehow," he says, "we're going to have to figure out how to meet a demand for a doubling of the world's food supply, when it's impossible to conceive of a doubling of the world's acreage under cultivation. And it is impossible, indeed, even to conceive of increases in productivity—using current technologies—that don't produce major issues for the sustainability of agriculture."

Those "major issues" preoccupy a growing number of economists, environmentalists, and other analysts concerned with agriculture. Given the widespread erosion of topsoil, the continued loss of genetic variety in the major crop species, the uncertain effects of long-term agrochemical use, and the chronic hunger that now haunts nearly 1 billion people, it would seem that a major paradigm shift in agriculture is long overdue. Yet Shapiro was anything but gloomy. Noting "the sense of excitement, energy, and confidence" that engulfed the room, he argued that "biotechnology represents a potentially sustainable solution to the issue of feeding people."

To its proponents, biotech is the key to that new agricultural paradigm. They envision crops genetically engineered to tolerate dry, low-nutrient, or salty soils—allowing some of the world's most degraded farmland to flourish once again. Crops that produce their own pesticides would reduce the need for toxic chemicals, and engineering for better nutrition would help the overfed as well as the hungry. In industry gatherings, biotech appears as some rare hybrid between corporate mega-opportunity and international social program.

The roots of this new paradigm were put down nearly 50 years ago, when James Watson and Francis Crick defined the structure of DNA, the giant molecule that makes up a cell's chromosomes. Once the structure of the genetic code was understood, researchers began looking for ways to isolate little snippets of DNA—particular genes—and manipulate them in various ways. In 1973, scientists managed to paste a gene from one microbe into another microbe of a different species; the result was the first artificial transfer of genetic information across the species boundary. In the early 1980s, several research teams—including one at Monsanto, then a multinational pesticide company—succeeded in splicing a bacterium gene into a petunia. The first "transgenic" plant was born.

Such plants represented a quantum leap in crop breeding: the fact that a plant could not interbreed with a bacterium was no longer an obstacle to using the microbe's genes in crop design. Theoretically, at least, the world's entire store of genetic wealth became available to plant breeders, and the bio-tech labs were quick to test the new possibilities. Among the early creations was a tomato armed with a flounder gene to enhance frost resistance and with a rebuilt tomato gene to retard spoilage....

Transgenic crops are no longer just a laboratory phenomenon. Since 1986, 25,000 transgenic field trials have been conducted worldwide—a full 10,000 of these just in the last two years. More than 60 different crops—ranging from corn to strawberries, from apples to potatoes—have been engineered. From 2 million hectares in 1996, the global area planted in transgenics jumped to 27.8 million hectares in 1998. That's nearly a fifteenfold increase in just two years.

In 1992, China planted out a tobacco variety engineered to resist viruses and became the first nation to grow transgenic crops for commercial use. Farmers in the United States sowed their first commercial crop in 1994; their counterparts in Argentina, Australia, Canada, and Mexico followed suit in 1996. By 1998, nine nations were growing transgenics for market and that number is expected to reach 20 to 25 by 2000.

Ag biotech is now a global phenomenon, but it remains powerfully concentrated in several ways:

In terms of where transgenics are planted. Three-quarters of transgenic cropland is in the United States. More than a third of the U.S. soybean crop last year was transgenic, as was nearly one-quarter of the corn and one-fifth of the cotton. The only other countries with a substantial transgenic harvest are Argentina and Canada: over half of the 1998 Argentine soybean crop was transgenic, as was over half of the Canadian canola crop. These three nations account for 99 percent of global transgenic crop area. (Most countries have been slow to adopt transgenics because of public concern over possible risks to ecological and human health.)

In terms of which crops are in production. While many crops have been engineered, only a very few are cultivated in appreciable quantities. Soybeans account for 52 percent of global transgenic area, corn for another 30 percent. Cotton—almost entirely on U.S. soil—and canola in Canada cover most of the rest.

In terms of which traits are in commercial use. Most of the transgenic harvest has been engineered for "input traits" intended to replace or accommodate the standard chemical "inputs" of large-scale agriculture, especially insecticides and herbicides. Worldwide, nearly 30 percent of transgenic cropland is planted in varieties designed to produce an insect-killing toxin, and almost all of the rest is in crops engineered to resist herbicides....

These two types of crops—the insecticidal and the herbicide-resistant varieties—are biotech's first large-scale commercial ventures. They provide the first real opportunity to test the industry's claims to be engineering a new agricultural paradigm.

The Bugs

The only insecticidal transgenics currently in commercial use are "Bt crops." Grown on nearly 8 million hectares worldwide in 1998, these plants have been equipped with a gene from the soil organism *Bacillus thuringiensis* (Bt), which produces a substance that is deadly to certain insects.

The idea behind Bt crops is to free conventional agriculture from the highly toxic synthetic pesticides that have defined pest control since World War II. Shapiro, for instance, speaks of Monsanto's Bt cotton as a way of substituting "information encoded in a gene in a cotton plant for airplanes flying over cotton fields and spraying toxic chemicals on them." ... At least in the short term, Bt varieties have allowed farmers to cut their spraying of insecticide-intensive crops, like cotton and potato. In 1998, for instance, the typical Bt cotton grower in Mississippi sprayed only once for tobacco budworm and cotton bollworm—the insects targeted by Bt—while non-Bt growers averaged five sprayings.

Farmers are buying into this approach in a big way. Bt crops have had some of the highest adoption rates that the seed industry has ever seen for new varieties. In the United States, just a few years after commercialization, nearly 25 percent of the corn crop and 20 percent of the cotton crop is Bt. In some counties in the southeastern states, the adoption rate of Bt cotton has reached 70 percent. The big draw for farmers is a lowering of production costs from reduced insecticide spraying, although the savings is partly offset by the more expensive seed. Some farmers also report that Bt crops are doing a better job of pest control than conventional spraying, although the crops must still be sprayed for pests that are unaffected by Bt. (Bt is toxic primarily to members of the Lepidoptera, the butterfly and moth family, and the Coleoptera, the beetle family.)

Unfortunately, there is a systemic problem in the background that will almost certainly erode these gains: pesticide resistance. Modern pest management tends to be very narrowly focused; the idea, essentially, is that when faced with a problematic pest, you should look for a chemical to kill it. The result has been a continual toughening of the pests, which has rendered successive generations of chemicals useless. After more than 50 years of this evolutionary rivalry, there is abundant evidence that pests of all sorts—insects, weeds, or pathogens—will develop resistance to just about any chemical that humans throw at them.

The Bt transgenics basically just replace an insecticide that is sprayed on the crop with one that is packaged inside it. The technique may be more sophisticated but the strategy remains the same: aim the chemical at the pest. Some entomologists are predicting that, without comprehensive strategies to prevent it, pest resistance to Bt could appear in the field within three to five years of widespread use, rendering the crops ineffective. Widespread resistance to Bt would affect more than the transgenic crops, since Bt is also commonly used in conventional spraying. Farmers could find one of their most environmentally benign pesticides beginning to slip away.

In one respect, Bt crops are a throwback to the early days of synthetic pesticides, when farmers were encouraged to spray even if their crops didn't appear to need it. The Bt crops show a similar lack of discrimination: they are programmed to churn out toxin during the entire growing season, regardless of the level of infestation. This sort of prophylactic control greatly increases the likelihood of resistance because it tends to maximize exposure to the toxin—it's the plant equivalent of treating antibiotics like vitamins.

Agricultural entomologists now generally agree that Bt crops will have to be managed in a way that discourages resistance if the effectiveness of Bt is to be maintained. In the United States, the Environmental Protection Agency, which regulates the use of pesticides, now requires producers of Bt crops to develop "resistance management plans." This is a new step for the EPA, which has never required analogous plans from manufacturers of conventional pesticides.

The usual form of resistance management involves the creation of "refugia"— areas planted in a crop variety that isn't armed with the Bt gene. If the refugia are large enough, then a substantial proportion of the target pest population will never encounter the Bt toxin, and will not be under any selection pressure to develop resistance to it. Interbreeding between the refugia insects and the insects in the Bt fields should stall the development of resistance in the population as a whole, assuming the resistance gene is recessive.

The biotech companies themselves have been recommending that their customers plant refugia, although the recommendations generally fall short of what most resistance experts consider necessary. This is not surprising, of course, since there is an inherent inconsistency between the refugia idea and the inevitable interest on the part of the manufacturer in selling as much product as possible. An even greater obstacle may be the reactions of farmers themselves, since the refugia concept is counter-intuitive: farmers, who spend much of their lives trying to control pests, are being told that the best way to maintain a high yield is to leave substantial portions of their land vulnerable to pests. The impulse to plant smaller refugia—or to count someone else's land as part of one's own refugia—may prove irresistible. And the possibility of enforcing the planting of larger refugia seems remote, especially once Bt crops are deployed to hundreds of millions of small-scale farmers throughout the developing world....

According to Gary Barton, director of ag biotech communications at Monsanto, "products now in the pipeline which rely on different insecticidal toxins or multiple toxins could replace Bt crops in the event of widespread resistance."...

The result, according to Fred Gould, an entomologist at the University of North Carolina, would be "a crop with a series of silver bullet pest solutions." And each of these solutions, in Gould's view, would be highly vulnerable to pest resistance. This scenario does not differ essentially from the current one: in place of a pesticide treadmill, we would substitute a sort of gene treadmill. The arms race between farmers and pests would continue, but would include an additional biochemical dimension. Transgenic plants, designed to secrete increasingly potent combinations of pesticides, would vie with a host of increasingly resistant pests.

Figure 1

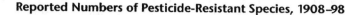

Reported Numbers of Pesticide-Resistant Species, 1908–98

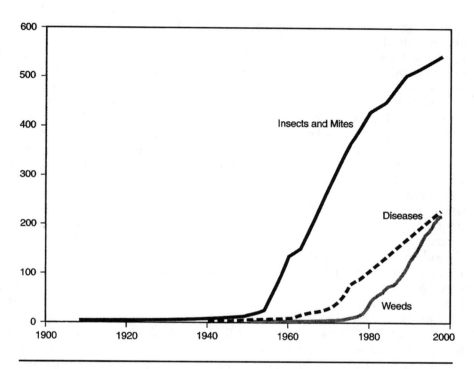

Source: *Worldwatch Institute,* Vital Signs 1999

The Weeds

The global transgenic harvest is currently dominated, not by Bt crops, but by herbicide-resistant crops (HRCs), which occupy 20 million hectares worldwide. HRCs are sold as part of a "technology package" comprised of HRC seed and the herbicide the crop is designed to resist. The two principal product lines are currently Monsanto's "Roundup Ready" crops—so-named because they tolerate Monsanto's best-selling herbicide, "Roundup" (glyphosate)—and AgrEvo's "Liberty Link" crops, which tolerate that company's "Liberty" herbicide (glufosinate).

It may sound contradictory, but one ostensible objective of HRCs is to reduce herbicide use. By designing crops that tolerate fairly high levels of exposure to a broad-spectrum herbicide (a chemical that is toxic to a wide range of plants), the companies are giving farmers the option of using a heavy, once-in-the-growing-season dousing with that herbicide, instead of the standard practice, which calls for a series of applications of several different compounds. It's not yet clear whether this new herbicide regime actually reduces the amount of material used, but its simplicity is attracting many farmers into the package.

Another potential benefit of HRCs is that they may allow for more "conservation tillage," farming techniques that reduce the need for plowing or even—under "no till" cultivation—eliminate it entirely. A primary reason for plowing is to break up the weeds, but because it exposes bare earth, plowing causes top soil erosion. Top soil is the capital upon which agriculture is built, so conserving soil is one of agriculture's primary responsibilities....

The bigger problem is that HRCs, like Bt crops, are really just an extension of the current pesticide paradigm. HRCs may permit a reduction in herbicide use over the short term, but obviously their widespread adoption would encourage herbicide dependency. In many parts of the developing world, where herbicides are not now common, the herbicide habit could mean substantial additional environmental stresses: herbicides are toxic to many soil organisms, they can pollute groundwater, and they may have long-term effects on both people and wildlife.

And of course, resistance will occur. Bob Hartzler, a weed scientist at Iowa State University, warns that if HRCs encourage reliance on just a few broad-spectrum herbicides, then resistance is likely to develop faster—and agriculture is likely to be more vulnerable to it....

In the U.S. Midwest, heavy use of Roundup on Roundup Ready soybeans is already encouraging weed species, like waterhemp, that are naturally resistant to that herbicide. (As Roundup suppresses the susceptible weeds, the resistant ones have more room to grow.) Thus far, the evolution of resistance in weed species that are susceptible to Roundup has been relatively rare, despite decades of use. The first reported case involved wild ryegrass in Australia, in 1995. But with increasing use, more such cases are all but inevitable—especially since Monsanto is on the verge of releasing Roundup Ready corn. Corn and soybeans are the classic crop rotation in the U.S. Midwest—corn is planted in one year, soy in the next. Roundup Ready varieties of both crops could subject vast areas of the U.S. "breadbasket" to an unremitting rain of that herbicide. As with the Bt crops, the early promise of HRCs is liable to be undercut by the very mentality that inspired them: the single-minded chemical pursuit of the pest.

Transgenes on the Loose

In 1997, just one year after its first commercial planting in Canada, a farmer reported—and DNA testing confirmed—that Roundup Ready canola had cross-pollinated with a related weed species growing in the field's margins, and produced an herbicide-tolerant descendant....

If a transgenic crop is capable of sexual reproduction (and they generally are), the leaking of "transgenes" is to some degree inevitable, if any close relatives are growing in the vicinity. This type of genetic pollution is not likely to be common in the industrialized countries, where most major crops have relatively few close relatives. But in the developing world—especially in regions where a major crop originated—the picture is very different. Such places are the "hot spots" of agricultural diversity: the cultivation of the ancient, traditional varieties—whether it's corn in Mexico or soybeans in China—often involves a subtle genetic interplay between cultivated forms of a species, wild forms, and related species that aren't cultivated at all. The possibilities for genetic pollution in such contexts are substantial.

Ordinary breeding creates some degree of genetic pollution too. But according to Allison Snow, an Ohio State University plant ecologist who studies transgene flow, biotech could amplify the process considerably because of the far more diverse array of genes it can press into service. Any traits that confer a substantial competitive advantage in the wild could be expected to spread widely. The Bt gene would presumably be an excellent candidate for this process, since its toxin affects so many insect species.

There's no way to predict what would happen if the Bt gene were to escape into a wild flora, but there's good reason to be concerned. John Losey, an entomologist at Cornell University, has been experimenting with Monarch butterflies, by raising their caterpillars on milkweed dusted with Bt-corn pollen. Losey found that nearly half of the insects raised on this fare died and the rest were stunted. (Caterpillars raised on milkweed dusted with ordinary corn pollen did fine.) According to Losey, "these levels of mortality are comparable to those you find with especially toxic insecticides." If the gene were to work a change that dramatic in a wild plant's toxicity, then it could trigger a cascade of second- and third-order ecological effects.

The potential for this kind of trouble is likely to grow, since a major interest in biotech product development is "trait-stacking"—combining several engineered genes in a single variety, as with the attempts to develop corn with multiple toxins. Monsanto's "stacked cotton"—Roundup Ready and Bt-producing—is already on the market in the United States....

In the agricultural hot spots, there is an important practical reason to be concerned about any resulting genetic pollution. Plant breeders depend on the genetic wealth of the hot spots to maintain the vigor of the major crops—and there's no realistic possibility of biotech rendering this natural wealth "obsolete." But it certainly is possible that foreign genes could upset the relationships between the local varieties and their wild relatives....

Toward a New Feudalism

The advent of transgenic crops raise serious social questions as well—beginning with ownership. All transgenic seed is patented, as are most nontransgenic commercial varieties. But beginning in the 1980s, the tendency in industrialized countries and in international law has been to permit increasingly broad agricultural patents—and not just on varieties but even on specific genes. Under the earlier, more limited patents, farmers could buy seed and use it in their own breeding; they could grow it out and save some of the resulting seed for the next year; they could even trade it for other seed. About the only thing they couldn't do was sell it outright. But under the broader patents, all of those activities are illegal; the purchaser is essentially just paying for one-time use of the germplasm.

The right to own genes is a relatively new phenomenon in world history and its effects on agriculture—and life in general—are still very uncertain. The biotech companies argue that ownership is essential for driving their industry: without exclusive rights to a product that costs hundreds of millions of dollars to develop, how will it be possible to attract investors? ... Val Giddings of the Biotechnology Industry Organization makes this case: "intellectual property rights allow us to harness genetic resources for commercial use, making biodiversity concretely more valuable."...

Patents are clearly an important ingredient in the industry's expansion. Global sales of transgenic crop products grew from $75 million in 1995 to $1.5 billion in 1998—a 20-fold increase. Sales are expected to hit $25 billion by 2010. And as the market has expanded, so has the scramble for patents. Recently, for example, the German agrochemical firm AgrEvo, the maker of "Liberty" herbicide, bought a Dutch biotech company called Plant Genetic Systems (PGS), which owned numerous wheat and corn patents. The patents were so highly valued that AgrEvo was willing to pay $730 million for the acquisition—$700 million more than PGS's annual sales....

This patent frenzy is contributing to an intense wave of consolidation within the industry.... Hoechst recently merged with one of its French counterparts, Rhône-Poulenc, to form Aventis, which is now the world's largest agrochemical firm and a major player in the biotech industry. On the other side of the Atlantic, Monsanto has spent nearly $8 billion since 1996 to purchase various seed companies. DuPont, a major competitor, has bought the world's largest seed company, Pioneer Hi-Bred. DuPont and Monsanto were minor players in the seed industry just a decade ago, but are now respectively the largest and second-largest seed companies in the world.

... Of the 56 transgenic products approved for commercial planting in 1998, 33 belonged to just four corporations: Monsanto, Aventis, Novartis, and DuPont. The first three of these companies control the transgenic seed market in the United States, which amounts to three-fourths of the global market....

But there is far more at stake here than the fortunes of the industry itself: patents and similar legal mechanisms may be giving companies additional control over farmers. As a way of securing their patent rights, biotech companies are requiring farmers to sign "seed contracts" when they purchase transgenic seed—a wholly new phenomenon in agriculture. The contracts may stipulate what brand of pesticides the farmer must use on the crop—a kind of legal cement for those crop-herbicide "technology packages." ...

The most troubling aspect of these contracts is the possible effect on seed saving—the ancient practice of reserving a certain amount of harvested seed for the next planting. In the developing world, some 1.4 billion farmers still rely almost exclusively on seed saving for their planting needs. As a widespread, low-tech form of breeding, seed saving is also critical to the husbandry of crop diversity, since farmers generally save seed from plants that have done best under local conditions. The contracts have little immediate relevance to seed saving in the developing world, since the practice there is employed largely by farmers who could not afford transgenic seed in the first place. But even in industrialized countries, seed saving is still common in certain areas and for certain crops, and Monsanto has already taken legal action against over 300 farmers for replanting proprietary seeds.

... The substitution of commercial for farm-saved seed has been a primary reason for the loss of genetic diversity in the agricultural hot spots. Hope Shand, research director for the Rural Advancement Foundation International (RAFI), a farmer advocacy group based in Winnipeg, Canada, regards the extension of patents in general as a means of reducing farmers to "bioserfs," who provide little more than land and labor to agribusiness.

... [B]eyond these control issues, there remains the basic question of biotech's potential for feeding the world's billions. Here too, the current trends are not very encouraging. At present, the industry has funneled its immense pool of investment into a limited range of products for which there are large, secured markets within the capital-intensive production systems of the First World. There is very little connection between that kind of research and the lives of the world's hungry. HRCs, for example, are not helpful to poor farmers who rely on manual labor to pull weeds because they couldn't possibly afford herbicides. (The immediate opportunities for biotech in the developing world are not the subsistence farmers, of course, but the larger operations, which are often producing for export rather than for local consumption.)

Just to get a sense of proportion on this subject, consider this comparison. The entire annual budget of the Consultative Group for International Agricultural Research (CGIAR), a consortium of international research centers that form the world's largest public-sector crop breeding effort, amounts to $400 million. The amount that Monsanto spent to develop Roundup Ready soybeans alone is estimated at $500 million. In such numbers, one can see a kind of financial disconnect. Per Pinstrup-Andersen, director of the International Food Policy Research Institute, the CGIAR's policy arm, puts it flatly: "the private sector will not develop crops to solve poor people's problems, because there is not enough money in it." The very nature of their affliction—poverty—makes hungry people poor customers for expensive technologies.

In addition to the financial obstacle, there is a biological obstacle that may limit the role of biotech as agricultural savior. The crop traits that would be most useful to subsistence farmers tend to be very complex. Miguel Altieri, an entomologist at the University of California at Berkeley, identifies the kind of products that would make sense in a subsistence context: "crop varieties responsive to low levels of soil fertility, crops tolerant of saline or drought conditions and other stresses of marginal lands, improved varieties that are not dependent on agro-chemical inputs for increased yields, varieties that are compatible with small, diverse, capital-poor farm settings." In HRCs and Bt crops, the engineering involves the insertion of a single gene. Most of the traits Altieri is talking about are probably governed by many genes, and for the present at least, that kind of complexity is far beyond the technology's reach.

Beyond the Techno-Fix

... On a 300-acre farm in Boone, Iowa—the heart of the U.S. corn belt—Dick Thompson rotates corn, soybeans, oats, wheat interplanted with clover, and a hay combination that includes an assortment of grasses and legumes. The pests that plague neighboring farmers—including the corn borer targeted by Bt corn—are generally a minor part of the picture on Thompson's farm. High crop diversity tends to reduce insect populations because insect pests are usually "specialists" on one particular crop. In a very diverse setting, no single pest is likely to be able to get the upper hand. Diversity also tends to shut out weeds, because complex cropping uses resources more efficiently than monocultures, so there's less left over for the weeds to consume.... Even without herbicides, Thompson's farm has been on conservation tillage for the last three decades.... [C]attle, a hog operation, and the nitrogen-fixing legumes provide the soil nutrients that most U.S. farmers buy in a bag. The soil organic matter content—the sentinel indicator of soil health—registers at 6 percent on Thompson's land, which is more than twice that of his neighbors.... Thompson's soybean and corn yields are well above the county average and even as the U.S. government continues to bail out indebted farmers, Thompson is making money. He profits both from his healthy soil and crops, and from the fact that his "input" costs—for chemical fertilizer, pesticides, and so forth—are almost nil.

In the activities of people like Herren and Thompson it is possible to see a very different kind of agricultural paradigm, which could move farming beyond the techno-fix approach that currently prevails. Known as agroecology, this paradigm recognizes the farm as an ecosystem—an agroecosystem—and employs ecological principles to improve productivity and build stability. The emphasis is on adapting farm design and practice to the ecological processes actually occurring in the fields and in the landscape that surrounds them. Agroecology aims to substitute detailed (and usually local) ecological knowledge for off-the-shelf and off-the-farm "magic bullet" solutions. The point is to treat the disease, rather than just the symptoms. Instead of engineering a corn variety that is toxic to corn rootworm, for example, an agroecologist would ask why there's a rootworm problem in the first place.

Where would biotech fit within such a paradigm? In the industry's current form, at least, it doesn't appear to fit very well at all. Biotech's first agricultural products are "derivative technologies," to use a term favored by Frederick Buttel, a rural sociologist at the University of Wisconsin. Buttel sees those products as "grafted onto an established trajectory, rather than defining or crystallizing a new one."

There is no question that biotech contains some real potential for agriculture, for instance as a supplement to conventional breeding or as a means of studying crop pathogens. But if the industry continues to follow its current trajectory, then biotech's likely contribution will be marginal at best and at worst, given the additional dimensions of ecological and social unpredictability—who knows? In any case, the biggest hope for agriculture is not something biochemists are going to find in a test tube. The biggest opportunities will be found in what farmers already know, or in what they can readily discover on their farms.

POSTSCRIPT

Is Genetic Engineering an Environmentally Sound Way to Increase Food Production?

The full report of the Royal Society of London et al. includes sections that deal with funding issues, capacity building, and intellectual property. The report, complete with references, can be found on the Internet at http://www.royalsoc.ac.uk/policy/index.html.

Kathleen Hart, in *Eating in the Dark* (Pantheon, 2002), expresses horror at the fact that "Frankenfood" is not labeled and that U.S. consumers are not as alarmed by genetically modified foods as European consumers are. The worries—and the scientific evidence to support them—are summarized by Kathryn Brown, in "Seeds of Concern," and Karen Hopkin, in "The Risks on the Table," *Scientific American* (April 2001). In the same issue, Sasha Nemecek poses the question "Does the World Need GM Foods?" to two prominent figures in the debate: Robert B. Horsch, vice president of the Monsanto Corporation and recipient of the 1998 National Medal of Technology for his work on modifying plant genes, says yes; Margaret Mellon, of the Union of Concerned Scientists, says no, adding that much more work needs to be done with regard to safety. The May 2002 U.S. General Accounting Office Report to Congressional Requesters, "Genetically Modified Foods: Experts View Regimen of Safety Tests as Adequate, but FDA's Evaluation Process Could Be Enhanced," urges more attention to verifying safety testing performed by biotechnology companies. Carl F. Jordan, in "Genetic Engineering, the Farm Crisis, and World Hunger," *Bioscience* (June 2002), says that a major problem is already apparent in the way agricultural biotechnology is widening the gap between the rich and the poor.

Charles Mann, in "Biotech Goes Wild," *Technology Review* (July/August 1999), discusses the continuing "lack of a rigorous regulatory framework to sort out the risks inherent in agricultural biotech." Margaret Kriz, in "Global Food Fight," *National Journal* (March 4, 2000), describes the January 2000 Montreal meeting, in which representatives of 130 countries reached "an agreement that requires biotechnology companies to ask permission before importing genetically altered seeds [and] forces food companies to clearly identify all commodity shipments that may contain genetically altered grain." Also see "Environmental Effects of Transgenic Plants: The Scope and Adequacy of Regulation," a report of the National Research Council's Committee on Environmental Impacts Associated With Commercialization of Transgenic Crops (National Academy Press, 2002).

Gregory Conko and C. S. Prakash, in "The Attack on Plant Biotechnology," in Ronald Bailey, ed., *Global Warming and Other Eco-Myths: How the Environmental Movement Uses False Science to Scare Us to Death* (Prima, 2002), say that genetically engineered crops have successfully increased yields and decreased pesticide usage, have not had notable bad environmental side effects, and will be essential for feeding the world's growing population. However, Monsanto has discontinued development of herbicide-tolerant wheat; reasons include marginal benefit and a threat that all U.S. wheat will be banned from European markets if GM wheat is grown in the United States; see Erik Stokstad, "Monsanto Pulls the Plug on Genetically Modified Wheat," *Science* (May 21, 2004). The UN Food and Agriculture Organization's 2004 annual report, *The State of Food and Agriculture 2003-2004*, FAO Agriculture Series No. 35 (Rome, 2004), maintains that the biggest problem with genetically engineered crops is that the technology has focused so far on crops of interest to large commercial firms. GM crops have not spread fast enough to small farmers, although where they have been introduced into developing countries, they have yielded economic gains and reduced the use of toxic chemicals. The report concludes that there have been no adverse health or environmental consequences so far. Continued safety will require more research and governmental regulation and monitoring. Jerry Cayford notes in "Breeding Sanity into the GM Food Debate," *Issues in Science and Technology* (Winter 2004) that the issue is one of social justice as much as it is one of science. Who will control the world's food supply? Which philosophy—democratic competition or technocratic monopoly—will prevail? Meanwhile, researchers are expanding the technology to turn crop plants into "bioreactors," GM crops that produce medically important proteins; see Deborah A. Fitzgerald, "Revving up the Green Express," *The Scientist* (July 14, 2003).

ISSUE 15

Are Marine Reserves Needed to Protect Global Fisheries?

YES: Robert R. Warner, from "Marine Protected Areas," Statement Before the Subcommittee on Fisheries Conservation, Wildlife and Oceans Committee on House Resources, United States House of Representatives (May 23, 2002)

NO: Sean Paige, from "Zoned to Extinction," *Reason* (October 2001)

ISSUE SUMMARY

YES: Professor of marine ecology Robert R. Warner argues that marine reserves, areas of the ocean completely protected from all extractive activities such as fishing, can be a useful tool for preserving ecosystems and restoring productive fisheries.

NO: Sean Paige, a fellow at the market-oriented Competitive Enterprise Institute, argues that marine reserves are based on immature and uncertain science and that they will have a direct and detrimental effect on commercial fishermen.

Carl Safina called attention to the poor state of the world's fisheries in "The World's Imperiled Fish," *Scientific American* (November 1995) and "Where Have All the Fishes Gone?" *Issues in Science and Technology* (Spring 1994). Expanding population, improved fishing technology, and growing demand had combined to drive down fish stocks around the world. Fishers going further from shore and deploying larger nets kept the catch growing, but the UN's Food and Agriculture Organization (FAO) had noted that the fisheries situation was already "globally non-sustainable, and major ecological and economic damage [was] already visible."

The UN declared 1998 the International Year of the Ocean. Kieran Mulvaney, in "A Sea of Troubles," *E Magazine* (January/February 1998), reported, "According to the United Nations Food and Agriculture Organization (FAO), an estimated 70 percent of global fish stocks are 'over-exploited,' 'fully exploited,' 'depleted' or recovering from prior over-exploitation. By 1992, FAO had recorded 16 major fishery species whose global catch had declined by more than 50 percent over the previous three decades—and in half of these, the collapse had begun after 1974." Ocean fishing is not sustainable, concluded Mulvaney.

Daniel Pauly and Reg Watson, in "Counting the Last Fish," *Scientific American* (July 2003), note that desirable fish tend to be top predators, such as tuna and cod. When the numbers of these fish decline due to overfishing, fishers shift their attention to fish lower on the food chain, and consumers see a change in what is available at the market. The cod are smaller, and monkfish and other once less-desirable fish join them on the crushed ice at the market. Such a change is an indicator of trouble in the marine ecosystem.

Responses to the situation have included government buyouts of fishing fleets and closures of fisheries such as the Canadian cod fishery. But the situation has not improved. David Helvarg, in "The Last Fish," *Earth Island Journal* (Spring 2003), concludes that about half of America's commercial seafood species are now overfished. Globally, the figure is still over 70 percent. And the North Atlantic contains only one third as much biomass of commercially valuable fish as it did in the 1950s. Helvarg recommends more buying out of excess fishing capacity; limiting the number of people allowed to enter the fishing industry; creating reserves; and perhaps most importantly, taking fisheries management out of the hands of people with a vested interest in the status quo.

In June 2000, the independent Pew Oceans Commission undertook the first national review of ocean policies in more than 30 years. Its report, *America's Living Oceans: Charting a Course for Sea Change* (Pew Oceans Commission, 2003), available at http://www.pewoceans.org/, notes that many commercially fished species are in decline; North Atlantic cod, haddock, and yellowtail flounder reached historic lows in 1989. The reasons include intense fishing pressure to feed demand for seafood; pollution; coastal development; fishing practices, such as bottom dragging, that destroy habitat; and fragmented ocean policy that makes it difficult to prevent or control the damage. The answers, the commission suggests, must have clear benefits for commercial fishing and include such practices as the no-fishing zones known as marine protected areas, which have been shown to restore habitat and fish populations.

In the following selections, Robert R. Warner argues that marine reserves, areas of the ocean completely protected from all extractive activities such as fishing, lead quickly to increased abundance and size of most species—especially exploited species—even outside the borders of the reserves. He insists that marine reserves can be a useful tool for preserving ecosystems and restoring productive fisheries. Sean Paige argues that marine reserves are based on immature and uncertain science and will have a direct and detrimental effect on commercial fishermen. The traditional species-based approach to conservation is based on the assumption that humans can both conserve and consume.

Robert R. Warner

 **YES**

Marine Protected Areas

Statement of Robert R. Warner, Professor of Marine Ecology, University of California, Santa Barbara

Before the Subcommittee on Fisheries Conservation, Wildlife and Oceans Committee on House Resources

United States House of Representatives

May 23, 2002

...We depend on ocean life in many ways, far beyond the 80 million metric tons of food that we draw from the sea each year. The ecologist Stuart Pimm recently estimated that marine ecosystem services have a value of $20 trillion, with most of that being provided from coastal ecosystems. Yet these ecosystems have been altered dramatically over the past decades—in some places, they have essentially collapsed. Many of the fisheries of the world are depleted, and the species we catch are getting smaller and further down the food chain. The problems of habitat alteration, pollution, aquaculture, exotic species, and climate change all converge on the species that make up ecosystems, and effects on one species can severely affect others. For example, in Hawaii, nutrient pollution fuels algal growth, and fishing removes the fishes that eat the algae, and corals die underneath the encroaching seaweeds. In every marine ecosystem that one of the NCEAS [National Center for Ecological Analysis and Synthesis] working groups investigated, there was clear evidence of fundamental change and loss of resources, and these losses are accelerating. Ecosystem health is often measured in terms of productivity and species diversity, and it is precisely these measures that are declining in many coastal habitats.

Entire marine ecosystems are affected by threats at many levels, and evaluating and responding to these threats in an integrated fashion is the challenge we currently face. Let me make this clear: there is a real need to shift our attention to ecosystems—based management of the marine environment, away from the confusing and often conflicting mass of single-species management plans. On the West coast, there are 88 species that generate more than $1 million a year in fisheries revenue. In New England, there are 41 such species, and in both areas invertebrates like urchins, squid, and lobsters are the most valuable resources. Multiple overlapping

From the United States House of Representatives, May 23, 2002.

single-species management plans can become cumbersome and difficult. A complementary approach to this problem is a scheme of ecosystem-based management.

Marine reserves, areas of the ocean completely protected from all extractive activities, can be a useful tool for ecosystem-based management. They cannot solve all of the problems of the coastal ocean, but they can stop habitat alteration and allow the recovery of depleted populations of several species at a time. Reserves are a place- and habitat-based approach to management, distinctly different from single-species management.

Because much of the sea is hidden from our view, and because the ocean is so vast, we have not been as aware of changes in marine ecosystems as we are of terrestrial changes. On land, many of the larger animals went extinct soon after humans arrived on the scene, and commercial hunting disappeared at the turn of the last century. In the sea, many of the large animals are rare but still present, and harvesting of wild animals continues at high levels. There is hope in this fact—it may be possible to restore marine ecosystems in some places to conditions approaching their former glory, because most of the key players are still present. This is a chance to do more than build a small monument to what existed before. We have a much more rewarding goal: rebuilding coastal ecosystems and recharging coastal fisheries. This is one of the few instances where we can combine benefit to both the extractive users and to the conservation community. It can be done.

The simplest question to ask is what happens when reserves are established. That is, can we document the effect of reserves on coastal ecosystems?

Documented Responses of Animals and Plants to Protection Inside Reserves

The overall coastal area currently under full protection in marine reserves is less than a fraction of one per cent. Although reserves are rare in the US, several have been the subject of careful study. The NCEAS working group summarized these studies and scores of other peer-reviewed reports of the responses of animals and plants to reserve protection around the world. The results were striking. Regardless of whether the reserve was in the tropics or in temperate waters, there was strong evidence that reserves function to increase the abundance and size of many species within their borders. On average, population sizes of animals nearly double, and the animals themselves average about 30% larger. This means that the biomass (or capacity for production) of these species showed a dramatic increase, at least doubling regardless of the location of the reserves.

Not surprisingly, it is exploited species that show the strongest positive response to protection, including species thought to be too mobile to benefit from reserve protection. But I want to stress that the changes seen inside reserves are ecosystem-level changes—not just the recovery of exploited species. For example, when reserves were established in New Zealand, the increase in lobsters resulted in a major decrease in sea urchins, the lobster's prey. This, in turn allowed kelp beds to flourish (because urchins eat kelp), and the overall productivity of the area has increased.

When year-round area closures were instituted on the Georges Bank to aid in the recovery of cod and other finfish, it was scallops that responded the most quickly, becoming unbelievably abundant inside the closed areas. Thus many species can be simultaneously affected by any particular closure.

Responses occurred in reserves of all sizes, and they appear rather quickly—reserves only two to four years old showed increased levels of animal abundance and size equivalent to reserves that had been established for decades.

As I mentioned previously, not all species increase inside reserves, but the great majority show a strong positive response. Neither will all species show a rapid response, especially those that are long-lived and slow-growing. However, the overwhelming result from over 20 years of studies is that species recover within reserve borders, becoming more numerous and larger. Although local conditions may affect the exact result in any particular place, the value of reserves in generating broad changes within their boundaries has been demonstrated in scores of well-documented studies in virtually all settings. This is good news for ecosystem-based management.

While reserves cannot stop pollution, prevent catastrophes, or slow the arrival of exotic invaders into marine ecosystems, they can help to withstand these threats simply because they contain larger populations and more species. Many studies have shown that healthy ecosystems are more resilient to chronic or acute threats, and species-rich ecosystems are more resistant to invasion.

Effects Outside of Reserve Borders

While the major role envisioned for reserves is the protection of habitats and ecosystems, there is added benefit if they export some of their population to surrounding areas. This function is particularly important when reserves are viewed as a fishery management tool, because this export could be used to replenish species subject to harvest in non-reserve areas.

The large variety of life histories, movement patterns, and time spent as a planktonic, drifting larva means that spillover will occur differently for different species. There are so few reserves established, and most of them are so small, that there have been relatively few studies done on spillover. Nevertheless, the evidence is compelling that reserves can recharge nearby areas.

Spillover can take two forms. The first is simple movement of adult animals out of reserves. Several studies have shown that numbers and sizes of species are greater in areas near reserve boundaries, and other studies have shown that the catches of fishers near reserves are higher than in other areas. Fishermen may not have read these studies, but they often know where the fish are, and this has led to concentrations of recreational and commercial fishing activity along reserve borders, an activity known as "fishing the line."

The other major potential contribution of marine reserves to fisheries is through larval export. Most marine species produce tiny young that drift in the water for days or weeks. We know that the rate of production of young by animals inside reserves can be tremendous—at the Edmonds Underwater Park in Washington, for example, it is estimated that the large lingcod there produce 20 times as many young than are produced in equivalent areas outside. But do some of these young make their way into the fishery? There has been little documentation of the effects of larval spillover, mostly because reserves are simply too small to have much effect. The Edmonds reserve, for example, is only 25 acres in extent, a tiny fraction of the area over which the larvae produced there could be expected to drift.

In one US example of a marine reserve large enough to have the potential to recharge fisheries through larval export, this apparently has occurred. On the Georges Bank, several large areas were set aside in 1994 to preserve cod and other groundfish, and as I have mentioned the strongest response so far has been in the fast-growing scallops. By 1998 scallops were 14 times more dense in the protected areas than outside, and dense settlement of young was predicted in downcurrent areas near the reserves. These areas are in fact now yielding higher catches than other areas, and overall revenues have increased from $91 million in 1995 to $123 million in 1999.

Reserve Size and Reserve Networks

A common perception is that conservation and fishery objectives for marine reserves are incompatible, and there will be inevitable conflict between these competing interests. That is certainly what appears to be happening at this point, but models of reserve function suggest that this need not be so. It is true that the larger the reserve, the more species will be able to complete their entire life cycles inside reserves. A reserve too small will not be self-sustaining because most larvae produced in it will be transported elsewhere, and thus a small reserve needs to be seeded from a fished area. Very large reserves, on the other hand, leave little area left to in which to fish.

Most single-species fisheries models of reserves suggest that the most substantial impacts on yields occur when between 20 and 50% of the area is set aside. The amount of area required in reserves varies, but few models show significant benefit at levels below 10%. The more depleted the fishery is on the outside, the more substantial the benefit from reserves.

Where does this leave conservation interests? To what extent can set-asides at this level work to rebuild ecosystems? Fortunately, the most recent scientific findings have suggested a solution: networks of smaller reserves. While these reserves may individually be too small for self-seeding, they are close enough together so that one reserve can seed another. In addition, networks can provide high amounts of spillover into fished areas because they have extensive borders, and networks can boost regional production of young as long as the aggregate area in reserves is sufficiently large.

Studies also suggest networks of reserves can provide additional protection against catastrophic loss (because we're not putting all of our eggs in one basket), and they may make reserve siting easier and more flexible because there are simply more options available.

Where to Put a Marine Reserve?

Recent scientific work on the criteria for siting marine reserves has emphasized that in any management area, there are many different reserve designs that might fit the biological needs of the protected community. That is, science can suggest a range of options that can then be evaluated for other criteria, like their social, economic, or political impact. This flexibility is good news for the process of establishing marine reserves, because it can include input from many different sectors of the community in forming the final decisions.

The most important criterion for designing reserves is to include representation of all habitat types within an area, preferably adjacent to one another, simply because many species use different habitats over the course of their lives. A common misconception is that reserves should be placed in the areas of best fishing. In fact, reserves should show the best response in areas that were formerly productive but are currently overfished—protection can allow these areas of proven potential to recover.

Conclusions

I realize that much of the regulatory process is constrained by mandated consideration of one species at a time. However, the solution to managing multiple threats to the oceans requires an integrated approach that includes the need to preserve intact marine ecosystems on a regional basis. Single species management is not sufficient for the future, especially since many fisheries already affect many different species through by-catch.

Marine reserves are one of the best tools we have to address management of entire marine ecosystems. While they are not the solution to every problem facing the coastal ocean, they can stem habitat destruction, alleviate the effects of local overfishing, simplify the simultaneous management of multiple species, and restore biodiversity within their borders. The healthier ecosystems inside reserves can be more resistant to threats from the outside, and more resilient in their recovery. A regional network of marine reserves may be the best solution for the broad enhancement of coastal ecosystems, with substantial contributions to biodiversity and recruitment of young both inside and outside their borders. While reserves are ideal tools for habitat protection and ecosystem preservation, they are best used as a complement to traditional fisheries management.

Zoned to Extinction

"**W**e've got water coming in!" the captain announced abruptly as he poked a flashlight through a torso-wide hatch to where the Detroit diesels fretfully thrumbled below.

Stunned by the statement, I did a quick assessment of the situation. I was on a small fishing boat a 20-mile swim from the nearest land, Fort Jefferson and the Dry Tortugas. It was the dead of night in seas notorious for their sharks and hull-ripping coral. We were taking on water and, what's more, our depth-finding fathometer was on the fritz, making our proximity to that coral a troubling mystery. Black waves were grabbing hold of the idle craft and pounding on it broadside.

A leaking boat in dark and dangerous seas wasn't my only reason for anxiety, however. The captain was a Key West fisherman—a "conch" in the local vernacular—named Harvey Watkins, whom I'd met the previous day. All I'd really heard about Harvey was that he had once done prison time and might be a little crazy. His mates, Christian and Ramon, were strangers to me as well. And after having watched everyone aboard (including myself) dip into the beer cooler with great frequency all day, my confidence in our collective reflexes and judgment was at ebb tide.

This was a bit more than I'd bargained for when I showed up at Watkins' modest Key West bungalow the day before. I was looking for a story about government regulations that were creating a new endangered species in the southern tip of Florida: commercial fishermen. Watkins, I had heard, was one of the best of a dying breed of "crawfishermen," the area's jargon for lobstermen. He worked the distant and dangerous waters around the Dry Tortugas, which would soon become the site of the nation's largest "no fishing" zone, a disturbing new approach to fisheries management being pushed by federal bureaucrats, environmentalists, and others. This zone represents the boldest step yet toward the creation of a national network of marine wilderness areas, mandated by an executive order, that may eventually blanket large areas of U.S. coastal waters, depriving both sport and commercial fishers of their most fertile fishing grounds.

The waves continued to pound and I desperately tried to assure myself that this couldn't really be happening. I recalled with bitter amusement that this type of participant-observer reporting is sometimes called "immersion journalism." The captain and his crew (who had hurriedly emerged from the berthing space in nothing but briefs and rubber boots) began moving with amazing efficiency and confidence, quickly clearing away gear from the rectangular coffin lid shielding the boat's flooding bowels.

With practiced precision, Watkins and his crew found and fixed a ruptured engine intake hose that was the source of the problem—one of a hundred acts of grace under pressure I saw the crew exhibit during what was, to them, just another routine haul out to sea. Technically, they're fishermen. But life at sea, I quickly began to appreciate, also requires them to be carpenters, diesel mechanics, electricians, butchers, chemists, medical corpsmen, and short-order cooks.

Spending even a couple of days in their world will make you understand that even the priciest lobster dinner is worth every penny. Quarters on Watkins' 48-foot *Fryde Conch* are spare and impossibly cramped. Life aboard is a constant battle against rust, mildew, and rot. The stench of bait clings even in a stiff breeze. The slop-covered deck pitches wildly, and working on it in high seas is a job for the Wallendas. There are a thousand ways to get slashed, crushed, snagged, speared, or dragged overboard, with no medical help—save for a bottle of Captain Morgan's stashed in the wheelhouse—within easy reach. The boat is a playground for tetanus. Scorpions nest in the rope coils. Competence and sound instincts are a must, because even minor mistakes invite major disaster. It's a difficult life, though men like Watkins love it and want no other.

Those are the unavoidable dangers of a fishing life. What really exercises fishermen like Watkins, he explained to me bitterly in our days together, are the threats from dry land: meddlesome government agencies and litigious environmental groups that have descended on this serene southern Florida archipelago with a vengeance, turning the island chain into arguably the most regulated stretch of real estate in the U.S. Few have been hit harder by the new regime than commercial fishermen, whose presence here has shrunken to a shadow of what it was. Tourists now search old Key West's waterfront in vain for its once-mighty commercial fishing fleet, the remnants of which have been banished to Stock Island—a working class, seedy appendage to Key West where few visitors venture. There, folks like Watkins watch the clock run out on their livelihood, done in by often arbitrary and capricious regulations.

A Sinking Feeling

All Harvey Watkins ever wanted to be was a fisherman. He's a wiry, frenetic man of 51, with hair the color and texture of steel wool and a mustache bleached by decades in the Florida sun. He wears aviator glasses with tinted lenses and speaks in staccato sentences, gesturing wildly and flicking cigarette ashes when he gets wound-up. His accent somehow mixes the bayou and the Bronx, but is neither. Once he gets to know you, you quickly become "bubba," "skipper," "bro," or "brother."

As a kid he learned to build wooden boats under the tutelage of a revered neighbor. By 14, he was taking a skiff out at night and on weekends, glorying in the heyday of Key West's commercial fishing boom. Along the way he has been a lobsterman, a stonecrabber, a long-liner, and, for a brief time in the late 1970s, a marijuana smuggler for Colombia's Escobar clan. The "import business" was unbelievably wide open then. The perks were good and the paydays filled garbage bags with cash. But Watkins, like boatloads of fellow islanders, wound up serving time at a federal work farm at Florida's Eglin Air Force Base, winning early release for saving the lives of two drowning men. After prison he returned to fishing, thinking his days of trouble with the government were through. No such luck.

Watkins has a lot in common with another famous Key West fisherman, Harry Morgan, the protagonist of Ernest Hemingway's 1932 novel *To Have and Have Not*. Morgan turns to running booze and guns across the Straits of Florida to Cuba when his failing charter boat business no longer can feed his family. "I don't know who made the laws but I know there ain't no law that you got to go hungry," is how Morgan sums up his philosophy. "I been sore a long time."

Incessant government intervention has made plenty of people in the Florida Keys as sore as Morgan. Certainly that's the case with Watkins and the dwindling commercial fishermen, whose catches and exploits once fattened the local economy and animated its lore. In today's Key West, they often feel like pariahs, shunted by government rules, elitist transplants, busy-body environmentalists, and the tourist economy. A mangrove-like tangle of at least 22 federal, state, and local regulatory agencies has taken root here, each agency trying to carve out a domain for itself, creating a picture-perfect prison and disillusioning iconoclastic locals who migrated here precisely to escape mainland hassles.

Watkins has resisted these changes in sometimes-unconventional ways, earning a reputation as a hell-raiser and gadfly. He's painted his modest bungalow eye-burning shades of blue, green, yellow, and red, as if to give passing, tourist-laden "conch trains" the middle finger. And there's a large painted sign propped against its front porch, in case somebody doesn't get the message. It reads: "The B.O.W.E.L. Movement: Butt Out Worthless Environmental Liars. No Marine Sanctuary!" The sign is a reminder of the decade-long fight over the establishment of the Florida Keys National Marine Sanctuary, which was created in 1990 in the face of overwhelming local resistance and which ushered in a new wave of rules and regulations. Watkins frequently paraded around town with the sign displayed in the bed of his pickup truck.

"My taxes are paying the salaries of the people that are trying their hardest to put me out of work," Watkins told me one night over beers in his back yard. "You shouldn't mess with the working people, bubba, cuz the working people are the backbone of America." Though he tries to maintain a sense of morbid humor about the continuous crisis of fishermen in the Keys, his anger and disillusionment creep in when talk turns to the government. "Stuff is building up out here," he says. "I can envision how somebody snaps."

Like most professional fishers, Watkins cherishes the independence and self-reliance the job affords, and has gone to great lengths to stay far from the madding crowd. When waters closer to shore got too crowded for his tastes, Watkins fished farther and farther out, eventually ending up in the Dry Tortugas, a

120-mile round-trip from Key West, where conflicts over territory and government hassles were fewer—at least until recently.

Working the Tortugas sets Watkins and his mates apart from most crawfishermen. The typical lobsterman generally traps in shallower waters close to shore, so they can be out and back, baiting and servicing their traps, in a single day. The risks are greater working in the Tortugas, but so are the potential rewards. Seas and shoals can be tricky, as hundreds of shattered galleons attest. Traps, floats, and gear get beaten to hell, as do captain and crew, who often work for 10 or 12 days at a time while being battered by 5- to 10-foot waves. But the crawfish are generally larger than those caught near shore, and in a good season, the Tortugas traps produce consistently while shore-huggers are pulling up empties.

Scales of Justice

Fishing the Tortugas has another disadvantage that couldn't possibly have occurred to Watkins when he gravitated there years ago: Because relatively few fishermen have the skills, motivation, or inclination to work this far out, it became a tempting target for government regulators looking for somewhere to place a 150 square mile no-fishing zone called the Tortugas Ecological Reserve. Environmentalists and federal regulators hope the reserve will serve as a model for similar no-fishing zones they want to establish along 20 percent of the U.S. coastline.

To proponents in government agencies, environmental groups, and academia, such "marine protected areas" (or MPAs) are the wave of the future for fisheries management—a maritime equivalent of federal wilderness areas that protect not just species but habitat. Greens have fixed like a barnacle on the idea of a "place-based" method for protecting marine life, much as they've pushed hard for the establishment of "critical habitats" for endangered species. Traditional, species-based fisheries management strategies are aimed at establishing a sustainable balance between the needs of fish and fishermen. The "place based" approach instead places marine wilderness areas off limits to all human meddling, which they believe will in time replenish the rest of the seas. For most MPA supporters, how fishermen fare as a result seems a matter of indifference.

For the fishermen being displaced by them, MPAs seem like a frivolous federal science fair project with potentially serious economic consequences—and yet another blow to the traditional Key West fishing life.

The no-fishing zone being created in the Tortugas this summer "is just the tip of the iceberg," according to one government fisheries official who asks not to be identified. "This is an idea that's not only being embraced at the federal level, but has taken root at the local level, too. A number of the people in the research community think as much as 20 percent of the U.S. coastline should be off limits to fishing. Before long, you may see another marine reserve sitting at the mouth of the Chesapeake Bay."

In addition to the 150 square mile Tortugas Reserve, which is administered by the National Oceanic and Atmospheric Administration's Florida Keys

National Marine Sanctuary, the Department of Interior plans to close nearly half the adjoining 100 square mile Dry Tortugas National Park to all "consumptive" activities.

This summer, the South Atlantic Fisheries Management Council, one of two federal panels that regulate fisheries in the Keys, began compiling a "shopping list" of potential new no-fishing zones from the Carolinas to Key West. The council is under mounting pressure from environmental organizations such as the Center for Marine Conservation and Earthjustice Legal Defense Fund. Already included on the list of proposed areas is "the Hump," a popular fishing spot off Islamorada in the upper Keys, and the Carysford Reef, where an existing no-take zone could be enlarged.

Even the city of Key Wes–in response to lobbying by the Center for Marine Conservation–recently considered a no-fishing zone extending 600 feet from the island's shoreline. In the end, the city shelved the notion, which would have effectively banned angling from the beach, from docks, and on flats close to shore.

In August, the South Atlantic Fisheries Management Council will be holding public hearings on the proposals for more fishing bans. The proposals may end up uniting commercial fishermen and sport fishing enthusiasts, who have often been at odds in the past. A number of media reports have suggested that there is support among commercial fishermen for the no-take zones, yet during a visit earlier this year, I failed to meet a single commercial fisherman in the Keys who supported the reserve concept or put much stock in the science behind it. According to most commercial fishermen, many of whom have completely lost faith in a public hearings process they see as a charade, the few fishermen who chose to participate in the government's public consultation process did so either in a vain effort to mitigate regulatory excesses or in the hope of avoiding being the next group targeted for cutbacks and closures.

Fishy Science

"It's clear that some environmental groups see MPAs as a panacea, but we think the science is still equivocal on the question of how these areas will affect overall health of the fishery," says Justin LeBlanc of the National Fisheries Institute, a trade association representing commercial fishing interests. "It doesn't take a scientist to understand that if you stop fishing in an area you're going to find more fish and larger fish in that area. But if you're making the argument that it's going to improve the fishery overall then I think the jury is still out and the science is more equivocal." LeBlanc calls the proposals to designate at least 20 percent of U.S. coastal areas as no-fishing zones "arbitrary" and "completely irrational."

Though temporary closures have been used as a fisheries management tool for decades–during spawning seasons, for instance–permanent closures over large areas are a relatively new idea on which hard scientific data either hasn't been collected or isn't conclusive, according to LeBlanc.

While there's little doubt that an unfished area will have more fish, the real issue is whether abundance of marine life in one closed-off area will translate into abundance elsewhere, improving the fishery overall. That question depends on a complicated and only partially understood set of variables, including the mating, spawning, and migratory patterns of myriad species; currents, weather, and tidal conditions; and the size, depth, and bottom composition of the zones.

Fishermen make the following, experience-based observations to challenge what they see as a windy academic theory. First, a lot hangs on the unique characteristics and habits of the fish you want to catch. For instance, territorial species that spend their lives on a reef aren't going to suddenly leave en masse for open water, where they can fall into the waiting nets of fishermen. And highly migratory species passing through seasonally aren't likely to settle down and retire to a fishing-free reef, escaping capture. As anyone who's ever put a hook in water knows, fish (like fishermen) tend to congregate in certain areas (reefs, wrecks, humps, walls, etc.) and are unevenly distributed throughout the seas. Making these crowded areas off limits to fishermen may affect their numbers, but likely will not alter the overall distribution of fish.

So one of the basic ideas behind MPAs is mistaken. When the Tortugas zone was first proposed as part of a Florida Keys National Marine Sanctuary management plan, it was billed as a "replenishment reserve," a name suggesting that it would somehow repopulate the seas. But that name was dropped after sanctuary officials were pressed on the specifics of how this "replenishment" would occur, an acknowledgment that the term and, by extension, the concept, was spurious.

Perhaps more important, the whole exercise skirts the most fundamental issue, according to many fishermen I spoke with. The question isn't whether you will have more fish if you stop fishing (common sense tells us that's so), but how you can have sustainable fish stocks and sustainable fishermen, too. The traditional, species-based approach to fisheries management, though it functions imperfectly, is at least based upon established procedures, benchmarks, methods, and decades of trial and error. It is also based on the assumption that we can adequately negotiate between our desire to eat fish and our desire to preserve species. No-take zones, by contrast, represent a radical departure, based on immature and uncertain science, which will have a direct and detrimental effect on the fishermen being displaced.

"It is a fad, but it is being effectively campaigned by national environmental groups that have gotten a lot of foundation money to promote this," says Ted Forsgren, spokesman for the Florida chapter of the Coastal Conservation Association, a group representing saltwater sport anglers. "The label 'marine protected areas' is false advertising, camouflaging what's really going on, which is the creation of no-fishing areas. And when they talk about prohibiting fishing in *just* 20 to 30 percent of the ocean, what they're not saying is that these are the areas that have all the fish."

Though President [George W.] Bush vowed that he would not be formulating policy based on environmentalist "fads," he seems to be giving in to this particular craze. He recently announced that he would not be reversing the last-minute Clinton administration executive order that called on federal agencies to develop a national system of MPAs. In June, Commerce Secretary Donald L. Evans took a trip to California, where a no-fishing zone is being proposed near the Channel Islands, over the objection of commercial fishermen. Evans not only signaled that the Bush administration would continue to implement Clinton's executive order, but announced $3 million for scientific research in support of the concept. Evans specifically lauded the Tortugas Reserve as a model that should be followed elsewhere, as has the president's brother, Florida Gov. Jeb Bush. Neither paid any mind to the concept's potentially devastating impact on Watkins and fishermen like him.

When the Tortugas no-take zone became law July 1, Harvey Watkins and at least two dozen Tortugas fishermen lost large parts of the fishing grounds it took them decades of hard experience to master. "They just took half of my livelihood," Watkins says. "I've got a lot of people here who count on me for support." Now Watkins will have to place more of his traps near shore ("Where I first started fishing 27 years ago," he says), inviting gear conflicts with boats already established in those areas, or even further out at sea, exposing himself and his crew to even greater hardships and dangers.

At night on the water, with the mates asleep, not a light on the horizon, and an anchorage at Fort Jefferson still a long hour away, Harvey Watkins' usual mood of manic defiance gradually yields to sadness and resignation as he ponders the government's apparent antipathy toward his chosen way of life.

"When I was a kid growing up in the '50s, I used to watch Westerns, cowboys and Indians, and I sided with the cowboys, like all the rest of the kids," he says, peering through the darkness for some sign of safe anchorage. "I've been reading lately about the Indians and how they were treated, and today I feel more like an Indian. Today, as an Indian, I say, 'To hell with the cowboys.'"

POSTSCRIPT

Are Marine Reserves Needed to Protect Global Fisheries?

Oran R. Young, in "Taking Stock: Management Pitfalls in Fisheries Science,"*Environment* (April 2003), notes that despite putting great effort into assessing fish stocks and managing fisheries for sustainable yield, marine fish stocks have continued to decline. This is partly the "result of the inability of managers to resist pressures from interest groups to set total allowable catches too high, even in the face of warnings from scientists about the dangers of triggering stock depletions. The problem also arises, however, from repeated failures on the part of analysts and policy makers to anticipate the collapse of major stocks or to grasp either the current condition or the reproductive dynamics of important stocks." He cautions against putting "blind faith in the validity of scientific assessments" and suggests more use of the *precautionary principle* (see Issue 1) despite the risk that this would set allowed catch levels lower than many would like. Lydia K. Bergen and Mark H. Carr, in "Establishing Marine Reserves," *Environment* (March 2003), favor the development of marine reserves. They review the Channel Islands reserve discussed by Warner, praising its incorporation of scientific data. Carl Safina, in "The Continued Danger of Overfishing," *Issues in Science and Technology* (Summer 2003), sees some grounds for optimism, for recovery plans have helped a number of fish species, but he remains concerned. He says, "I predict that over the next few years, consumer education will become the largest area of growth and change in the toolbox of ocean conservation strategy."

In April 2004, the U.S. Commission on Ocean Policy (http://ocean-commission.gov/) released a preliminary report calling for improved management systems "to handle mounting pollution, declining fish populations and coral reefs, and promising new industries such as aquaculture." In many ways it agreed with the Pew report, as mentioned in the introduction to this issue. See B. Harder, "Sea Change," *Science News* (April 24, 2004). Marine reserves are definitely gaining favor as part of the solution. John Temple Swing, in "What Future for the Oceans?" *Foreign Affairs* (September/October 2003), calls marine protected areas "one of the promising trends of the past two decades." See also "Marine Protected Areas," *Congressional Digest* (September 2003) and Sascha K. Hooker and Leah R. Gerber, "Marine Reserves as a Tool for Ecosystem-Based Management: The Potential Importance of Megafauna," *Bioscience* (January 2004). But Garry R. Russ and Angela C. Alcala, in "Marine Reserves: Long-Term Protection Is Required for Full Recovery of Predatory Fish Populations," *Oecologia* (March 2004), conclude that reserves cannot be viewed as temporary creations; reserve management systems must be designed to last generations in the face of increasing pressures for food and other resources. However, it is essential to engage the fishing community in the process of designing and establishing reserves. See Mark Helvey, "Seeking Consensus on Designing Marine Protected Areas: Keeping the Fishing Community Engaged," *Coastal Management* (April/June 2004).

While commercial fishers object to conservation measures such as marine reserves because they see them as affecting their livelihood, sports fishers also object to marine reserves. See Jerry Gibbs, "Freedom to Fish?" *Outdoor Life* (May 2004). Sports fishers, too, may need to be engaged in the process of designing and establishing reserves.

On the Internet . . .

The Pesticide Action Network North America

The Pesticide Action Network North America (PANNA) challenges the global proliferation of pesticides, defends basic rights to health and environmental quality, and works to insure the transition to a just and viable society.

http://www.panna.org/

e.hormone

e.hormone is hosted by the Center for Bioenvironmental Research at Tulane and Xavier Universities in New Orleans. The site provides accurate, timely information about environmental hormones and their impacts.

http://e.hormone.tulane.edu/

The Silicon Valley Toxics Coalition

The Silicon Valley Toxics Coalition (SVTC) was formed in 1982 to engage in research, advocacy, and organizing associated with environmental and human problems caused by the rapid growth of the high-tech electronics industry; to advance environmental sustainability and clean production in the industry; and to improve health, promote justice, and ensure democratic decision making for affected communities and workers in the United States and the world.

http://www.svtc.org/

The Nevada Nuclear Waste Task Force

The Nevada Nuclear Waste Task Force is a nonprofit organization fighting nuclear waste in Nevada and providing public information about Yucca Mountain.

http://www.nvantinuclear.org/

Superfund

The U.S. Environmental Protection Agency (EPA) provides a great deal of information on the Superfund program, including material on environmental justice.

http://www.epa.gov/superfund/

The Military Toxics Project

The mission of the Military Toxics Project (MTP) is to unite activists, organizations, and communities in the struggle to clean up military pollution, safeguard the transportation of hazardous materials, and develop and implement preventative solutions to the toxic and radioactive pollution caused by military activities. MTP's mission is based on mutual respect and justice for all peoples, free from any form of discrimination or bias.

http://www.miltoxproj.org/

The Center for Liquefied Natural Gas

The Center for Liquefied Natural Gas's (LNG) overall goals are to enhance public education and LNG acceptance by serving as a clearinghouse for LNG information; to promote efficient regulation for permitting, siting, as well as the building and operation of LNG facilities and infrastructure; and to work toward continued safe and secure operations, stressing industry guidelines and best practices.

http://www.lngfacts.org/

Federal Energy Regulatory Commission

The Federal Energy Regulatory Commission regulates and oversees energy industries in the economic and environmental interest of the American public. This page also provides information about LNG.

http://www.ferc.gov/industries/gas/indus-act/lng-what.asp

PART 5

Toxic Chemicals

A *great many of today's environmental issues have to do with industrial development, which expanded greatly during the twentieth century. Just since World War II, many thousands of synthetic chemicals—pesticides, plastics, and antibiotics—have flooded the environment. People have become dependent on the production and use of energy, particularly in the form of fossil fuels. Air and water pollution have become global problems. We have discovered that industrial processes generate huge amounts of waste, much of it toxic. And we have discovered that our actions may change the world for generations to come.*

Among other controversies, this section deals with two prominent ones concerning toxic chemicals and two concerning toxic wastes.

- Should DDT Be Banned Worldwide?

- Do Environmental Hormone Mimics Pose a Potentially Serious Health Threat?

- Is the Superfund Program Successfully Protecting the Environment from Hazardous Wastes?

- Should the United States Continue to Focus Plans for Permanent Nuclear Waste Disposal Exclusively at Yucca Mountain?

- Should the Military Be Exempt from Environmental Regulations?

- Is Addition Federal Oversight Needed for the Construction of LNG Import Facilities?

ISSUE 16

Should DDT Be Banned Worldwide?

YES: Anne Platt McGinn, from "Malaria, Mosquitoes, and DDT," *World Watch* (May/June 2002)

NO: Alexander Gourevitch, "Better Living Through Chemistry," *Washington Monthly* (March 2003)

ISSUE SUMMARY

YES: Anne Platt McGinn, a senior researcher at the Worldwatch Institute, argues that although DDT is still used to fight malaria, there are other, more effective and less environmentally harmful methods. She maintains that DDT should be banned or reserved for emergency use.

NO: Alexander Gourevitch, an *American Prospect* writing fellow, argues that, properly used, DDT is not as dangerous as its reputation insists and that it remains the cheapest and most effective way to combat malaria.

DDT is a crucial element in the story of environmentalism. The chemical was first synthesized in 1874. Swiss entomologist Paul Mueller was the first to notice that DDT has insecticidal properties, which, it was quickly realized, implied that the chemical could save human lives. It had long been known that more soldiers died during wars because of disease than because of enemy fire. During World War I, for example, some 5 million lives were lost to typhus, a disease carried by body lice. DDT was first deployed during World War II to halt a typhus epidemic in Naples, Italy. It was a dramatic success, and DDT was soon used routinely as a dust for soldiers and civilians. During and after the war, DDT was also deployed successfully against the mosquitoes that carry malaria and other diseases. In the United States cases of malaria fell from 120,000 in 1934 to 72 in 1960, and cases of yellow fever dropped from 100,000 in 1878 to none. In 1948 Mueller received the Nobel Prize for medicine and physiology because DDT had saved so many civilian lives.

DDT was by no means the first pesticide. But its predecessors—arsenic, strychnine, cyanide, copper sulfate, and nicotine—were all markedly toxic to humans. DDT was not only more effective as an insecticide, it was also less hazardous to users. It is therefore not surprising that DDT was seen as a beneficial substance. It was soon applied routinely to agricultural crops and used to control mosquito populations in American suburbs. However, insects quickly became resistant to the insecticide. (In any population of insects, some will be more resistant than others; when the insecticide kills the more vulnerable members of the population, the resistant ones are left to breed and multiply. This is an example of natural selection.) In *Silent Spring* (Houghton Mifflin, 1962), marine scientist Rachel Carson demonstrated that DDT was concentrated in the food chain and affected the reproduction of predators such as hawks and eagles. In 1972 the U.S. Environmental Protection Agency banned almost all uses of DDT (it could still be used to protect public health). Other developed countries soon banned it as well, but developing nations, especially those in the tropics, saw it as an essential tool for fighting diseases such as malaria. Roger Bate, director of Africa Fighting Malaria, argues in "A Case of the DDTs," *National Review* (May 14, 2001) that DDT remains the cheapest and most effective way to combat malaria and that it should remain available for use.

It soon became apparent that DDT is by no means the only pesticide or organic toxin with environmental effects. As a result, on May 24, 2001, the United States joined 90 other nations in signing the Stockholm Convention on Persistent Organic Pollutants (POPs). This treaty aims to eliminate from use the entire class of chemicals to which DDT belongs, beginning with the "dirty dozen," pesticides DDT, aldrin, dieldrin, endrin, chlordane, heptachlor, mirex, and toxaphene, and the industrial chemicals polychlorinated biphenyls (PCBs), hexachlorobenzene (HCB), dioxins, and furans. Since then, 59 countries, not including the United States and the European Union (EU), have formally ratified the treaty. It took effect in May 2004. Fiona Proffitt, in "U.N. Convention Targets Dirty Dozen Chemicals," *Science* (May 21, 2004), notes, "About 25 countries will be allowed to continue using DDT against malaria-spreading mosquitoes until a viable alternative is found."

In the following selection, Anne Platt McGinn, granting that malaria remains a serious problem in the developing nations of the tropics, especially Africa, contends that although DDT is still used to fight malaria in these nations, it is far less effective than it used to be. She argues that the environmental effects are also serious concerns and that DDT should be banned or reserved for emergency use. In the second selection, Alexander Gourevitch argues that, properly used, DDT is not as dangerous as its reputation insists. He contends that it remains the cheapest and most effective way to combat malaria and that environmentalist opposition to its use in Africa threatens the lives of millions.

Anne Platt McGinn
 **YES**

Malaria, Mosquitoes, and DDT

This year, like every other year within the past couple of decades, uncountable trillions of mosquitoes will inject malaria parasites into human blood streams billions of times. Some 300 to 500 million full-blown cases of malaria will result, and between 1 and 3 million people will die, most of them pregnant women and children. That's the official figure, anyway, but it's likely to be a substantial underestimate, since most malaria deaths are not formally registered, and many are likely to have escaped the estimators. Very roughly, the malaria death toll rivals that of AIDS, which now kills about 3 million people annually.

But unlike AIDS, malaria is a low-priority killer. Despite the deaths, and the fact that roughly 2.5 billion people (40 percent of the world's population) are at risk of contracting the disease, malaria is a relatively low public health priority on the international scene. Malaria rarely makes the news. And international funding for malaria research currently comes to a mere $150 million annually. Just by way of comparison, that's only about 5 percent of the $2.8 billion that the U.S. government alone is considering for AIDS research in fiscal year 2003.

The low priority assigned to malaria would be at least easier to understand, though no less mistaken, if the threat were static. Unfortunately it is not. It is true that the geographic range of the disease has contracted substantially since the mid-20th century, but over the past couple of decades, malaria has been gathering strength. Virtually all areas where the disease is endemic have seen drug-resistant strains of the parasites emerge—a development that is almost certainly boosting death rates. In countries as various as Armenia, Afghanistan, and Sierra Leone, the lack or deterioration of basic infrastructure has created a wealth of new breeding sites for the mosquitoes that spread the disease. The rapidly expanding slums of many tropical cities also lack such infrastructure; poor sanitation and crowding have primed these places as well for outbreaks—even though malaria has up to now been regarded as predominantly a rural disease.

What has current policy to offer in the face of these threats? The medical arsenal is limited; there are only about a dozen antimalarial drugs commonly in use, and there is significant malaria resistance to most of them. In the absence of a reliable way to kill the parasites, policy has tended to focus on killing the mosquitoes that bear them. And that has led to an abundant use of synthetic pesticides, including one of the oldest and most dangerous: dichlorodiphenyl trichloroethane, or DDT.

DDT is no longer used or manufactured in most of the world, but because it does not break down readily, it is still one of the most commonly detected pesticides in the milk of nursing mothers. DDT is also one of the "dirty dozen" chemicals included in the 2001 Stockholm Convention on Persistent Organic Pollutants [POPs]. The signatories to the "POPs Treaty" essentially agreed to ban all uses of DDT except as a last resort against disease-bearing mosquitoes. Unfortunately, however, DDT is still a routine option in 19 countries, most of them in Africa. (Only 11 of these countries have thus far signed the treaty.) Among the signatory countries, 31—slightly fewer than one-third—have given notice that they are reserving the right to use DDT against malaria. On the face of it, such use may seem unavoidable, but there are good reasons for thinking that progress against the disease is compatible with *reductions* in DDT use.

❧❀❧

Malaria is caused by four protozoan parasite species in the genus *Plasmodium*. These parasites are spread exclusively by certain mosquitoes in the genus *Anopheles*. An infection begins when a parasite-laden female mosquito settles onto someone's skin and pierces a capillary to take her blood meal. The parasite, in a form called the *sporozoite*, moves with the mosquito's saliva into the human bloodstream. About 10 percent of the mosquito's lode of sporozoites is likely to be injected during a meal, leaving plenty for the next bite. Unless the victim has some immunity to malaria—normally as a result of previous exposure—most sporozoites are likely to evade the body's immune system and make their way to the liver, a process that takes less than an hour. There they invade the liver cells and multiply asexually for about two weeks. By this time, the original several dozen sporozoites have become millions of *merozoites*—the form the parasite takes when it emerges from the liver and moves back into the blood to invade the body's red blood cells. Within the red blood cells, the merozoites go through another cycle of asexual reproduction, after which the cells burst and release millions of additional merozoites, which invade yet more red blood cells. The high fever and chills associated with malaria are the result of this stage, which tends to occur in pulses. If enough red blood cells are destroyed in one of these pulses, the result is convulsions, difficulty in breathing, coma, and death.

As the parasite multiplies inside the red blood cells, it produces not just more merozoites, but also *gametocytes*, which are capable of sexual reproduction. This occurs when the parasite moves back into the mosquitoes; even as they inject sporozoites, biting mosquitoes may ingest gametocytes if they are feeding on a person who is already infected. The gametocytes reproduce in the insect's gut and the resulting eggs move into the gut cells. Eventually, more sporozoites emerge from the gut and penetrate the mosquito's salivary glands, where they await a chance to enter another human bloodstream, to begin the cycle again.

Of the roughly 380 mosquito species in the genus *Anopheles*, about 60 are able to transmit malaria to people. These malaria vectors are widespread throughout the tropics and warm temperate zones, and they are very efficient at spreading the disease. Malaria is highly contagious, as is apparent from a measurement that epidemiologists call the "basic reproduction number," or BRN. The BRN indicates, on average, how many new cases a single infected person is likely to cause. For example, among the nonvectored diseases (those in which the pathogen travels directly from person to person without an intermediary like a mosquito), measles is one of the most contagious. The BRN for measles is 12 to 14, meaning that someone with measles is likely to infect 12 to 14 other people. (Luckily, there's an inherent limit in this process: as a pathogen spreads through any particular area, it will encounter fewer and fewer susceptible people who aren't already sick, and the outbreak will eventually subside.) HIV/ AIDS is on the other end of the scale: it's deadly, but it burns through a population slowly. Its BRN is just above 1, the minimum necessary for the pathogen's survival. With malaria, the BRN varies considerably, depending on such factors as which mosquito species are present in an area and what the temperatures are. (Warmer is worse, since the parasites mature more quickly.) But malaria can have a BRN in excess of 100: over an adult life that may last about a week, a single, malaria-laden mosquito could conceivably infect more than 100 people.

Seven Years, Seven Months

"Malaria" comes from the Italian "mal'aria." For centuries, European physicians had attributed the disease to "bad air." Apart from a tradition of associating bad air with swamps—a useful prejudice, given the amount of mosquito habitat in swamps—early medicine was largely ineffective against the disease. It wasn't until 1897 that the British physician Ronald Ross proved that mosquitoes carry malaria.

The practical implications of Ross's discovery did not go unnoticed. For example, the U.S. administration of Theodore Roosevelt recognized malaria and yellow fever (another mosquito-vectored disease) as perhaps the most serious obstacles to the construction of the Panama Canal. This was hardly a surprising conclusion, since the earlier and unsuccessful French attempt to build the canal—an effort that predated Ross's discovery—is thought to have lost between 10,000 and 20,000 workers to disease. So the American workers draped their water supplies and living quarters with mosquito netting, attempted to fill in or drain swamps, installed sewers, poured oil into standing water, and conducted mosquito-swatting campaigns. And it worked: the incidence of malaria declined. In 1906, 80 percent of the workers had the disease; by 1913, a year before the Canal was completed, only 7 percent did. Malaria could be suppressed, it seemed, with a great deal of mosquito netting, and by eliminating as much mosquito habitat as possible. But the labor involved in that effort could be enormous.

That is why DDT proved so appealing. In 1939, the Swiss chemist Paul Müller discovered that this chemical was a potent pesticide. DDT was first used during World War II, as a delousing agent. Later on, areas in southern Europe, North Africa, and Asia were fogged with DDT, to clear malaria-laden mosquitoes from the paths of invading Allied troops. DDT was cheap and it seemed to be harmless to anything other than insects. It was also long-lasting: most other insecticides lost their potency in a few days, but in the early years of its use, the effects of a single dose of DDT could last for up to six months. In 1948, Müller won a Nobel Prize for his work and DDT was hailed as a chemical miracle.

A decade later, DDT had inspired another kind of war—a general assault on malaria. The "Global Malaria Eradication Program," launched in 1955, became one of the first major undertakings of the newly created World Health Organization [WHO]. Some 65 nations enlisted in the cause. Funding for DDT factories was donated to poor countries and production of the insecticide climbed.

The malaria eradication strategy was not to kill every single mosquito, but to suppress their populations and shorten the lifespans of any survivors, so that the parasite would not have time to develop within them. If the mosquitoes could be kept down long enough, the parasites would eventually disappear from the human population. In any particular area, the process was expected to take three years—time enough for all infected people either to recover or die. After that, a resurgence of mosquitoes would be merely an annoyance, rather than a threat. And initially, the strategy seemed to be working. It proved especially effective on islands—relatively small areas insulated from reinfestation. Taiwan, Jamaica, and Sardinia were soon declared malaria-free and have remained so to this day. By 1961, arguably the year at which the program had peak momentum, malaria had been eliminated or dramatically reduced in 37 countries.

One year later, Rachel Carson published *Silent Spring*, her landmark study of the ecological damage caused by the widespread use of DDT and other pesticides. Like other organochlorine pesticides, DDT bioaccumulates. It's fat soluble, so when an animal ingests it—by browsing contaminated vegetation, for example—the chemical tends to concentrate in its fat, instead of being excreted. When another animal eats that animal, it is likely to absorb the prey's burden of DDT. This process leads to an increasing concentration of DDT in the higher links of the food chain. And since DDT has a high chronic toxicity—that is, long-term exposure is likely to cause various physiological abnormalities—this bioaccumulation has profound implications for both ecological and human health.

With the miseries of malaria in full view, the managers of the eradication campaign didn't worry much about the toxicity of DDT, but they were greatly concerned about another aspect of the pesticide's effects: resistance. Continual exposure to an insecticide tends to "breed" insect populations that are at least partially immune to the poison. Resistance to DDT had been reported as early as 1946. The campaign managers knew that in mosquitoes, regular exposure to DDT tended to produce widespread resistance in four to seven years. Since it took three years to clear malaria from a human population, that didn't leave a lot of leeway for the eradication effort. As it turned out, the logistics simply couldn't be made to work in large, heavily infested areas with high human populations, poor housing and roads, and generally minimal infrastructure. In 1969, the campaign was abandoned. Today, DDT resistance is widespread in *Anopheles*, as is resistance to many more recent pesticides.

Undoubtedly, the campaign saved millions of lives, and it did clear malaria from some areas. But its broadest legacy has been of much more dubious value. It engendered the idea of DDT as a first resort against mosquitoes and it established the unstable dynamic of DDT resistance in *Anopheles* populations. In mosquitoes, the genetic mechanism that confers resistance to DDT does not usually come at any great competitive "cost"—that is, when no DDT is being sprayed, the resistant mosquitoes may do just about as well as nonresistant mosquitoes. So once a population acquires resistance, the trait is not likely to disappear even if DDT isn't used for years. If DDT is reapplied to such a population, widespread resistance will reappear very rapidly. The rule of thumb among entomologists is that you may get seven years of resistance-free use the first time around, but you only get about seven months the second time. Even that limited respite, however, is enough to make the chemical an attractive option as an emergency measure—or to keep it in the arsenals of bureaucracies committed to its use.

Malaria Taxes

In December 2000, the POPs Treaty negotiators convened in Johannesburg, South Africa, even though, by an unfortunate coincidence, South Africa had suffered a potentially embarrassing setback earlier that year in its own POPs policies. In 1996, South Africa had switched its mosquito control programs from DDT to a less persistent group of pesticides known as pyrethroids. The move seemed solid and supportable at the time, since years of DDT use had greatly reduced *Anopheles* populations and largely eliminated one of the most troublesome local vectors, the appropriately named *A. funestus* ("funestus" means deadly). South Africa seemed to have beaten the DDT habit: the chemical had been used to achieve a worthwhile objective; it had then been discarded. And the plan worked—until a year before the POPs summit, when malaria infections rose to 61,000 cases, a level not seen in decades. *A. funestus* reappeared as well, in KwaZulu-Natal, and in a form resistant to pyrethroids. In early 2000, DDT was reintroduced, in an indoor spraying program. (This is now a standard way of using DDT for mosquito control; the pesticide is usually applied only to walls, where mosquitoes alight to rest.) By the middle of the year, the number of infections had dropped by half.

Initially, the spraying program was criticized, but what reasonable alternative was there? This is said to be the African predicament, and yet the South African situation is hardly representative of sub-Saharan Africa as a whole.

Malaria is considered endemic in 105 countries throughout the tropics and warm temperate zones, but by far the worst region for the disease is sub-Saharan Africa. The deadliest of the four parasite species, *Plasmodium falciparum*, is widespread throughout this region, as is one of the world's most effective malaria vectors, *Anopheles gambiae*. Nearly half the population of sub-Saharan Africa is at risk of infection, and in much of eastern and central Africa, and pockets of west Africa, it would be difficult to find anyone who has not been exposed to the parasites. Some 90 percent of the world's malaria infections and deaths occur in sub-Saharan Africa, and the disease now accounts for 30 percent of African childhood mortality. It is true that malaria is a grave problem in many parts of the world, but the African experience is misery on a very different order of magnitude. The average Tanzanian suffers more infective bites each *night* than the average Thai or Vietnamese does in a year.

As a broad social burden, malaria is thought to cost Africa between $3 billion and $12 billion annually. According to one economic analysis, if the disease had been eradicated in 1965, Africa's GDP would now be 35 percent higher than it currently is. Africa was also the gaping hole in the global eradication program: the WHO planners thought there was little they could do on the continent and limited efforts to Ethiopia, Zimbabwe, and South Africa, where eradication was thought to be feasible.

But even though the campaign largely passed Africa by, DDT has not. Many African countries have used DDT for mosquito control in indoor spraying programs, but the primary use of DDT on the continent has been as an agricultural insecticide. Consequently, in parts of west Africa especially, DDT resistance is now widespread in *A. gambiae*. But even if *A. gambiae* were not resistant, a full-bore campaign to suppress it would probably accomplish little, because this mosquito is so efficient at transmitting malaria. Unlike most *Anopheles* species, *A. gambiae* specializes in human blood, so even a small population would keep the disease in circulation. One way to get a sense for this problem is to consider the "transmission index"—the threshold number of mosquito bites necessary to perpetuate the disease. In Africa, the index overall is 1 bite per person per month. That's all that's necessary to keep malaria in circulation. In India, by comparison, the TI is 10 bites per person per month.

And yet Africa is not a lost cause—it's simply that the key to progress does not lie in the general suppression of mosquito populations. Instead of spraying, the most promising African programs rely primarily on "bednets"— mosquito netting that is treated with an insecticide, usually a pyrethroid, and that is suspended over a person's bed. Bednets can't eliminate malaria, but they can "deflect" much of the burden. Because *Anopheles* species generally feed in the evening and at night, a bednet can radically reduce the number of infective bites a person receives. Such a person would probably still be infected from time to time, but would usually be able to lead a normal life.

In effect, therefore, bednets can substantially reduce the disease. Trials in the use of bednets for children have shown a decline in malaria-induced mortality by 25 to 40 percent. Infection levels and the incidence of severe anemia also declined. In Kenya, a recent study has shown that pregnant women who use bednets tend to give birth to healthier babies. In parts of Chad, Mali, Burkina Faso, and Senegal, bednets are becoming standard household items. In the tiny west African nation of The Gambia, somewhere between 50 and 80 percent of the population has bednets.

Bednets are hardly a panacea. They have to be used properly and retreated with insecticide occasionally. And there is still the problem of insecticide resistance, although the nets themselves are hardly likely to be the main cause of it. (Pyrethroids are used extensively in agriculture as well.) Nevertheless, bednets can help transform malaria from a chronic disaster to a manageable public health problem—something a healthcare system can cope with.

So it's unfortunate that in much of central and southern Africa, the nets are a rarity. It's even more unfortunate that, in 28 African countries, they're taxed or subject to import tariffs. Most of the people in these countries would have trouble paying for a net even without the tax. This problem was addressed in the May 2000 "Abuja Declaration," a summit agreement on infectious diseases signed by 44 African countries. The Declaration included a pledge to do away with "malaria taxes." At last count, 13 countries have actually acted on the pledge, although in some cases only by reducing rather than eliminating the taxes. Since the Declaration was signed, an estimated 2 to 5 million Africans have died from malaria.

This failure to follow through with the Abuja Declaration casts the interest in DDT in a rather poor light. Of the 31 POPs treaty signatories that have reserved the right to use DDT, 21 are in Africa. Of those 21, 10 are apparently still taxing or imposing tariffs on bednets. (Among the African countries that have *not* signed the POPs treaty, some are almost certainly both using DDT and taxing bednets, but the exact number is difficult to ascertain because the status of DDT use is not always clear.) It is true that a case can be made for the use of DDT in situations like the one in South Africa in 1999—an infrequent flare-up in a context that lends itself to control. But the routine use of DDT against malaria is an exercise in toxic futility, especially when it's pursued at the expense of a superior and far more benign technology.

Learning to Live With the Mosquitoes

A group of French researchers recently announced some very encouraging results for a new anti-malarial drug known as G25. The drug was given to infected aotus monkeys, and it appears to have cleared the parasites from their systems. Although extensive testing will be necessary before it is known whether the drug can be safely given to people, these results have raised the hope of a cure for the disease.

Of course, it would be wonderful if G25, or some other new drug, lives up to that promise. But even in the absence of a cure, there are opportunities for progress that may one day make the current incidence of malaria look like some dark age horror. Many of these opportunities have been incorporated into an initiative that began in 1998, called the Roll Back Malaria (RBM) campaign, a collaborative effort between WHO, the World Bank, UNICEF, and the UNDP [United Nations Development Programme]. In contrast to the earlier WHO eradication program, RBM grew out of joint efforts between WHO and various African governments specifically to address African malaria. RBM focuses on household- and community-level intervention and it emphasizes apparently modest changes that could yield major progress. Below are four "operating principles" that are, in one way or another, implicit in RBM or likely to reinforce its progress.

1. Do away with all taxes and tariffs on bednets, on pesticides intended for treating bednets, and on antimalarial drugs. Failure to act on this front certainly undercuts claims for the necessity of DDT; it may also undercut claims for antimalaria foreign aid.

2. Emphasize appropriate technologies. Where, for example, the need for mud to replaster walls is creating lots of pothole sized cavities near houses—cavities that fill with water and then with mosquito larvae—it makes more sense to help people improve their housing maintenance than it does to set up a program for squirting pesticide into every pothole. To be "appropriate," a technology has to be both affordable and culturally acceptable. Improving home maintenance should pass this test; so should bednets. And of course there are many other possibilities. In Kenya, for example, a research institution called the International Center for Insect Physiology and Ecology has identified at least a dozen native east African plants that repel *Anopheles gambiae* in lab tests. Some of these plants could be important additions to household gardens.

3. Use existing networks whenever possible, instead of building new ones. In Tanzania, for example, an established healthcare program (UNICEF's Integrated Management of Childhood Illness Program) now dispenses antimalarial drugs—and instruction on how to use them. The UNICEF program was already operating, so it was simple and cheap to add the malaria component. Reported instances of severe malaria and anemia in infants have declined, apparently as a result. In Zambia, the government is planning to use health and prenatal clinics as the network for a coupon system that subsidizes bednets for the poor. Qualifying patients would pick up coupons at the clinics and redeem them at stores for the nets.

4. Assume that sound policy will involve action on many fronts. Malaria is not just a health problem—it's a social problem, an economic problem, an environmental problem, an agricultural problem, an urban planning problem. Health officials alone cannot possibly just make it go away. When the disease flares up, there is a strong and understandable temptation to strap on the spray equipment and douse the mosquitoes. But if this approach actually worked, we wouldn't be in this situation today. Arguably the biggest opportunity for progress against the disease lies, not in our capacity for chemical innovation, but in our capacity for *organizational innovation*—in our ability to build an awareness of the threat across a broad range of policy activities. For example, when government officials are considering loans to irrigation projects, they should be asking: has the potential for malaria been addressed? When foreign donors are designing antipoverty programs, they should be asking: do people need bednets? Routine inquiries of this sort could go a vast distance to reducing the disease.

Where is the DDT in all of this? There isn't any, and that's the point. We now have half a century of evidence that routine use of DDT simply will not prevail against the mosquitoes. Most countries have already absorbed this lesson, and banned the chemical or relegated it to emergency only status. Now the RBM campaign and associated efforts are showing that the frequency and intensity of those emergencies can be reduced through systematic attention to the chronic aspects of the disease. There is less and less justification for DDT, and the futility of using it as a matter of routine is becoming increasingly apparent: in order to control a disease, why should we poison our soils, our waters, and ourselves?

NO  Alexander Gourevitch

Better Living Through Chemistry

By most measures, Uganda is one of the success stories of African development. Under President Yoweri Museveni, a former guerrilla and darling of the international donor community, Uganda has achieved GDP [gross domestic product] growth of 6 percent per year, gradually expanding political freedoms, and a measure of peace. In a continent beset by poverty and political disorder, Uganda has slowly become a model for how to get things right. Even the AIDS epidemic, which infected an estimated 20 percent of Uganda's population a decade ago, seems to be under control.

But a more old-fashioned plague has come back to haunt Uganda: malaria. The mosquito-borne illness costs Uganda more than $347 million a year. Today, up to 40 percent of the country's outpatient care goes to people thus infected. Total infections are so numerous that the government doesn't even try to track them; but last year, 80,000 people died of the disease, half of them children under the age of five.

So last December [2202], at a convention of regional health ministers held in Kampala, Jim Muhwezi, an army officer and member of parliament who today serves as Uganda's minister of health, announced the launch of a new campaign against the epidemic, using Dichloro-diphenyl-trichloroethane, or DDT. To Muhwezi, DDT—a pesticide widely proscribed in Europe and banned in the United States since 1972—was a cheap, effective weapon against malaria for a poor country with minimal public health resources. And in South Africa, the recent reintroduction of DDT spraying had reduced malaria rates by 75 percent over two years. "Instead of sitting back and watching our people die of malaria and lose in economic terms," he proclaimed, "an all-out war against the disease must be waged."

But Muhwezi encountered opposition almost immediately. After his announcement, Andrew Sisson, a USAID official attending the Kampala convention, told one session that in the United States DDT had been found to "cause environmental problems," according to Muhwezi. A member of Uganda's parliament warned Muhwezi that Europe and the United States might ban imports of Uganda's fish and agricultural exports, a fear shared by local environmentalists, according to the *Nairobi East African*, Kenya's leading daily. Since USAID prefers to fund bednets as a solution to Uganda's mosquito problem, Muhwezi is unsure if he'll be able to obtain international assistance to fund a DDT-based malarial eradication project. "We hope they'll come along. But if they don't, we'll do it alone."

Until recently, one might have considered Uganda's to be a tragic but unavoidable tradeoff—deprive many of an uncontaminated natural environment, or save few from malaria. In much of the world, after all, the popular conception of DDT is of a dangerous and toxic chemical that pollutes water and poisons the food chain; in the United States, DDT is remembered as the pesticide that helped put bald eagles on the endangered species list. But a growing body of scientific evidence suggests that the popular conception is wrong. Older studies on the effects of DDT have been called into question, and newer ones militate against the notion that DDT is inherently dangerous. For the kind of use Muhwezi has in mind, in fact, DDT may not be dangerous at all.

The stakes are high. Uganda is but one of many African countries suffering from malaria epidemics. Africa already accounts for 90 percent of the 2 million deaths and 300 million infections around the world each year, and it costs the continent 1.3 percent in annual growth per year, according to the economist Jeffrey Sachs. Mosquitoes are increasingly resistant to the main insecticide put into use to replace DDT; and the parasite that causes the disease has, in recent years, become increasingly resistant to the cheapest and most common medical treatment, a drug called chloroquine. Yet most international aid agencies, development agencies, and lending institutions have moved away from funding spraying projects in general, and DDT use specifically. Without assistance, African governments cannot afford spraying programs, leaving them bereft of a safe, effective, and cheap defense. Which means that aid agencies and governments opposed to DDT use may end up costing Africa millions of needless deaths.

Bad Medicine?

DDT first came to the United States after the late 1930s, when Dr. Paul Muller, a chemist with the Swiss firm J.R. Geigy, found that minuscule amounts of DDT killed just about every insect he could find. Slow to break down, a single application of DDT remained toxic for up to a year, which made spraying programs much easier to administer, especially in remote locations. It was cheap to produce, easy to ship, and did not require the extensive safety gear of other insecticides. And remarkably, even when mosquitoes developed resistance to the toxicity of DDT, it still acted as a repellent and irritant, driving nocturnal mosquitoes out of homes before they had a chance to bite. (This mechanism was discovered later—originally, it was just toxicity and safety that attracted people to DDT.)

Impressed, the U.S. military deployed DDT in 1942 to fight a third front against diseases like malaria, dengue, and typhus, which until then had seriously impaired U.S. fighting forces, especially in Italy and the Pacific Theater. Army personnel sprayed soldiers, dusted beachheads, and even deloused concentration camp survivors with DDT. After the war, DDT came into use for commercial and public health purposes. Farmers used DDT to protect cash crops like cotton, corn, and apples from a wide variety of agricultural pests. Around the same time, the U.S. government launched an ambitious DDT-centered malaria eradication project which by the early '60s had virtually eliminated malaria from Southern Europe, the Caribbean, and parts of East and South Asia. (In India, for example, annual deaths went from 800,000 to zero.) At the time, DDT was thought to be such an effective and useful substance that in 1948, Muller received a Nobel Prize in medicine. "To only a few chemicals does man owe as great a debt as to DDT," declared the National Academy of Sciences in a report in 1970. "In little more than two decades, DDT has prevented 500 million human deaths, due to malaria."

But by then, the tide had begun to turn against DDT. During the 1960s, reports began to emerge of increasing resistance to the drug among insects, probably sparked by its widespread use in agriculture. At the same time, case-detection followed by medical treatment began to emerge as the new model for malaria control. (By 1979, the World Health Organization [WHO] had formally endorsed this approach over that of preemptive insecticide spraying.) Most important, however, was the publication of Rachel Carson's book *Silent Spring* in 1962.

Carson's book was a lyrical broadside against synthetic chemicals in general, but against DDT in particular. She noted that as DDT seeped into the ground and ran off into streams, worms and fish stored it in their fatty tissues. Over time, songbirds like the robin and other prized avians, including bald eagles and peregrine falcons, ingested enough contaminated prey that they died of DDT poisoning. If they didn't die outright, Carson warned, studies also showed that DDT prevented reproduction by thinning eggshells. Carson also trumpeted studies of rats which suggested DDT was a liver carcinogen, and gathered anecdotal evidence of harm in human beings, like a farmer whose bone marrow wasted away after repeatedly inhaling a mixture of DDT and benzene hexachloride he used to spray his fields.

Silent Spring practically launched the modern environmental movement. The Environmental Defense Fund cut its teeth in national politics raising public alarm over—and bringing lawsuits against—DDT use, which in turn pushed the recently created Environmental Protection Agency [EPA] to hold a series of hearings on DDT. The critics were so successful that, although the administrative judge presiding over the hearings concluded that "DDT is not a carcinogenic hazard to man... DDT is not a mutagenic or teratogenic hazard to man," the EPA banned it anyway in 1972. (Chemical companies, of course, were more than happy to supply the less practical, more expensive alternatives.) The U.S. ban was a turning point; soon after, anti-DDT sentiment went global. Environmental organizations campaigned against its use abroad, wealthy countries began to restrict funding for DDT projects, and the World Health Organization shifted away from promoting it for public health uses. By 2000, a group of environmental activists, led by the World Wildlife Fund, was promoting a U.N. "persistent organic pollutants" treaty known as the Stockholm Convention, which would have banned DDT worldwide for all uses. Only at the last minute was an exemption added for public health use.

Reconsidering a Rogue Agent

But over the years, mainstream scientific opinion has absolved DDT of many of its supposed sins. Indeed, the Stockholm Convention partially backfired because it brought to light a slew of studies and literature reviews which contradicted the conventional wisdom on DDT. Like nearly any chemical, DDT is harmful in high enough doses. But when it comes to the kinds of uses once permitted in the United States and abroad, there's simply no solid scientific evidence that exposure to DDT causes cancer or is otherwise harmful to human beings.

Not a single study linking DDT exposure to human toxicity has ever been replicated. In 1993, Mary Wolff, an associate professor at Mount Sinai Medical Center, published a small study linking DDT exposure to breast cancer. But numerous follow-up studies with human subjects—including one large five-study review comparing 1,400 women with breast cancer to an equivalent number of controls—found no evidence for the link. David Hunter, an epidemiologist at Harvard University who ran one of the follow-up studies, says of the breast cancer connection, "the studies have really put that idea to rest." Similarly, various studies have contradicted initial concerns that DDT might cause myeloma, hepatic cancer, or non-Hodgkins lymphoma.

Other reports over the years postulating human toxicity in DDT exposure turned out to be cases of correlation without causation. In its heyday, for instance, DDT was mixed with a variety of dangerous chemicals, sometimes petroleum derivatives. In every anecdote of death or human harm by DDT that Carson related, the chemical had been dissolved in some other, highly toxic, substance, such as fuel oil, petroleum distillate, benzene hexachloride, or methylated naphthalenes. Such "mixtures with other chemicals or solvents," a 2000 review article in the medical journal *The Lancet* noted, were responsible for many of the reported deaths from DDT and for other problems like dermatitis. But even these dangers do not extend to public health use, where DDT is dissolved in water and sprayed as a thin film.

That's not to say that DDT is harmless. Matthew Longnecker studied American women who had lived during the period of high DDT use and suggested that high levels of DDT in the bloodstream of pregnant women might cause pre-term delivery and low birthweight, for instance. But public health use doses—two grams per square meter of wall sprayed indoors at most every six months—aren't likely to produce those concentrations. Since DDT is not absorbed through the skin, spraying DDT in houses is unlikely to expose pregnant women—or any one else—to amounts great enough to pose a danger. And scant evidence suggests DDT gets into the environment in significant amounts when sprayed indoors. According to a WHO report in 2000, "The targeted application of insecticides to indoor walls... greatly reduces dispersion of the chemicals into the environment. For this reason, the environmental risks from such targeted measures [are] considered minimal."

Agricultural use, on the other hand, is very different, amounting to literally tons of the chemical sprayed outdoors every few weeks. But almost nobody who supports using DDT to combat malaria wants to see it come back into use as an agricultural pesticide. The ideal pesticide is one that will stay on the crop but break down and virtually disappear by the time of harvest. DDT, on the other hand, is persistent and takes a long time to break down—which is why it tends to accumulate in the environment over time. "Even though there's no evidence right now that it's harmful to human beings, there's no sense in taking the risk of using it when other pesticides, better-suited for agricultural use, are available," says Donald Roberts, an expert on tropical health at the Uniformed Services University of the Health Sciences in Bethesda, Md. DDT is also so cheap that, when it was legal, farmers often used it well in excess of the officially prescribed amounts.

Yet environmental activists resist distinguishing between the agricultural and public health uses of DDT. Richard Liroff, a spokesman for the World Wildlife Fund, says, "We hang most of our argument" against DDT spraying on studies like Longnecker's. But the clear benefits of DDT use would seem to outweigh the potential dangers. Malaria, after all, also causes low birthweight in newborns (and mental retardation in infants). And while DDT may prove to have as-yet-unknown side effects, malaria has a well-known, direct effect: It kills millions of people a year.

Environmental Mea Culpa

But although prevailing scientific opinion favors the use of DDT in antimalarial campaigns, international aid agencies still take their cues from environmental groups. Roll Back Malaria (RBM), a WHO-sponsored consortium of aid agencies, international institutions, and NGOs [nongovernmental organizations], has a 40-page action plan for reducing countries' reliance on DDT, with the goal of eventually eliminating its use for public health purposes. And the international donors who fund most anti-malaria campaigns usually follow RBM's technical guidelines. "Bottom line is, [RBM] favors the ultimate elimination of DDT from the malaria toolbox," says Dr. John Paul Clark, a former RBM adviser with expertise in DDT, although he concedes that "there are a number of countries that are not economically or epidemiologically ready to make that switch at this time."

While few organizations have a *de jure* ban on DDT projects, very few have actually put money behind them. No international aid agency will fund DDT use. The World Bank is currently funding a malaria control project in Eritrea on the condition that the country not use DDT. The recently formed Global Environmental Facility has donated money to projects in both Africa and South America, likewise with the intent of weaning recipient nations off DDT. In an emailed statement, USAID's malaria team informed me that its "activities are focused to reduce reliance on the pesticide DDT." They are "emphasizing prevention, medical intervention, and mosquito nets dipped in pyrethroid." Richard Tren, head of a group called Africa Fighting Malaria, says that the international aid agencies of Sweden, the United Kingdom, Norway, Japan, and Germany have all told him they would not fund DDT projects, nor will UNICEF. And lacking the resources to develop domestic programs on their own, most African countries bend to the requirements of these international funders. (South Africa is one of the few African countries wealthy enough to fund its own program.)

Those alternatives that aid agencies will fund are either less effective, more expensive, harder to administer, or inadequate on their own. "Eco-friendly" approaches like mosquito repellent trees or mosquito-larvae-eating fish have been tried in East Africa, where the malaria epidemic is particularly bad, but with little success. The pesticide pyrethroid was originally developed as a biodegradable DDT alternative, but mosquito resistance throughout Africa is rendering it increasingly useless. Other substitute pesticides, like car-bamates and organophosphates, have turned out to be no more safe or effec-tive than DDT, and most lack DDT's ability to repel mosquitoes even after they build up resistance to it. DDT is at least four times less expensive than the cheapest alternative—even though it is only still produced by one factory in India—and requires less frequent spraying. Both are significant advantages in poor African countries with minimal infrastructure, where every dollar not spent bringing malaria under control can be used for other public health pri-orities, such as supplying clean water. "DDT is long-acting, the alternatives are not," says Donald Roberts. "DDT is cheap, the alternatives are not. End of story."

Amir Attaran, a former WHO expert on malaria, once supported funding alternative pesticides, but South Africa's experience changed his mind. "If South Africa can't get by without DDT, it's pretty much as if to say that nobody can," says Attaran. "They really tried to phase this stuff out, and had the budget to afford the alternatives... They tried and failed." (South Africa had switched from DDT to pyrethroid in the mid-1990s, but switched back in 2000 when the mosquitoes became resistant to the pyrethroid, causing malaria cases to skyrocket). There has also been a move to use insecticide treated bednets, particularly in East Africa, but few believe that bednets alone can address the problem. "All large-scale programs that have been successful have been based on insecticide control," notes Brian Sharp, Director of the Malaria Research Program for South Africa's Medical Research Council and director of their spraying program. "I don't believe we should discriminate against [bednets or DDT]... One has to practice integrated vector control... [But] DDT is an important tool in this fight."

It is difficult to get a clear answer from aid agencies why they won't fund DDT. They may be hesitant because they receive contradictory guidance: National DDT bans conflict with WHO guidelines saying it's safe and effec-tive, which in turn conflict with Roll Back Malaria's blueprint for phasing out DDT. Nobody seems to want to stick his or her neck out to clarify things. Most importantly, already-underfunded Western aid agencies are concerned about a backlash if they did fund DDT, since doing so might well provoke the lingering fear of DDT among the citizens of wealthier countries. Several experts told me that they are specifically afraid of tangling with the environ-mental lobby. When Attaran circulated a letter two years ago protesting a total ban on DDT, the head of Roll Back Malaria excoriated him for undermining RBM's relations with environmental groups. Attaran, formerly a lawyer for the Sierra Club; thinks the environmentalists should correct the misperceptions they have perpetuated. They should do what "the pharmaceutical [companies] did on access to AIDS medicine in Africa. They did a mea culpa. The environmentalists need to do the same thing."

"It Shouldn't Be This Hard"

That's not to say that all anti-malaria aid dollars should go to DDT. It makes sense to balance between funding available measures and investing in new ones. Presently, however, too little money goes to DDT at a time when few effective tools are available. Local conditions will determine the best course of action; in some places, DDT may be less effective than others, and funding should be adequate and open enough for countries to experiment with what's right for them. Brian Sharp, for instance, argues that money could go to a rotational spraying program, under which DDT would be rotated with other insecticides to prevent the development of resistance among mosquitoes and extending the effectiveness of non-DDT alternatives. Yet opposition to DDT has undercut even that compromise.

An environmentalist mea culpa would be a start, but in the United States, at least, nothing short of congressional hearings or an executive order from the Bush administration is likely to spur USAID to change its ways. The most direct approach would be a reconsideration of the EPA ban on DDT, with an explicit mandate to use some of the foreign aid budget for DDT spraying should countries ask for it. USAID's current goals for Uganda are for at least 60 percent of the country's population to have access to drugs and bednets; Jim Muhwezi, for one, would like USAID to set its goals higher. But it's not easy to get them to listen, especially with poor African countries trying to curry favor with aid agencies. Indeed, neighboring Tanzania has stayed away from DDT because, among other things, it is too "controversial," according to Alex Mwita, the program manager for Tanzania's National Malaria Control Programme. "You have to remove the myths that people have in their minds that it is not a good chemical." It shouldn't be this hard. African governments know what they need to do to control malaria—they just need the money. Like Mwita, Brian Sharp says he's waiting for the West to get over its "misguided opposition to DDT." So is Africa.

POSTSCRIPT

Should DDT Be Banned Worldwide?

Gourevitch comes close to accusing environmentalists of condemning DDT on the basis of politics or ideology rather than of science. Angela Logomasini comes even closer in "Chemical Warfare: Ideological Environmentalism's Quixotic Campaign Against Synthetic Chemicals," in Ronald Bailey, ed., *Global Warming and Other Eco-Myths: How the Environmental Movement Uses False Science to Scare Us to Death* (Prima, 2002). Her admission that public health demands have softened some environmentalists' resistance to the use of DDT points to a basic truth about environmental debates: over and over again, they come down to what we should do first: Should we meet human needs regardless of whether or not species die and air and water are contaminated? Or should we protect species, air, water, and other aspects of the environment even if some human needs must go unmet? What if this means endangering the lives of children? In the debate over DDT, the human needs are clear, for insect-borne diseases have killed and continue to kill a great many people. Yet the environmental needs are also clear. The question is one of choosing priorities and balancing risks. See John Danley, "Balancing Risks: Mosquitoes, Malaria, Morality, and DDT," *Business and Society Review* (Spring 2002). See also Jon Cohen, "Mothers' Malaria Appears to Enhance Spread of AIDS Virus," *Science* (November 21, 2003).

Mosquitoes can be controlled in various ways: Swamps can be drained (which carries its own environmental price), and other breeding opportunities can be eliminated. Fish can be introduced to eat mosquito larvae. And mosquito nets can be used to keep the insects away from people. But these (and other) alternatives do not mean that there does not remain a place for chemical pesticides. In "Pesticides and Public Health: Integrated Methods of Mosquito Management," *Emerging Infectious Diseases* (January–February 2001), Robert I. Rose, an arthropod biotechnologist with the Animal and Plant Health Inspection Service of the U.S. Department of Agriculture, says, "Pesticides have a role in public health as part of sustainable integrated mosquito management. Other components of such management include surveillance, source reduction or prevention, biological control, repellents, traps, and pesticide-resistance management." "The most effective programs today rely on a range of tools," says McGinn in "Combating Malaria," *State of the World 2003* (W. W. Norton, 2003). Still, as Gourevitch notes, some countries see DDT as essential. For more on Uganda, see Charles Wendo, "Uganda Considers DDT to Protect Homes From Malaria," *The Lancet* (April 24, 2004). Indonesia makes its case in "Bring Back DDT," *Far Eastern Economic Review* (March 4, 2004). Tina Rosenberg attempts to speak for all in "What the World Needs Now Is DDT," *The New York Times Magazine* (April 11, 2004).

Researchers have long sought a vaccine against malaria, but the parasite has demonstrated a persistent talent for evading all attempts to arm the immune system against it. The difficulties are covered by Thomas L. Richie and Allan Saul in "Progress and Challenges for Malaria Vaccines," *Nature* (February 7, 2002). In March 2002 a new vaccine was reported to be effective in animals; see Michael Greer, "Malaria Vaccine Based on Parasite Protein Effective in Animals," *Vaccine Weekly* (March 13 & 20, 2002). A newer approach is to develop genetically engineered (transgenic) mosquitoes that either cannot support the malaria parasite or cannot infect humans with it; see Jane Bradbury, "Transgenic Mosquitoes Bring Malarial Control Closer," *The Lancet* (May 25, 2002).

It is worth stressing that malaria is only one of several mosquito-borne diseases that pose threats to public health. Two others are yellow fever and dengue. A recent arrival to the United States is West Nile virus, which mosquitoes can transfer from birds to humans. However, West Nile virus is far less fatal than malaria, yellow fever, or dengue, and a vaccine is in development. See Dwight G. Smith, "A New Disease in the New World," *The World & I* (February 2002) and Michelle Mueller, "The Buzz on West Nile Virus," *Current Health 2* (April/May 2002).

It is also worth stressing that global warming means climate changes that may increase the geographic range of disease-carrying mosquitoes. Many climate researchers are concerned that malaria, yellow fever, and other now mostly tropical and subtropical diseases may return to temperate-zone nations and even spread into areas where they have never been known.

ISSUE 17

Do Environmental Hormone Mimics Pose a Potentially Serious Health Threat?

YES: Michele L. Trankina, from "The Hazards of Environmental Estrogens," *The World & I* (October 2001)

NO: Michael Gough, from "Endocrine Disrupters, Politics, Pesticides, the Cost of Food and Health," *Cato Institute* (December 15, 1997)

ISSUE SUMMARY

YES: Professor of biological sciences Michele L. Trankina argues that a great many synthetic chemicals behave like estrogen, alter the reproductive functioning of wildlife, and may have serious health effects—including cancer—on humans.

NO: Michael Gough, a biologist and expert on risk assessment and environmental policy, argues that only "junk science" supports the hazards of environmental estrogens.

Following World War II there was an exponential growth in the industrial use and marketing of synthetic chemicals. These chemicals, known as "xeno-biotics," were used in numerous products, including solvents, pesticides, refrigerants, coolants, and raw materials for plastics. This resulted in increasing environmental contamination. Many of these chemicals, such as DDT, PCBs, and dioxins, proved to be highly resistant to degradation in the environment; they accumulated in wildlife and were serious contaminants of lakes and estuaries. Carried by winds and ocean currents, these chemicals were soon detected in samples taken from the most remote regions of the planet, far from their points of introduction into the ecosphere.

Until very recently most efforts to assess the potential toxicity of synthetic chemicals to bio-organisms, including human beings, focused almost exclusively on their possible role as carcinogens. This was because of legitimate public concern about rising cancer rates and the belief that cancer causation was the most likely outcome of exposure to low levels of synthetic chemicals.

Some environmental scientists urged public health officials to give serious consideration to other possible health effects of xenobiotics. They were generally ignored because of limited funding and the common belief that toxic effects other than cancer required larger exposures than usually resulted from environmental contamination.

In the late 1980s Theo Colborn, a research scientist for the World Wildlife Fund who was then working on a study of pollution in the Great Lakes, began linking together the results of a growing series of isolated studies. Researchers in the Great Lakes region, as well as in Florida, the West Coast, and Northern Europe, had observed widespread evidence of serious and frequently lethal physiological problems involving abnormal reproductive development, unusual sexual behavior, and neurological problems exhibited by a diverse group of animal species, including fish, reptiles, amphibians, birds, and marine mammals. Through Colborn's insights, communications among these researchers, and further studies, a hypothesis was developed that all of these wildlife problems were manifestations of abnormal estrogenic activity. The causative agents were identified as more than 50 synthetic chemical compounds that have been shown in laboratory studies to either mimic the action or disrupt the normal function of the powerful estrogenic hormones responsible for female sexual development and many other biological functions.

Concern that human exposure to these ubiquitous environmental contaminants may have serious health repercussions was heightened by a widely publicized European research study, which concluded that male sperm counts had decreased by 50 percent over the past several decades (a result that is disputed by other researchers) and that testicular cancer rates have tripled. Some scientists have also proposed a link between breast cancer and estrogen disrupters.

In response to the mounting scientific evidence that environmental estrogens may be a serious health threat, the U.S. Congress passed legislation requiring that all pesticides be screened for estrogenic activity and that the Environmental Protection Agency (EPA) develop procedures for detecting environmental estrogenic contaminants in drinking water supplies; see the EPA's Endocrine Disruptor Screening Program Web site at http://www. epa.gov/scipoly/oscpendo/index.htm. Government-sponsored studies of synthetic endocrine disrupters and other hormone mimics are also under way in the United Kingdom and in Germany.

In the following selections, Michele L. Trankina argues that a great many synthetic chemicals behave like estrogen, alter the reproductive functioning of wildlife, and may have serious health effects—including cancer—on humans. She insists that regulatory agencies must minimize public exposure. Michael Gough argues that only "junk science" supports the hazards of environmental estrogens. Expensive testing and regulatory programs can only drive up the cost of food, he says, which will make it harder for the poor to afford fresh fruits and vegetables. Furthermore, health protection will not be increased.

Michele L. Trankina

 **YES**

The Hazards of Environmental Estrogens

What do Barbie dolls, food wrap, and spermicides have in common? And what do they have to do with low sperm counts, precocious puberty, and breast cancer? "Everything," say those who support the notion that hormone mimics are disrupting everything from fish gender to human fertility. "Nothing," counter others who regard the connection as trumped up, alarmist chemophobia. The controversy swirls around the significance of a number of substances that behave like estrogens and appear to be practically everywhere—from plastic toys to topical sunscreens.

Estrogens are a group of hormones produced in both the female ovaries and male testes, with larger amounts made in females than in males. They are particularly influential during puberty, menstruation, and pregnancy, but they also help regulate the growth of bones, skin, and other organs and tissues.

Over the past 10 years, many synthetic compounds and plant products present in the environment have been found to affect hormonal functions in various ways. Those that have estrogenic activity have been labeled as environmental estrogens, ecoestrogens, estrogen mimics, or xenoestrogens (*xenos* means foreign). Some arise as artifacts during the manufacture of plastics and other synthetic materials. Others are metabolites (breakdown products) generated from pesticides or steroid hormones used to stimulate growth in livestock. Ecoestrogens that are produced naturally by plants are called phytoestrogens (*phyton* means plant).

Many of these estrogen mimics bind to estrogen receptors (within specialized cells) with roughly the same affinity as estrogen itself, setting up the potential to wreak havoc on reproductive anatomy and physiology. They have therefore been labeled as disruptors of endocrine function.

Bizarre Changes in Reproductive Systems

Heightened concern about estrogen-mimicking substances arose when several nonmammalian vertebrates began to exhibit bizarre changes in reproductive anatomy and fertility. Evidence that something was amiss came serendipitously in 1994, from observations by reproductive physiologist Louis Guillette of the University of Florida. In the process of studying the decline of alligator populations at Lake Apopka, Florida, Guillette and coworkers noticed that many male alligators had smaller penises than normal. In addition, females superovulated, with multiple nuclei in some of the surplus ova. Closer scrutiny linked these findings to a massive spill of DDT (dichloro-diphenyl-trichloroethane) into Lake Apopka in 1980. Guillette concluded that the declining alligator population was related to the effects of DDT exposure on the animals' reproductive systems.

Although DDT was banned for use in the United States in the early 1970s, it continues to be manufactured in this country and marketed abroad, where it is sprayed on produce that is then sold in U.S. stores. The principal metabolite derived from DDT is called DDE, a xenoestrogen that lingers in fat deposits in the human body for decades. Historically, there have been reports of estrogen-mimicking effects in various fish species, especially in the Great Lakes, where residual concentrations of DDT and PCBs (polychlorinated biphenyls) are high. These effects include feminization and hermaphroditism in males. In fact, fish serve as barometers of the effects of xenoestrogen contamination in bodies of water. An index of exposure is the presence of vitellogenin, a protein specific to egg yolk, in the blood of male fish. Normally, only females produce vitellogenin in their livers, upon stimulation by estrogen from the ovaries.

Ecoestrogens are further concentrated in animals higher up in the food chain. In the Great Lakes region, birds including male herring gulls, terns, and bald eagles exhibit hermaphroditic changes after feeding on contaminated fish. Increased embryo mortality among these avians has also been observed. In addition, evidence from Florida links sterility in male and female panthers to their predation on animals exposed to pesticides with estrogenic activity.

The harmful effects of DDT have also been observed in rodents. Female rodents treated with high concentrations of DDT become predisposed to mammary tumors, while males tend to develop testicular cancer. These observations raise the question of whether pharmacological (that is, low) doses of substances with estrogenic activity translate into physiological effects. Usually they do not, but chronic exposure may be enough to trigger such effects.

Dangers to Humans

If xenoestrogens cause such dramatic reproductive effects in vertebrates, including mammals, what might be the consequences for humans? Nearly a decade ago, Frederick vom Saal, a developmental biologist at the University of Missouri Columbia, cautioned that mammalian reproductive mechanisms are similar enough to warrant concern about the effects of hormone disruptors on humans.

A 1993 article in *Lancet* looked at decreasing sperm counts in men in the United States and 20 other countries and correlated these decreases with the growing concentration of environmental estrogens. The authors—Niels Skakkenbaek, a Danish reproductive endocrinologist, and Richard Sharpe, of the British Medical Research Council Reproductive Biology Unit in Scotland—performed a meta-analysis of 61 sperm-count studies published between 1938 and 1990 to make their connection.

"Nonbelievers" state that this interpretation is contrived. Others suggest alternative explanations. For instance, the negative effects on sperm counts could have resulted from simultaneous increases in the incidence of venereal diseases. Besides, there are known differences in steroid hormone metabolism between lower vertebrates (including nonprimate mammals) and primates, so one cannot always extrapolate from the former group to the latter.

Even so, the incidence of testicular cancer, which typically affects young men in their 20s and 30s, has increased worldwide. Between 1979 and 1991, over 1,100 new cases were reported in England and Wales—a 55 percent increase over previous rates. In Denmark, the rate of testicular cancer increased by 300 percent from 1945 to 1990. Intrauterine exposure to xenoestrogens during testicular development is thought to be the cause.

Supporting evidence comes from Michigan, where accidental contamination of cattle feed with PCBs in 1973 resulted in high concentrations in the breast milk of women who consumed tainted beef. Their sons exhibited defective genitalia. Furthermore, observers in England have noted increased incidences of cryptorchidism (undescended testes), which results in permanent sterility if left untreated, and hypospadias (urethral orifice on the underside of the penis instead of at the tip). The one area of agreement between those who attribute these effects to ecoestrogens and those who deny such a connection is that more research is needed.

Perhaps one of the most disturbing current trends is the alarming increase in breast cancer incidence. Fifty years ago, the risk rate was 1 woman in 20; today it is 1 in 8. Numerous studies have implicated xenoestrogens as the responsible agents. For instance, high concentrations of pesticides, especially DDT, have been found in the breast tissue of breast cancer patients on Long Island. In addition, it has long been known that under certain conditions, estrogen from any source can be tumor promoting, and that most breast cancer cell types have estrogen receptors.

Precocious Puberty

If that were not enough, the growing number of estrogen mimics in the environment has been linked to early puberty in girls. The normal, average age of onset is between 12 and 13. A recent study of 17,000 girls in the United States indicated that 7 percent of white and 27 percent of black girls exhibited physical signs of puberty by age seven. For 10-year-old girls, the percentages increased to 68 and 95, respectively. Studies from the United Kingdom, Canada, and New Zealand have shown similar changes in the age of puberty onset.

It is difficult, however, to elucidate the exact mechanisms that underlie these trends toward precocious puberty. One explanation, which applies especially to cases in the United States, points to the increasing number of children who are overweight or obese as a result of high-calorie diets and lack of regular exercise. Physiologically, an enhanced amount of body fat implies reproductive readiness and signals the onset of puberty in both boys and girls.

For girls, more body fat ensures that there is enough stored energy to support pregnancy and lactation. Young women with low percentages of body fat caused by heavy exercise, sports, or eating disorders usually do not experience the onset of menses until their body composition reflects adequate fat mass.

Many who study the phenomenon of premature puberty attribute it to environmental estrogens in plastics and secondhand exposure through the meat and milk of animals treated with steroid hormones. An alarming increase in the numbers of girls experiencing precocious puberty occurred in the 1970s and '80s in Puerto Rico. Among other effects, breast development occurred in girls as young as one year. Premature puberty was traced to consumption of beef, pork, and dairy products containing high concentrations of estrogen.

Another study from Puerto Rico revealed higher concentrations of phthalate—a xenoestrogen present in certain plastics—in girls who showed signs of early puberty, compared with controls.

It may be that excess body fat and exposure to estrogenic substances operate in concert to hasten puberty. Body fat is one site of endogenous estrogen synthesis. Exposure to environmental estrogens may add just enough exogenous hormone to exert the synergistic effect necessary to bring on puberty, much like the last drop of water that causes the bucket to overflow.

Although most xenoestrogens produce detrimental effects, at least one subgroup—the phytoestrogens—includes substances that can have beneficial effects. Phytoestrogens are generally weaker than natural estrogens. They are found in various foods, such as flax seeds, soybeans and other legumes, some herbs, and many fruits and vegetables. Some studies suggest that soy products may offer protection against certain cancers, including breast, prostate, uterus, and colon cancers. On the other hand, high doses of certain phytoestrogens— such as coumestrol (in sunflower seeds and alfalfa sprouts)—have been found to adversely affect the fertility and reproductive cycles of animals.

Unlike artificially produced xenoestrogens, phytoestrogens are generally not stored in the body but are readily metabolized and excreted. Their health effects should be evaluated on a case-by-case basis, considering such factors as the individual's age, medical and family history, and potential interactions with medications or supplements.

Plastics, Plastics Everywhere

It has been estimated that perhaps 100,000 synthetic chemicals are registered for commercial use in the world today and 1,000 new ones are formulated every year. While many are toxic and carcinogenic, little is known about the chronic effects of the majority of them. And there is growing concern about their potential hormone-disrupting effects. The problem of exposure is complicated by numerous carrier routes, including air, food, water, and consumer products.

Consider certain synthetics that turn up in familiar places, including food and consumer goods. Such seemingly inert products as plastic soda and water bottles, baby bottles, food wrap, Styrofoam, many toys, cosmetics, sunscreens, and even spermicides either contain or break down to yield xenoestrogens. In addition, environmental estrogens are among the byproducts created by such processes as the incineration of biological materials or industrial waste and chlorine bleaching of paper products.

In April 1999, Consumers Union confirmed information previously reported by the Food and Drug Administration regarding 95 percent of baby bottles sold in the United States. The bottles, made of a hard plastic known as polycarbonate, leach out the synthetic estrogen named bisphenol-A, especially when heated or scratched. Studies verifying the estrogenic activity of bisphenol-A were published in *Nature* in the 1930s, but it did not arouse much concern then. A 1993 report published in *Endocrinology* showed that bisphenol-A produced estrogenic effects in a culture of human breast cancer cells.

Additional studies published by vom Saal in 1997 and '98 have shown that bisphenol-A stimulates precocious puberty and obesity in mice. Others have detected leaching of bisphenol-A from polycarbonate products such as plastic tableware, water cooler jugs, and the inside coatings of certain cans (used for some canned foods) and bottle tops. Autoclaving in the canning process causes bisphenol-A to migrate into the liquid in cans.

Spokespersons from polycarbonate manufacturers have stated that they cannot replicate vom Saal's results, but he counters that industry researchers have not done the experiments correctly.

DEHA (di-[2-ethylhexyl] adipate) is a liquid plasticizer added to some plastic food wraps made of polyvinyl chloride (PVC). Scientific studies have shown that the fat-soluble DEHA can migrate into foods, especially luncheon meats, cheese, and other products with a high fat content. For a 45-pound child eating cheese wrapped in such plastic, the limit of safe intake is 1.5 ounces by European standards or 2.5 ounces by Environmental Protection Agency criteria. Studies conducted by Consumers Union indicate that DEHA leaching from commercial plastic wraps is eight times higher than European directives allow. Fortunately, alternative wraps—such as Handi-Wrap™ and Glad Cling Wrap—are made of polyethylene; chemicals in them do not appear to leach into foods.

Barbie dolls manufactured in the 1950s and '60s are made of PVC containing a stabilizer that degrades to a sticky, estrogen-mimicking residue and accumulates on the dolls' bodies. This phenomenon was noted by Danish museum officials in August 2000. Yvonne Shashoua, an expert in materials preservation at the National Museum of Denmark, warns that young children who play with older Barbie and Ken dolls expose themselves to this estrogenic chemical, and she suggests enclosing the dolls with xenoestrogen-free plastic wrap. Storing the dolls in a cool, dark place also helps prevent the harmful stabilizer from oozing out. Who would have thought that these models of glamour would become health hazards?

Another consumer nightmare is a group of chemical plasticizers known as phthalates. They have been associated with various problems, including testicular depletion of zinc, a necessary nutrient for spermatogenesis. Zinc deficiency results in sperm death and consequent infertility. Many products that previously contained phthalates have been reformulated to eliminate them, but phthalates continue to be present in vinyl flooring, medical tubing and bags, adhesives, infants' toys, and ink used to print on food wrap made of plastic and cardboard. They have been detected in fat-soluble foods such as infant formula, cheese, margarine, and chips.

Some environmental estrogens can be found in unusual places, such as contraceptive products containing the spermicide nonoxynol-9. This chemical degrades to nonylphenol, a xenoestrogen shown to stimulate breast cancer cells. Nonylphenol and other alkylphenols have been detected in human umbilical cords, plastic test tubes, and industrial detergents.

Various substances added to lotions (including sunscreens) and cosmetics serve as preservatives. Some of them, members of a chemical family called parabens, were shown to be estrogen mimics by a study at Brunel University in the United Kingdom. The researchers warned that "the safety of these chemicals should be reassessed with particular attention being paid to .... levels of systemic exposure to humans." Officials of the European Cosmetic Toiletry and Perfumery Association dismissed the Brunel study as "irrelevant," on the grounds that parabens do not enter the systemic circulation. But they ignored the possibility of transdermal introduction.

Additional questions have been raised about the safety—in terms of xenoestrogen content—of recycled materials, especially plastics and paper. Because it is unlikely that a moratorium on chemical synthesis will occur anytime soon, such questions will continue to surface until the public is satisfied that regulatory agencies are doing all they can to minimize exposure. Fortunately, organizations such as the National Institutes of Health, the National Academy of Sciences, the Environmental Protection Agency, the Centers for Disease Control, many universities, and other institutions are involved in efforts to monitor and minimize the effects of environmental estrogens on wildlife and humans.

NO

Michael Gough

Endocrine Disrupters, Politics, Pesticides, the Cost of Food and Health

Environmentalists and politicians and federal regulators have added environmental estrogens or endocrine disrupters to the "concerns" or scares that dictate "environmental health policy." That policy, from its beginning, has been based on ideology, not on science. To provide some veneer to the ideology, its proponents have spawned bad science and junk science that claims chemicals in the environment are a major cause of human illness. There is no substance to the claims, but the current policies threaten to cost billions of dollars in wasted estrogen testing programs and to drive some substantial proportion of pesticides from the market.

Rachel Carson, conjuring up a cancer-free, pre-industrial Garden of Eden launched the biggest environmental scare of all in the 1960s. She charged that modern industrial chemicals in the environment caused human cancers. It mattered not at all to her or to her readers that cancers are found in every society, pre-industrial and modern. What mattered were opinions of people such as Umberto Safiotti of the National Cancer Institute, who wrote:

> I consider cancer as a social disease, largely caused by external agents which are derived from our technology, conditioned by our societal lifestyle and whose control is dependent on societal actions and policies.

When Saffioti said "societal actions and policies," he meant government regulations.

By 1968, environmental groups and individuals—including some scientists—appeared on TV and on the floors of the House and Senate to say, over and over again, "The environment causes 90 percent of cancers." They didn't have to say "environment" meant pollution from modern industry and chemicals—especially pesticides—everyone already knew that. Saffioti and others had told them.

In the 1970s, the National Cancer Institute [NCI] released reports that blamed elevated rates for all kinds of cancers on chemicals in the workplace or in the environment. The institute did not have evidence to link those exposures to cancer. It didn't exist then, and it doesn't, except for a limited number of high exposures in the workplace, exist now. So what? The reports were gobbled up by the press, politicians, and the public.

In our ignorance of what causes most cancers, the "90 percent" misstatement provided great hope. If the carcinogenic agents in the environment could be identified and eliminated, cancer rates should drop. NCI scientists said so, and they said success was just around the corner if animal tests were used to identify carcinogens. Congress responded. It created the Environmental Protection Agency [EPA] and the Occupational Safety and Health Administration [OSHA]. Both agencies have lots of tasks, but both place an emphasis on controlling exposures to carcinogens. Congress passed and amended law after law. The Clean Air Act, the Clean Water Act, the Safe Drinking Water Act, amendments to the Fungicide, Insecticide, and Rodenticide Act—the euphonious FIFRA—the Toxic Substances Control Act, and the Resource Recovery and Control Act poured forth from Capitol Hill.

And in return, EPA and OSHA, to justify their existence, generated scare after scare. They are aided by all kinds of people eager for explanations about their health problems or for government grants and contracts for research or other work or for money to compensate for health effects or other problems that could be blamed on chemicals.

In 1978, we had the occupational exposure scare. Astoundingly, according to a government report, six workplace substances caused 38 percent of all the cancers in the United States. It was nonsense, of course, and many scientists ridiculed the report, but the government never retracted it. The government scientists who contributed to it never repudiated it.

At about the same time, wastes disposed in Love Canal near Niagara Fall, NY, spewed liquids and gases into a residential community. The chemicals were blamed as the cause of cancers, birth defects, miscarriages, skin diseases, you name it. None of it was true, but waste sites around the country were routinely identified as "another Love Canal" or a "Love Canal in the making," and Congress gave the nation the Superfund Law. Since its passage, Superfund has enriched lawyers and provided secure employment to thousands who wear moon suits and dig up, burn, and rebury wastes, and done nothing for the nation's health. For those who doubt the importance of politics in the environmental health saga, it's worth recalling that every state had two waste sites on the first list of sites slated for priority cleanup under Superfund.

By the 1980s, EPA was chucking out the scares. We had the 2,4,5-T scare, the dioxin scare, the 2,4-D scare, the asbestos in schools scare, the radon in homes scare, the Alar scare, the EMF scare. I've left some out, but the common thread that linked the scares together was cancer. Each scare prompted investigations by affected industries and non-government scientists. Each scare fell apart, revealed as a house of cards jerry-rigged from bad science, worse interpretations of the science, and terrible policy.

In fact, by the late 1970s, there was ample evidence that the much-talked about "cancer epidemic" and the 90 percent statement were simply wrong. Cancer rates were not increasing and rates for some cancers were higher in industrial countries and rates for others were higher in non-industrial countries. The U.S. fell in the middle of countries when ranked by cancer rates. Sure, there are some carcinogenic substances in the nation's workplaces, but the best estimates are that they cause four percent or less of all cancers, and the percentage is decreasing because the biggest occupational threat, asbestos, is gone. Environmental exposures might cause two or three percent of cancer—on the outside—and they might cause much less.

The research into causes of cancer—not stories designed to bolster the chemicals cause cancer myth—did reveal that there are preventable causes of cancer. Not smoking is a good idea, as is eating lots of fruits and vegetables, not gaining too much weight, restricting the number of sexual partners, and, for those who are fair-skinned, being careful about sun exposures. It's not a lot different from what your mother or grandmother told you.

The government can take a nanny role in urging us to behave, but that's not where the big bucks are. The big bucks are in regulation, and regulation doesn't seem to have much to do with cancer.

In any case, cancer death rates began to fall in 1990, they've fallen since, and the fall is growing steeper. Maybe that information is blunting the cancer scare. I somehow doubt it. I think that the public has become numb to the cancer scare or that it fatalistically accepts the notion that "everything causes cancer." In any case, the environmentalists and the regulators needed a new scare.

The collapse of the cancer scare wasn't good news to everyone. Government bureaucrats and scientists in the anti-carcinogen offices and programs at EPA and elsewhere have secure jobs. Congress easily finds the will to write laws establishing environmental protection activities, but it lacks the will or patience to examine those activities to see if they've accomplished anything. And, let's face it, Congress doesn't eliminate established programs. But the growth of programs slows, and money can become scarce, and that can squeeze researchers who depend on EPA grants and contracts to fund their often senseless surveys and testing programs. Moreover, the fading of scares doesn't benefit environmental organizations that utter shrill cries about scares and coming calamities in their campaigns for contributions.

Here's an example of just how disappointed some people can be that cancer isn't on the rise. Dr. Theo Colborn, a wildlife biologist working for the Conservation Foundation in the late 1980s, was convinced that the chemicals in the Great Lakes were causing human cancer. She set out to prove it by reviewing the available literature about cancer rates in that region. She couldn't. In fact, she found that the rates for some cancers in the Great Lakes region were lower than the rates for the same cancers in other parts of the United States and Canada....

Failing to find cancer slowed her down but didn't stop her. She knew that those chemicals were causing something. All she had to do was find it.

And find it, she did. She collected every paper that described any abnormality in wildlife that live on or around the Great Lakes, and concluded that synthetic chemicals were mimicking the effects of hormones. They were causing every problem in the literature, whether it was homosexual behavior among gulls, crossed bills in other birds, cancer in fish, or increases or decreases in any wildlife population.

The chemicals that have those activities were called "environmental estrogens" or "endocrine disrupters." There was no more evidence to link them to every abnormality in wildlife than there had been in the 1960s to link every human cancer to chemicals. The absence of evidence wasn't much of a problem. Colborn and her colleagues believed that chemicals were the culprit, and the press and much of the public, nutured on the idea that chemicals were bad, didn't require evidence.

Even so, Colborn had a problem that EPA faced in its early days. Soon after EPA was established, the agency leaders realized that protecting wildlife and the environment might be a good thing, but that Congress might not decide to lavish funds on such activities. They were sure, however, that Congress would throw money at programs that were going to protect human health from environmental risks. Whether Colborn knew that history or not, she apparently realized that any real splash for endocrine disrupters depended on tying them to human health effects.

Using the same techniques she'd used to catalogue the adverse effects of endocrine disrupters on wildlife, she reviewed the literature about human health effects that some way or another might be related to disruption of hormone activity. The list was long, including cancers, birth defects, and learning disabilities, but the big hitter on the list was decreased sperm counts. According to Colborn and others' analyses of sperm counts made in different parts of the world under different conditions of nutrition and stress and at different time periods, sperm counts had decreased by 50 percent in the post–World War II period.

If there's anything that catches the attention of Congress, it's risks to males. Congress banned leaded gasoline after EPA released a report that said atmospheric lead was a cause of heart attacks in middle-aged men. The reported decrease in sperm counts leaped up for attention, and attention it got. Congressional hearings were held, magazine articles were written, experts opined about endocrine disrupters and sexual dysfunctions.

And then it fell apart. Scientists found large geographical variations in sperm counts that have not changed over time. Those geographical variations and poor study designs accounted for the reported decrease. That scare went away, but endocrine disrupters were here to stay.

Well-organized and affluent women's groups are convinced that breast cancer is unusually common on Long Island.... We know that obesity, estrogen replacement therapy, and late child-bearing or no child-bearing, all of which are more common in affluent women, are associated with breast cancer. Nevertheless, from the very beginning, environmental chemicals have been singled out as the cause of the breast cancer excess. The insecticide aldicarb, which is very resistant to degradation was blamed, but subsequent studies failed to confirm the link. A well-publicized study found a link between DDT and breast cancer, but larger, follow-up studies failed to confirm it. But there're lots of environmental chemicals, and no evidence is required to justify a suggestion of a link between the chemicals and cancer.

Senator Al D'Amato is from Long Island, and he shares his constituents' concerns. During a hearing about the Clean Water Act, Senator D'Amato heard testimony by Dr. Anna Soto from Tufts University about her "E-Screen." According to Dr. Soto, her quick laboratory test could identify chemicals that behave as environmental estrogens or endocrine disrupters for $500 a chemical. Since environmental estrogens seem to some people to be a likely cause of breast cancer, Soto's test appeared to be a real bargain.

Senator D'Amato pushed for an amendment to require E-Screen testing of chemicals that are regulated under the Clean Water Act, but he was unsuccessful. Later in 1995, a senior Senate staffer, Jimmy Powell, took the E-Screen amendment to a very junior Senate staffer and told her to incorporate it into the Safe Drinking Water Act as an "administrative amendment." She did, it passed the Senate, and, for the first time, there was a legislative requirement for endocrine testing.

In the spring of 1996, House Committees were considering legislation to amend the Safe Drinking Water Act and new legislation related to pesticides in food. Aware of the Senate's action, some members of the House committees were eager to include endocrine testing in their legislation, but there was resistance as well. Chemical companies viewed the imposition of yet another test as certain to be an expense, unlikely to cost as little as $500 a chemical, and bound to raise new concerns about chemicals that would require far more extensive tests and research to understand or discount.

Furthermore, so far as food was concerned, there was a general conviction that all the safety factors built into the testing of pesticides and other chemicals that might end up in food provided adequate margins of protection. That conviction was shattered by rumors that reached the House in May 1996. According to the rumors, Dr. John McLachlan and his colleagues at Tulane University had shown that mixtures of pesticides and other environmental chemicals such as PCBs were far more potent in activating estrogen receptors, the first step in estrogen modulation of biochemical pathways than were single chemicals. In the most extreme case, two chemicals at concentrations considered safe by all conventional toxicity tests were 1600 times as potent as estrogen receptor activators as either chemical by itself. The powerful synergy raised new alarms.

In May, everyone concerned about pesticides knew that EPA had a draft of the Tulane paper, and EPA staff were drifting around House offices, but they refused to answer questions about the Tulane results. The silence signaled the expected significance of the paper. A month later, in June, the paper appeared in *Science*. It was a big deal. *Science* ran a news article about the research with a picture of the Tulane researchers. It also ran an editorial by a scientist from the National Institutes of Health who offered some theoretical explanations for how combinations of pesticides at very low levels could affect cells and activate the estrogen receptors. *The New York Times, The Washington Post*, other major newspapers, and newsmagazines and TV reported the news. If there was ever any doubt that FQPA [Food Quality Protection Act] would require tests for endocrine activity, the flurry of news about the Tulane results erased them.

While the House was drafting the FQPA, Dr. Lynn Goldman, an assistant administrator at EPA, established a committee called the "Endocrine Disrupter Screening and Testing Advisory Committee" (EDSTAC) [which is now] considering tests for all of the 70,000 chemicals that it estimates are present in commerce, and it's not limiting its recommendations to tests for estrogenic activity. It's adding tests for testosterone and thyroid hormone activity as well as for anti-estrogenic, anti-testosterone, and anti-thyroid activity. The relatively simple E-Screen, which is run on cultured cells, is to be supplemented by some whole animal tests. Tests on single compounds will have to be complemented by tests of mixtures of compounds. The FQPA requires that "valid" tests be used. None of the tests being considered by EDSTAC has been validated; many of them have never been done.

EDSTAC's estimate of 70,000 chemicals in commerce is on the high side—some of those chemicals are used in such small amounts and under such controlled conditions that there's no exposure to them. Dr. Dan Byrd has estimated that 50,000 is a more realistic number. He's also looked at the price lists from commercial testing laboratories to see how much they would charge for a battery of tests something like EDSTAC is considering. Some of the tests haven't been developed, but assuming they can be, Dr. Byrd estimates testing each chemical will cost between $100,000 and $200,000. The total cost would be between $5 and $10 billion....

The Tulane results played some major role in the passage of FQPA, the focus on endocrine disrupters, and Dr. Goldman's establishment of EDSTAC. The Tulane results are wrong. Several groups of scientists tried to replicate the Tulane results. None could. At first, Dr. McLachlan insisted his results were correct. He said that the experiments he reported required expertise and finesse and suggested that the scientists who couldn't repeat his findings were at fault, essentially incompetent. That changed. In July 1997, just 13 months after he published his report, he threw in the towel, acknowledging that neither his laboratory nor anyone else had been able to produce the results that had created such a stir.

Whether the initial results were caused by a series of mistakes or a willful desire to show, once and for all, that environmental chemicals, especially pesticides are bad, bad, bad, we don't know. We do know that the results were wrong.

No matter, EPA now assumes as a matter of policy that synergy occurs. Good science, repeatable science that showed the reported synergy didn't occur has been brushed aside. In its place, we have bad science or junk science. If the Tulane results were the products of honest mistakes, they're bad science; if they flowed from ideology, they're junk science. The effect is the same, but the reasons are different.

The estrogenic disrupter testing under FQPA is going to cost a lot of money and cause a lot of mischief. But the effects of that testing are off somewhere in the future. More immediately, a combination of ideology-driven science and congressional misreading of that science threatens to drive between 50 and 80 percent of all pesticides from the market.

In 1993, a committee of the National Research Council spun together the facts that childhood developmental takes place at specific times as an infant matures into a toddler and then a child, that infants, toddlers, and children eat, proportionally, far larger amounts of foods such as apple juice and apple sauce and orange juice than do adults, and that pesticides can be present in those processed foods. From those three observations, the committee concluded that an additional safety factor should be included in setting acceptable levels for pesticides in those foods. Left out from the analysis was any evidence that current exposures cause any harm to any infants, toddlers, or children. No matter.

Most people who worry about pesticides expected EPA and the Food and Drug Administration to react to the NRC recommendation by reducing the allowable levels of pesticides on foods that are destined for consumption by children. Maybe they would have. We'll never know. In the FQPA, Congress directed that a new ten-fold safety factor be incorporated into the evaluation of the risks from pesticides.

Safety factors are a fundamental part in the evaluation of pesticide risks. Pesticides are tested in laboratory animals to determine what concentrations to cause effects on the nervous, digestive, endocrine, and other systems. At some, sufficiently low dose that varies from pesticide to pesticide, the chemical does not cause those adverse effects. That dose, called the "No Observed Adverse Effect Level" (or NOAEL), is then divided by 100 to set the acceptable daily limit for human ingestion of the chemical. The FQPA requires division by another factor of 10, so the acceptable daily limit will be the NOAEL divided by 1000 instead of 100. Acceptable limits will be ten-fold less.

Dr. Byrd has estimated that up to 80 percent of all currently permitted uses of pesticides would be eliminated by an across the board application of the 1000-fold safety factor. He cites another toxicologist who estimates that 50 percent of all pesticides would be eliminated from the market. The extent to which these draconian reductions will be forced remains to be seen, but pesticide manufacturers and users can look forward to a period of even-greater limbo as EPA sorts through it new responsibilities and decides how to implement FQPA.

There's no convincing evidence that pesticides in food contribute to cancer causation and none that they cause other adverse health effects. Restrictions on pesticides in food will not have a demonstrable effect on human health. On the other hand, the estrogen testing program and the new safety factor will drive pesticide costs up and pesticide availability down.

Some manufacturers may lose profitable product lines; some may even lose their businesses. Farmers will pay more. They will pass those costs onto middlemen and processors, who, in turn, will pass them onto consumers. Increases in the costs of fruits and vegetables won't change the food purchasing habits of the middle class, but they may and probably will affect the purchases of the poor. The poor are already at greater risks because of poor diets, and the increased costs can be expected to further decrease their consumption of fresh fruits and vegetables.

POSTSCRIPT

Do Environmental Hormone Mimics Pose a Potentially Serious Health Threat?

Stephen H. Safe's "Environmental and Dietary Estrogens and Human Health: Is There a Problem?" *Environmental Health Perspectives* (April 1995) is often cited to support the contention that there is no causative link between environmental estrogens and human health problems. He draws a cautious conclusion, calling the link "implausible" and "unproven." Gough's belief that the battle against environmental estrogens is motivated by environmentalist ideology rather than facts is repeated by Angela Logomasini in "Chemical Warfare: Ideological Environmentalism's Quixotic Campaign Against Synthetic Chemicals," in Ronald Bailey, ed., *Global Warming and Other Eco-Myths: How the Environmental Movement Uses False Science to Scare Us to Death*, (Prima, 2002). Some caution is certainly warranted, for the complex and variable manner by which different compounds with estrogenic properties may affect organisms makes projections from animal effects to human effects risky.

Sheldon Krimsky, in "Hormone Disruptors: A Clue to Understanding the Environmental Cause of Disease," *Environment* (June 2001) summarizes the evidence that many chemicals released to the environment affect—both singly and in combination or synergistically—the endocrine systems of animals and humans and may threaten human health with cancers, reproductive anomalies, and neurological effects. He cautions that the regulatory machinery is likely to move very slowly, adding that we cannot wait for scientific certainty about the hazards before we act. See also M. Gochfeld, "Why Epidemiology of Endocrine Disruptors Warrants the Precautionary Principle," *Pure & Applied Chemistry* (December 1, 2003).

Theo Colborn, a senior scientist with the World Wildlife Fund, first drew public attention to the potential problems of environmental estrogens with the book *Our Stolen Future* (Dutton, 1996), coauthored by Dianne Dumanoski and John Peterson Myers. Colborn clearly believes that the problem is real; she finds the evidence that extensive damage is being done to wildlife by synthetic estrogenic chemicals convincing and thinks it likely that humans are experiencing similar health problems. Recent data reinforce her and Krimsky's points; see Rebecca Renner, "Human Estrogens Linked to Endocrine Disruption," *Environmental Science and Technology* (January 1, 1998) and Ted Schettler et al., *Generations at Risk: Reproductive Health and the Environment* (MIT Press, 1999). In 1999 the National Research Council published *Hormonally Active Agents in the Environment* (National Academy Press), in which the council's Committee on Hormonally Active Agents in the Environment reports on its evaluation of the scientific evidence pertaining to endocrine disruptors. The

National Environmental Health Association has called for more research and product testing; see Ginger L. Gist, "National Environmental Health Association Position on Endocrine Disrupters," *Journal of Environmental Health* (January–February 1998). And in 2002 legislation was introduced in Congress to provide funds for expanded research efforts; see Neil Franz, "Bill Floated to Boost Endocrine Funding," *Chemical Week* (May 22, 2002).

Elisabete Silva, Nissanka Rajapakse, and Andreas Kortenkamp, in "Something From 'Nothing'—Eight Weak Estrogenic Chemicals Combined at Concentrations Below NOECs Produce Significant Mixture Effects," *Environmental Science and Technology* (April 2002), find synergistic effects of exactly the kind dismissed by Gough. Also, after reviewing the evidence, the U.S. National Toxicology Program found that low-dose effects had been demonstrated in animals (see Ronald Melnick et al., "Summary of the National Toxicology Program's Report of the Endocrine Disruptors Low-Dose Peer Review," *Environmental Health Perspectives* [April 2002]). The evidence seems to favor the view that environmental hormone mimics have potentially serious effects.

ISSUE 18

Is the Superfund Program Successfully Protecting the Environment From Hazardous Wastes?

YES: Robert H. Harris, Jay Vandeven, and Mike Tilchin, from "Superfund Matures Gracefully," *Issues in Science & Technology* (Summer 2003)

NO: Margot Roosevelt, from "The Tragedy of Tar Creek," *Time* (April 26, 2004)

ISSUE SUMMARY

YES: Environmental consultants Robert H. Harris, Jay Vandeven, and Mike Tilchin argue that although the Superfund program still has room for improvement, it has made great progress in risk assessment and treatment technologies.

NO: Journalist Margot Roosevelt argues that because one-quarter of Americans live near Superfund sites, and sites like Tar Creek, Oklahoma, remain hazardous, Superfund's work is clearly not getting done.

The potentially disastrous consequences of improper hazardous waste disposal burst upon the consciousness of the American public in the late 1970s. The problem was dramatized by the evacuation of dozens of residents of Niagara Falls, New York, whose health was being threatened by chemicals leaking from the abandoned Love Canal, which was used for many years as an industrial waste dump. Awakened to the dangers posed by chemical dumping, numerous communities bordering on industrial manufacturing areas across the country began to discover and report local sites where chemicals had been disposed of in open lagoons or were leaking from disintegrating steel drums. Such esoteric chemical names as dioxins and PCBs have become part of the common lexicon, and numerous local citizens' groups have been mobilized to prevent human exposure to these and other toxins.

The expansion of the industrial use of synthetic chemicals following World War II resulted in the need to dispose of vast quantities of wastes laden with organic and inorganic chemical toxins. For the most part, industry adopted a casual attitude toward this problem and, in the absence of regulatory restraint, chose the least expensive means available. Little attention was paid to the ultimate fate of chemicals that could seep into surface water or groundwater. Scientists have estimated that less than 10 percent of the waste was disposed of in an environmentally sound manner.

The magnitude of the problem is truly mind-boggling: Over 275 million tons of hazardous waste is produced in the United States each year; as many as 10,000 dump sites may pose a serious threat to public health, according to the federal Office of Technology Assessment; and other government estimates indicate that more than 350,000 waste sites may ultimately require corrective action at a cost that could easily exceed $500 billion.

Congressional response to the hazardous waste threat is embodied in two complex legislative initiatives. The Resource Conservation and Recovery Act (RCRA) of 1976 mandated action by the Environmental Protection Agency (EPA) to create "cradle to grave" oversight of newly generated waste, and the Comprehensive Environmental Response, Compensation, and Liability Act of 1980 (CERCLA), commonly called "Superfund," gave the EPA broad authority to clean up existing hazardous waste sites. The implementation of this legislation has been severely criticized by environmental organizations, citizens' groups, and members of Congress who have accused the EPA of foot-dragging and a variety of politically motivated improprieties. Less than 20 percent of the original $1.6 billion Superfund allocation was actually spent on waste cleanup.

Amendments designed to close RCRA loopholes were enacted in 1984, and the Superfund Amendments and Reauthorization Act (SARA) added $8.6 billion to a strengthened cleanup effort in 1986 and an additional $5.1 billion in 1990. While acknowledging some improvement, both environmental and industrial policy analysts remain very critical about the way that both RCRA and Superfund/SARA are being implemented. Efforts to reauthorize and modify both of these hazardous waste laws have been stalled in Congress since the early 1990s. But the work went on. The Superfund program continued to identify hazardous waste sites that warranted cleanup and to clean up sites. In 2004, it even declared that the infamous Love Canal site was finally safe.

In the following selections, Robert H. Harris, Jay Vandeven, and Mike Tilchin argue that the Superfund program has had to struggle with unexpectedly large cleanup tasks such as mines, harbors, and the ruins of the World Trade Center in New York; is subject to greater resource demands than ever before; and still has room for improvement. But, they say, the program has made great progress in risk assessment and treatment technologies. Margot Roosevelt argues that because one-quarter of Americans live near Superfund sites, and sites like Tar Creek remain hazardous, Superfund's work is not getting done.

Robert H. Harris, Jay Vandeven, and Mike Tilchin

 **YES**

Superfund Matures Gracefully

Superfund, one of the main programs used by the Environmental Protection Agency (EPA) to clean up serious, often abandoned, hazardous waste sites, has been improved considerably in recent years. Notably, progress has been made in two important areas: the development of risk assessments that are scientifically valid yet flexible, and the development and implementation of better treatment technologies.

The 1986 Superfund Amendments and Reauthorization Act (SARA) provided a broad refocus to the program. The act included an explicit preference for the selection of remediation technologies that "permanently and significantly reduce the volume, toxicity, or mobility of hazardous substances." SARA also required the revision of the National Contingency Plan (NCP) that sets out EPA's rules and guidance for site characterization, risk assessment, and remedy selection.

The NCP specifies the levels of risk to human health that are allowable at Superfund sites. However, "potentially responsible parties"—companies or other entities that may be forced to help pay for the cleanup—have often challenged the risk assessment methods used as scientifically flawed, resulting in remedies that are unnecessary and too costly. Since SARA was enacted, fundamental changes have evolved in the policies and science that EPA embraces in evaluating health risks at Superfund sites, and these changes have in turn affected which remedies are most often selected. Among the changes are three that collectively can have a profound impact on the selected remedy and attendant costs: EPA's development of land use guidance, its development of guidance on "principal threats," and the NCP requirement for the evaluation of "short-term effectiveness."

Before EPA's issuance in 1995 of land use guidance for evaluating the potential future public health risks at Superfund sites, its risk assessments usually would assume a future residential use scenario at a site, however unrealistic that assumption might be. This scenario would often result in the need for costly soil and waste removal remedies necessary to protect against hypothetical risks, such as those to children playing in contaminated soil or drinking contaminated ground water, even at sites where future residential use was highly improbable. The revised land use guidance provided a basis for selecting more realistic future use scenarios, with projected exposure patterns that may allow for less costly remedies.

Potentially responsible parties also complained that there was little room to tailor remedies to the magnitude of cancer risk at a site, and that the same costly remedies would be chosen for sites where the cancer risks may differ by several orders of magnitude. However, EPA's guidance on principal threats essentially established a risk-based hierarchy for remedy selection. For example, if cancer risks at a site exceed 1 in 1,000, then treatment or waste removal or both might be required. Sites that posed a lower lifetime cancer risk could be managed in other ways, such as by prohibiting the installation of drinking water wells, which likely would be far less expensive than intrusive remedies.

Revisions to the NCP in 1990 not only codified provisions required by the 1986 Superfund amendments, but also refined EPA's evolving remedy-selection criteria. For example, these revisions require an explicit consideration of the short-term effectiveness of a remedy, including the health and safety risks to the public and to workers associated with remedy implementation. EPA had learned by bitter experience that to ignore implementation risks, such as those associated with vapor and dust emissions during the excavation of wastes, could lead to the selection of remedies that proved costly and created unacceptable risks.

Although these changes in risk assessment procedures have brought greater rationality to the evaluation of Superfund sites, EPA still usually insists on the use of hypothetical exposure factors (for example, the length of time that someone may come in contact with the site) that may overstate risks. The agency has been slow in embracing other methodologies, such as probabilistic exposure analysis, that might offer more accurate assessments. Thus, some remedies are still fashioned on risk analyses that overstate risk.

Technological Evolution

Cleanup efforts in Superfund's early years were dominated by containment and excavation-and-disposal remedies. But over the years, cooperative work by government, industry, and academia have led to the development and implementation of improved treatment technologies.

The period from the mid-1980s to the early 1990s was marked by a dramatic increase in the use of source control treatment, reflecting the preference expressed in SARA for "permanent solutions and alternative treatment technologies or resource recovery technologies to the maximum extent practicable." Two types of source control technologies that have been widely used are incineration and soil vapor extraction. Although the use of incineration decreased during the 1990s because of cost and other factors, soil vapor extraction remains a proven technology at Superfund sites.

Just as early source control remedies relied on containment or excavation and disposal offsite, the presumptive remedy for groundwater contamination has historically been "pump and treat." It became widely recognized in the early 1990s that conventional pump-and-treat technologies had significant limitations, including relatively high costs. What emerged to fill the gap was an approach called "monitored natural attenuation" (MNA), which makes use of a variety of technologies, such as biodegradation, dispersion, dilution, absorption, and volatilization. As the name suggests, monitoring the effectiveness of the process is a key element of this technology. And although cleanup times still may be on the order of years, there is evidence that MNA can achieve comparable results in comparable periods and at significantly lower costs than conventional pump-and-treat systems. EPA has taken an active role in promoting this technology, and its use has increased dramatically in recent years.

As suggested by the MNA example, what may prove an even more formidable challenge than selecting specific remedies is the post-remedy implementation phase—that is, the monitoring and evaluation that will be required during coming decades to ensure that the remedy chosen is continuing to protect human health and the environment. Far too few resources have been devoted to this task, which will require not only monitoring and maintaining the physical integrity of the technology used and ensuring the continued viability of institutional controls, but also evaluating and responding to the developing science regarding chemical detection and toxicity.

Coming Challenges

In recent years, the rate at which waste sites are being added to the National Priorities List (NPL) has been decreasing dramatically as compared with earlier years. In fiscal years 1983 to 1991, EPA placed an average of 135 sites on the NPL annually. The rate dropped to an average of 27 sites per year between 1992 and 2001. Although many factors have contributed to this trend, three stand out:

- There was a finite group of truly troublesome sites before Superfund's passage, and after a few years most of those were identified.
- The program's enforcement authority has had a profound impact on how wastes are managed, significantly reducing, although not eliminating, the types of waste management practices that result in the creation of Superfund sites.
- A range of alternative cleanup programs, such as voluntary cleanup programs and those for brownfields, have evolved at both the federal and state levels. No longer is Superfund the only path for cleaning up sites.

But such programmatic changes are about more than just site numbers. In 1988, most NPL sites were in the investigation stage, and the program was widely criticized as being too much about studies and not enough about cleanup. Superfund is now a program predominantly focused on the design and construction of cleanup remedies.

This shift reflects the natural progress of sites through the Superfund pipeline, the changes in NPL listing activity, and a deliberate emphasis on achieving "construction completion," which is the primary measure of achievement for the program as established under the Government Performance and Results Act. It is a truism in regulatory matters that what gets done is what gets measured, and Superfund is no exception.

In the late 1990s, many observers believed that the demands on Superfund were declining and that it would be completed sometime in the middle of the first decade of the new century. But this is not proving to be true. Although expenditures have not been changing dramatically over time, the resource demands on the program are greater today than ever before.

Few people would have predicted, for example, that among the biggest technical and resource challenges facing Superfund at this date would be the cleanup of hard-rock mining sites and of large volumes of sediments from contaminated waterways and ports. These sites tend to be very costly to clean up, with the driver behind these great costs weighted more toward the protection of natural resources than of human health. In mapping the future course of the program, Congress and EPA must address the question of whether Superfund is the most appropriate program for cleaning up these types of sites.

There are other uncertainties, as well. The substantial role that Superfund has played in emergency response in the aftermath of 9/11, the response to the anthrax attacks of October 2001, and the program's role in the recovery of debris from the crash of the space shuttle Columbia were all totally unforeseeable. Although many valuable lessons have been learned over the past 20 years of the program, there remain substantial opportunities for improvement as well as considerable uncertainty about the kinds of environmental problems Superfund will tackle in the coming decade.

NO

Margot Roosevelt

The Tragedy of Tar Creek

To get a better view of the situation, John Sparkman guns his flame-red truck up a massive pile of gravel. From the summit, a lifeless brown wasteland stretches to the horizon, like a scene from a science-fiction movie. Mountains of mine tailings, some as tall as 13-story buildings, others as wide as four football fields, loom over streets, homes, churches and schools. Dust, laced with lead, cadmium and other poisonous metals, blows off the man-made hills and 800 acres of dry settling ponds. "It gets in your teeth," says Sparkman, head of a local citizens' group. "It cakes in your ears and hair. It's like we've been environmentally raped."

Hyperbole? Drive through the desolate towns around Picher, Okla., and you might think differently. This is eco-assault on an epic scale. The prairie here in the northeast corner of the state is punctured with 480 open mine shafts and 30,000 drill holes. Little League fields have been built over an immense underground cavity that could collapse at any time. Acid mine waste flushes into drinking wells. When the water rises in Tar Creek, which runs through the site, a neon-orange scum oozes onto the roadside. Wild onions, a regional delicacy tossed into scrambled eggs, are saturated with cadmium—which may explain, local doctors say, why three different kidney dialysis centers have opened here to serve a population of only 30,000.

But the grimmest legacy of a century of intensive lead and zinc mining are the "lead heads," or "chat rats," as the kids who grew up around here are known. As toddlers, they played in sandboxes of chat—the powdery output of mills after ore is extracted from rock. As preteens, they rode their bikes across the gravel mounds and swam in lime-green sinkholes. Their parents used mine tailings to make driveways and foundations, never thinking that contaminated dust might blow through the heating ducts of their ranch houses. In the past decade, studies have shown that up to 38% of local children have had high levels of lead in their blood—an exposure that can cause permanent neurological damage and learning disabilities. "Our kids hit a brick wall," says Kim Pace, principal of the Picher-Cardin Elementary School. "Their eyes skip and jump. It takes them 100 repetitions to learn a sound."

At her kitchen table, Evona Moss helps her son Michael, 10, with his homework. Michael grew up across the street from a chat pile, and at one point the third-grader's lead levels measured 40% above the Centers for

From *Time Magazine*, April 26, 2004, pages 42, 44, and 47. Copyright © 2004 by Time Inc. Reprinted by permission.

Disease Control's danger level. He repeated kindergarten. "I used to think he was lazy," says his mother, "but he tries so hard. One minute he knows the words, and a half-hour later he doesn't. Every night he kneels down and prays to be a better reader."

It wasn't supposed to be like this. In 1980, Congress passed the Comprehensive Environmental Response, Compensation and Liability Act—commonly known as the Superfund law—one of the boldest environmental statutes in U.S. history. It was a law designed to fit all circumstances. It covered existing plants whose owners could be forced to clean up their dumps. It covered polluted sites long since abandoned by their owners: defunct factories, refineries and mines. Even when companies followed the standard, if dubious, practices of the day—dumping toxic waste in rivers, burying it in leaky drums or just leaving it, as in Oklahoma, to blow in the wind—they would be held accountable. And if they refused to clean up their messes, the Environmental Protection Agency (EPA) would do so for them and charge treble damages for its trouble. In the event that the perpetrators had disappeared or gone out of business, a general tax on polluting industries—a "Superfund"— would pay to fix the damage.

But today Superfund is a program under siege, plagued by partisan politics, industry stonewalling and bureaucratic inertia. The U.S. government has spent $27 billion on the effort and forced individual polluters to spend an additional $21 billion. Love Canal, the deadly dump in New York State that spurred the law's passage, has been capped with a layer of clay, and the EPA proposed last month to take it off the list. So far, 278 sites have been delisted. But there are thousands more out there. According to the General Accounting Office (GAO), 1 out of 4 Americans still lives within four miles of a Superfund site—many of them killing fields saturated with cancer-causing chemicals and other toxins.

The GAO reports that the program's budget fell 35% in inflation-adjusted dollars over the past decade. And environmentalists say that Bush appointees are slowing the pace of cleanups and failing to list potential new sites. According to the EPA's inspector general, 29 projects in 17 states were underfunded last year. The Administration, charges New Jersey Senator Frank Lautenberg, a Democrat, has "allowed—deliberately—these sites to rot where they are."

Tar Creek is a case in point. Two decades after it was targeted on the very first Superfund priority list, the 40-sq.-mi. site is worse off than ever. Early on, the government confined its effort to the polluted creek, without looking at chat piles, soil, air quality or the danger of subsidence. Was it a lack of knowledge of the danger, as EPA claims? Or industry influence, as environmentalists charge? Whatever the reason, federal attorneys settled with mining companies for pennies on the dollar. Now, after fruitless efforts to contain 28 billion gal. of acid mine water, contamination is spreading across a vast watershed. And although the EPA trucked out toxic dirt from about 2,000 homes and schools, Tar Creek's children still show elevated lead levels at six times the national average.

Administration officials say they are cleaning up the nation's 1,240 highest-priority sites as fast as they can. But that will be harder, since the multibillion-dollar industry-paid trust fund, set aside for abandoned sites such as Tar Creek, ran dry in October [2003]. The fund was supplied by taxes on the purchase of toxic chemicals and petroleum and on corporate profits above $2 million. But the Republican-led Congress allowed the fees to expire in 1995. Bush is the first President to oppose the levies, and last month Lautenberg and other Senate Democrats lost a narrow vote to reinstate them. In protest, the Sierra Club aired "Make Polluters Pay" TV ads in Pennsylvania, Florida and Michigan—all swing states. And on April 15, tax day, activists in 25 states picketed post offices to object. "We went from polluters paying to citizens paying," says Oklahoma environmentalist Earl Hatley. "Now EPA doesn't have the money for megasites like Tar Creek."

Meanwhile, Superfund defenders in Washington are bracing for a new battle: a Bush-appointed advisory committee, which they claim is heavily stacked with corporate members, issued a report [recently] that pushes for administrative changes. "It is a wonky thing," says Julie Wolk of the Public Interest Research Group. "But it could dramatically weaken the program." Companies want to limit liability and shift responsibility to the states, where rules are more flexible. Federal standards are "rigid and extreme," says Michael Steinberg of the Superfund Settlements Project, an industry group that includes General Electric, DuPont and IBM. "Groundwater must meet standards for tap water, even though at many of these sites no one drinks it. Soil at many sites must be clean enough so people could play in it. The costs exceed the benefits."

With the EPA's clout slackening, private attorneys are moving in. At Tar Creek, lawyers are suing seven mining companies on behalf of scores of lead-exposed children. A separate suit demanding a cleanup was filed by the Quapaw Indians, whose land was leased for the mines. And environmentalist Robert F. Kennedy Jr. has joined a class action to force companies to relocate the population of two polluted towns, Picher and Cardin. Court papers suggest that mining executives knew as early as the 1930s that the contaminated dust was dangerous but sought to, in their words, "dissuade" the government from intervening. A mining-company lawyer says the charge is based on "out-of-context reading" of historical documents.

Just how dangerous that dust might be is still a matter of dispute. Doctors at the Harvard School of Public Health have begun extensive studies in Tar Creek, not just of lead exposure but also of the cocktail mix of lead, manganese, cadmium and other metals that interact in unknown ways. "We're looking at four generations of poisoning," says Rebecca Jim of the L.E.A.D. agency, a local group. Meanwhile, parents like Evona Moss wonder what else the toxic brew might have done. Did it cause her obesity and bad teeth? Is it responsible for the malformation of her daughter's shins? Does her baby's asthma come from the chat? Her nephew's cancer? No one knows because no one has done careful, long-term studies.

Tar Creek is an extreme case. But like Tolstoy's unhappy families, every Superfund site is tragic and contentious in its own way. In Libby, Mont., a massive mine blanketed the town with asbestos dust, killing at least 215 people and sickening 1,100 more with cancer and lung disease—yet cleanup funds have been cut so sharply that it could take 10 to 15 years to finish the job. In Coeur d'Alene, Idaho, miners dumped 60 million tons of toxic metals into waterways, but state officials are fighting a Superfund cleanup, fearing a stigma that might hurt tourism. In New York, General Electric, which contaminated 40 miles of the Hudson River with cancer-causing PCBs, has hired high-profile attorney Laurence Tribe to convince federal courts that the Superfund law is unconstitutional. And in New Jersey, where the rabbits frolicking around the Chemical Insecticide Corp. plant once grew green-tinged fur, cleanup funds were restored only after locals sent green plush bunnies to members of Congress.

At Tar Creek, many residents have given up hope. Even the EPA, which has spent $107 million at the site, isn't sure if it can ever be repaired. "We don't have an off-the-shelf remedy," says EPA Superfund official Randy Deitz. "What do you do with the enormous chat piles? When does cleanup become impracticable? We have limited resources." In a show of no-confidence, the Oklahoma legislature last week passed a $5 million buyout for all families with children under 6. John Sparkman, who heads the Tar Creek Steering Committee, a group of buyout supporters, veers between cynicism and despair. "They think we're poor white trash," he says bitterly, driving past Picher's boarded-up storefronts. "The votes here don't affect any federal election—so why bother? We've agitated till we can't agitate anymore." Meanwhile, at Tar Creek, the toxic dust keeps blowing in the wind.

POSTSCRIPT

Is the Superfund Program Successfully Protecting the Environment From Hazardous Wastes?

Superfund cleanups, when those responsible for contaminated sites could not pay or could not be found, were to have been funded by taxes on industry (e.g., the Crude Oil Tax, the Chemical Feedstock Tax, the Toxic Chemicals Importation Tax, and the Corporate Environmental Income Tax). These taxes expired in 1995, and Congress has so far refused to reauthorize them; the Senate voted down the latest attempt in March 2004. The program exhausted its funds in September 2003 and is now running on government revenues. See Kara Sissell, "Senate Votes Down Bill to Reinstate Superfund Tax," *Chemical Week* (March 17, 2004) and "Superfund Program: Current Status and Future Fiscal Challenges," Report to the Chairman, Subcommittee on Oversight of Government Management, the Federal Workforce, and the District of Columbia, Committee on Governmental Affairs, U.S. Senate (GAO, July 2003). Activist groups, such as the Public Interest Research Group (PIRG), have issued calls for the Bush administration to reinstate funding without delay. But, says PIRG, "The Bush administration opposes reauthorization of the polluter pays taxes, supports a steep increase in the amount paid by taxpayers, and has dramatically slowed down the pace of cleanups at the nation's worst toxic waste sites." The Competitive Enterprise Institute objects that such taxes are an assault on consumer pocketbooks, as is the Comprehensive Environmental Response, Compensation, and Liability Act's (CERCLA's) "joint and several liability" clause, which can make minor contributors to toxics sites liable for large cleanup costs even when they acted according to all laws and regulations in force at the time.

Meanwhile, the hazardous waste problem takes new forms. Even in the 1990s, an increasingly popular method of disposing of hazardous wastes was to ship them from the United States to "dumping grounds" in developing countries. Iwonna Rummel-Bulska's "The Basel Convention: A Global Approach for the Management of Hazardous Wastes," *Environmental Policy and Law* (vol. 24, no. 1, 1994) describes an international treaty designed to prevent or at least limit such waste dumping. But eight years later, in February 2002, the Basel Action Network and the Silicon Valley Toxics Coalition published *Exporting Harm: The High-Tech Trashing of Asia*. This lengthy report documents the shipping of electronics wastes, including defunct personal computers, monitors, and televisions, as well as circuit boards and other products rich in lead, beryllium, cadmium, mercury, and other toxic materials. Some 50 to 80 percent of the "e-waste" collected for recycling in the western United States is shipped to destinations such as China, India, and Pakistan, where recycling and disposal methods lead to widespread human and environmental contamination. An updated version of this report, *Poison PCs and Toxic TVs: E-Waste Tsunami to Roll Across the US: Are We Prepared?* was released in February 2004 and available at `http://www.svtc.org/cleancc/pubs/poisonpc2004.htm`.

Among the solutions that have been urged to address the hazardous waste problem are "take-back" and "remanufacturing" practices. Gary A. Davis and Catherine A. Wilt, in "Extended Product Responsibility," *Environment* (September 1997), urge such solutions as crucial to the minimization of waste and describe how they are becoming more common in Europe. After *Exporting Harm* was published and drew considerable attention from the press, some industry representatives hastened to emphasize such practices as Hewlett Packard's recycling of printer ink cartridges. See Doug Bartholomew's "Beyond the Grave," *Industry Week* (March 1, 2002), which also stressed the need to minimize waste by intelligent design. The Institute of Industrial Engineers published in its journal *IIE Solutions* Brian K. Thorn's and Philip Rogerson's "Take It Back" (April 2002), which stressed the importance of designing for reuse or remanufacturing. Anthony Brabazon and Samuel Idowu, in "Costing the Earth" *Financial Management (CIMA)* (May 2001), note that "take-back schemes may [both] provide opportunities to build goodwill and [help] companies to use resources more efficiently."

ISSUE 19

Should the United States Continue to Focus Plans for Permanent Nuclear Waste Disposal Exclusively at Yucca Mountain?

YES: Spencer Abraham, from *Recommendation by the Secretary of Energy Regarding the Suitability of the Yucca Mountain Site for a Repository Under the Nuclear Waste Policy Act of 1982* (February 2002)

NO: Gar Smith, from "A Gift to Terrorists?" *Earth Island Journal* (Winter 2002–2003)

ISSUE SUMMARY

YES: Secretary of Energy Spencer Abraham argues that the Yucca Mountain, Nevada, nuclear waste disposal site is suitable technically and scientifically and that its development serves the U.S. national interest in numerous ways.

NO: Environmentalist writer Gar Smith argues that transporting nuclear waste to Yucca Mountain will expose millions of Americans to risks from accidents and terrorists.

Nuclear waste is generated when uranium and plutonium atoms are split to make energy in nuclear power plants, when uranium and plutonium are purified to make nuclear weapons, and when radioactive isotopes that are useful in medical diagnosis and treatment are made and used. These wastes are radioactive, meaning that as they break down they emit radiation of several kinds. Those that break down fastest are most radioactive and are said to have a short half-life (the time needed for half the material to break down). Uranium-238, the most common isotope of uranium, has a half-life of 4.5 billion years and is not very radioactive at all. Plutonium-239 (which is used in bombs) has one of 24,000 years and is radioactive enough to be quite hazardous to humans.

According to the U.S. Department of Energy (DOE), high-level waste includes spent reactor fuel (the current amount of which is 52,000 tons) and waste from weapons production (91 million gallons). Transuranic waste includes clothing, equipment, and other materials contaminated with pluto-nium and other radioactive materials (11.3 million cubic feet, some of which has been buried in the Waste Isolation Pilot Plant salt cavern in New Mex-ico). Low- and mixed-level waste includes waste from hospitals and research labs, remnants of decommissioned nuclear plants, and air filters (472 million cubic feet). The high-level waste is the most hazardous and poses the most severe disposal problems. In general, experts say, such materials must be kept away from people and other living things—with no possibility of contaminat-ing air, water (including ground water), or soil—for 10 half-lives.

Since the beginning of the nuclear age in the 1940s, nuclear waste has been accumulating. A sense of urgency about finding a place to put the waste where it would not threaten humans or ecosystems for a quarter-mil-lion years or more has also developed. The 1982 Nuclear Waste Policy Act called for locating candidate disposal sites for high-level wastes and choos-ing one by 1998. Since the people of the states containing candidate sites were unhappy about it, and since many of the investigated sites were less than ideal for various reasons, the schedule proved impossible to meet. In 1987 Congress attempted to settle the matter by designating Yucca Moun-tain, Nevada, as the one site to be intensively studied and developed. It was scheduled to be opened for use in 2010. Risk assessment expert D. Warner North, in "Unresolved Problems of Radioactive Waste: Motivation for a New Paradigm," *Physics Today* (June 1997), asserted that the technical and political problems related to nuclear waste disposal remained formidable and that a new approach was needed. Luther J. Carter and Thomas H. Pig-ford, in "Getting Yucca Mountain Right," *The Bulletin of the Atomic Scien-tists* (March/April 1998), wrote that those formidable problems could be defeated, given technical and congressional attention, and that the Yucca Mountain strategy was both sensible and realistic. However, problems have continued to plague the project; see Chuck McCutcheon, "High-Level Acri-mony in Nuclear Storage Standoff," *Congressional Quarterly Weekly Report* (September 25, 1999) and Sean Paige, "The Fight at the End of the Tunnel," *Insight on the News* (November 15, 1999). Jon Christensen, in "Nuclear Rou-lette," *Mother Jones* (September/October 2001), argues that a more basic problem is that estimates of Yucca Mountain's long-term safety are based on probabilistic computer models that are too uncertain to trust.

In February 2002 Secretary of Energy Spencer Abraham recommended to the president that the United States go ahead with development of the Yucca Mountain site. His report, which is excerpted in the following selec-tion, makes the points that a disposal site is necessary, that Yucca Mountain has been thoroughly studied, and that moving ahead with the site best serves "our energy future, our national security, our economy, our environment, and safety." Abraham further argues that objections to the site are not serious enough to stop the project. In the second selection, Gar Smith argues that in addition to any risks associated with Yucca Mountain itself, transporting large quantities of nuclear waste to that disposal site will expose millions of Americans to risks from accidents and terrorists.

Spencer Abraham  **YES**

Recommendation by the Secretary of Energy Regarding the Suitability of the Yucca Mountain Site for a Repository Under the Nuclear Waste Policy Act of 1982

Introduction

For more than half a century, since nuclear science helped us win World War II and ring in the Atomic Age, scientists have known that the Nation would need a secure, permanent facility in which to dispose of radioactive wastes. Twenty years ago, when Congress adopted the Nuclear Waste Policy Act of 1982 (NWPA or "the ACT"), it recognized the overwhelming consensus in the scientific community that the best option for such a facility would be a deep underground repository. Fifteen years ago, Congress directed the Secretary of Energy to investigate and recommend to the President whether such a repository could be located safely at Yucca Mountain, Nevada. Since then, our country has spent billions of dollars and millions of hours of research endeavoring to answer this question. I have carefully reviewed the product of this study. In my judgment, it constitutes sound science and shows that a safe repository can be sited there. I also believe that compelling national interests counsel in favor of proceeding with this project. Accordingly, consistent with my responsibilities under the NWPA, today I am recommending that Yucca Mountain be developed as the site for an underground repository for spent fuel and other radioactive wastes.

The first consideration in my decision was whether the Yucca Mountain site will safeguard the health and safety of the people, in Nevada and across the country, and will be effective in containing a minimum risk the material it is designed to hold. Substantial evidence shows that it will. Yucca Mountain is far and away the most thoroughly researched site of its kind in the world. It is a geologically stable site, in a closed groundwater basin, isolated on thousands of acres of Federal land, and farther from any metropolitan area than

From Spencer Abraham, *Recommendation by the Secretary of Energy Regarding the Suitability of the Yucca Mountain Site for a Repository Under the Nuclear Waste Policy Act of 1982* (February 2002). Notes omitted.

the great majority of less secure, temporary nuclear waste storage sites that exist in the country today.

This point bears emphasis. We are not confronting a hypothetical problem. We have a staggering amount of radioactive waste in this country—nearly 100,000,000 gallons of high-level nuclear waste and more than 40,000 metric tons of spent nuclear fuel with more created every day. Our choice is not between, on the one hand, a disposal site with costs and risks held to a minimum, and, on the other, a magic disposal system with no costs or risks at all. Instead, the real choice is between a single secure site, deep under the ground at Yucca Mountain, or making do with what we have now or some variant of it—131 aging surface sites, scattered across 39 states. Every one of those sites was built on the assumption that it would be temporary. As time goes by, every one is closer to the limit of its safe life span. And every one is at least a potential security risk—safe for today, but a question mark in decades to come.

The Yucca Mountain facility is important to achieving a number of our national goals. It will promote our energy security, our national security, and safety in our homeland. It will help strengthen our economy and help us clean up the environment.

The benefits of nuclear power are with us every day. Twenty percent of our country's electricity comes from nuclear energy. To put it another way, the "average" home operates on nuclear-generated electricity for almost five hours a day. A government with a complacent, kick-the-can-down-the-road nuclear waste disposal policy will sooner or later have to ask its citizens which five hours of electricity they would care to do without.

Regions that produce steel, automobiles, and durable goods rely in particular on nuclear power, which reduces the air pollution associated with fossil fuels—greenhouse gases, solid particulate matter, smog, and acid rain. But environmental concerns extend further. Most commercial spent fuel storage facilities are near large populations centers; in fact, more than 161 million Americans live within 75 miles of these facilities. These storage sites also tend to be near rivers, lakes, and seacoasts. Should a radioactive release occur from one of these older, less robust facilities, it could contaminate any of 20 major waterways, including the Mississippi River. Over 30 million Americans are served by these potentially at-risk water sources.

Our national security interests are likewise at stake. Forty percent of our warships, including many of the most strategic vessels in our Navy, are powered by nuclear fuel, which eventually becomes spent fuel. At the same time, the end of the Cold War has brought the welcome challenge to our Nation of disposing of surplus weapons-grade plutonium as part of the process of decommissioning our nuclear weapons. Regardless of whether this material is turned into reactor fuel or otherwise treated, an underground repository is an indispensable component in any plan for its complete disposition. An affirmative decision on Yucca Mountain is also likely to affect other nations' weapons decommissioning, since their willingness to proceed will depend on being satisfied that we are doing so. Moving forward with the repository will contribute to our global efforts to stem the proliferation of nuclear weapons in other ways, since it will encourage nations with weaker controls over their own materials to follow a similar path of permanent, underground disposal, thereby making it more difficult for these materials to fall into the wrong hands. By moving forward with Yucca Mountain, we will show leadership, set out a roadmap, and encourage other nations to follow it.

There will be those who say the problem of nuclear waste disposal generally, and Yucca Mountain in particular, needs more study. In fact, both issues have been studied for more than twice the amount of time it took to plan and complete the moon landing. My Recommendation today is consistent with the conclusion of the National Research Council of the National Academy of Sciences—a conclusion reached, not last week or last month, but 12 years ago. The Council noted "a worldwide scientific consensus that deep geological disposal, the approach being followed by the United States, is the best option for disposing of high-level radioactive waste." Likewise, a broad spectrum of experts agrees that we now have enough information, including more than 20 years of researching Yucca Mountain specifically, to support a conclusion that such a repository can be safely located there.

Nonetheless, should this site designation ultimately become effective, considerable additional study lies ahead. Before an ounce of spent fuel or radioactive waste could be sent to Yucca Mountain, indeed even before construction of the permanent facilities for emplacement of waste could begin there, the Department of Energy (DOE or "the Department") will be required to submit an application to the independent Nuclear Regulatory Commission (NRC). There, DOE would be required to make its case through a formal review process that will include public hearings and is expected to last at least three years. Only after that, if the license were granted, could construction begin. The DOE would also have to obtain an additional operating license, supported by evidence that public health and safety will be preserved, before any waste could actually be received.

In short, even if the Yucca Mountain Recommendation were accepted today, an estimated minimum of eight more years lies ahead before the site would become operational.

We have seen decades of study, and properly so for a decision of this importance, one with significant consequences for so many of our citizens. As necessary, many more years of study will be undertaken. But it is past time to stop sacrificing that which is forward-looking and prudent on the altar of a *status quo* we know ultimately will fail us. The *status quo* is not the best we can do for our energy future, our national security, our economy, our environment, and safety—and we are less safe every day as the clock runs down on dozens of older, temporary sites.

I recommend the deep underground site at Yucca Mountain, Nevada, for development as our Nation's first permanent facility for disposing of high-level nuclear waste.

Background

History of the Yucca Mountain Project and the Nuclear Waste Policy Act

The need for a secure facility in which to dispose of radioactive wastes has been known in this country at least since World War II. As early as 1957, a National Academy of Sciences report to the Atomic Energy Commission suggested burying radioactive waste in geologic formations. Beginning in the 1970s, the United States and other countries evaluated many options for the safe and permanent disposal of radioactive waste, including deep seabed disposal, remote island siting, dry cask storage, disposal in the polar ice sheets, transmutation, and rocketing waste into orbit around the sun. After analyzing these options, disposal in a mined geologic repository emerged as the preferred long-term environmental solution for the management of these wastes. Congress recognized this consensus 20 years ago when it passed the Nuclear Waste Policy Act of 1982.

In the Act, Congress created a Federal obligation to accept civilian spent nuclear fuel and dispose of it in a geologic facility. Congress also designated the agencies responsible for implementing this policy and specified their roles. The Department of Energy must characterize, site, design, build, and manage a Federal waste repository. The Environmental Protection Agency (EPA) must set the public health standards for it. The Nuclear Regulatory Commission must license its construction, operation, and closure.

The Department of Energy began studying Yucca Mountain almost a quarter century ago. Even before Congress adopted the NWPA, the Department had begun national site screening research as part of the National Waste Terminal Storage program, which included examination of Federal sites that had previously been used for defense-related activities and were already potentially contaminated. Yucca Mountain was one such location, on and adjacent to the Nevada Test Site, which was then under construction. Work began on the Yucca Mountain site in 1978. When the NWPA was passed, the Department was studying more than 25 sites around the country as potential repositories. The Act provided for the siting and development of two; Yucca Mountain was one of nine sites under consideration for the first repository program.

Following the provisions of the Act and the Department's siting Guidelines, the Department prepared draft environmental assessments for the nine sites. Final environmental assessments were prepared for five of these, including Yucca Mountain. In 1986, the Department compared and ranked the sites under construction for characterization. It did this by using a multi-attribute methodology—an accepted, formal scientific method used to help decision makers compare, on an equivalent basis, the many components that make up a complex decision. When all the components of the ranking decision were considered together, taking account of both preclosure and post-closure concerns, Yucca Mountain was the top-ranked site. The Department examined a variety of ways of combining the components of the ranking scheme; this only confirmed the conclusion that Yucca Mountain came out in first place. The EPA also looked at the performance of a repository in unsaturated tuff. The EPA noted that in its modeling in support of development of the standards, unsaturated tuff was one of the two geologic media that appeared most capable of limiting releases of radionuclides in a manner that keeps expected doses to individuals low.

In 1986, Secretary of Energy Herrington found three sites to be suitable for site characterization, and recommended the three, including Yucca Mountain, to President Reagan for detailed site characterization. The Secretary also made a preliminary finding, based on Guidelines that did not require site characterization, that the three sites were suitable for development as repositories.

The next year, Congress amended the NWPA, and selected Yucca Mountain as the single site to be characterized. It simultaneously directed the Department to cease activities at all other potential sites. Although it has been suggested that Congress's decision was made for purely political reasons, the record described above reveals that the Yucca Mountain site consistently ranked at or near the top of the sites evaluated well before Congress's action.

As previously noted, the National Research Council of the National Academy of Sciences concluded in 1990 (and reiterated [recently]) that there is "a worldwide scientific consensus that deep geological disposal, the approach being followed by the United States, is the best option for disposing of high-level radioactive waste." Today, many national and international scientific experts and nuclear waste management professionals agree with DOE that there exists sufficient information to support a national decision on designation of the Yucca Mountain site.

The Nuclear Waste Policy Act and the Responsibilities of the Department of Energy and the Secretary

Congress assigned to the Secretary of Energy the primary responsibility for implementing the national policy of developing a deep underground repository. The Secretary must determine whether to initiate the next step laid out in the NWPA—a recommendation to designate Yucca Mountain as the site for development as a permanent disposal facility.... Briefly, I first must determine whether Yucca Mountain is in fact technically and scientifically suitable to be a repository. A favorable suitability determination is indispensable for a positive recommendation of the site to the President. Under additional criteria I have adopted above and beyond the statutory requirements, I have also sought to determine whether, when other relevant considerations are taken into account, recommending it is in the overall national interest and, if so, whether there are countervailing arguments so strong that I should nonetheless decline to make the Recommendation.

The Act contemplates several important stages in evaluating the site before a Secretarial recommendation is in order. It directs the Secretary to develop a site characterization plan, one that will help guide test programs for the collection of data to be used in evaluating the site. It directs the Secretary to conduct such characterization studies as may be necessary to evaluate the site's suitability. And it directs the Secretary to hold hearings in the vicinity of the prospective site to inform the residents and receive their comments. It is at the completion of these stages that the Act directs the Secretary, if he finds the site suitable, to determine whether to recommend it to the President for development as a permanent repository.

If the Secretary recommends to the President that Yucca Mountain be developed, he must include with the Recommendation, and make available to the public, a comprehensive statement of the basis for his determination. If at any time the Secretary determines that Yucca Mountain is not a suitable site, he must report to Congress within six months his recommendations for further action to assure safe, permanent disposal of spent nuclear fuel and high-level radioactive waste.

Following a Recommendation by the Secretary, the President may recommend the Yucca Mountain site to Congress "if ... [he] considers [it] qualified for application for a construction authorization...." If the President submits a recommendation to Congress, he must also submit a copy of the statement setting forth the basis for the Secretary's Recommendation.

A Presidential recommendation takes effect 60 days after submission unless Nevada forwards a notice of disapproval to the Congress. If Nevada submits such a notice, Congress has a limited time during which it may nevertheless give effect to the President's recommendation by passing, under expedited procedures, a joint resolution of siting approval. If the President's recommendation takes effect, the Act directs the Secretary to submit to the NRC a construction license application.

The NWPA by its terms contemplated that the entire process of siting, licensing, and constructing a repository would have been completed more than four years ago, by January 31, 1998. Accordingly, it required the Department to enter into contracts to begin accepting waste for disposal by that date.

Decision

The Recommendation

After over 20 years of research and billions of dollars of carefully planned and reviewed scientific field work, the Department has found that a repository at Yucca Mountain brings together the location, natural barriers, and design elements most likely to protect the health and safety of the public, including those Americans living in the immediate vicinity, now and long into the future. It is therefore suitable, within the meaning of the NWPA, for development as a permanent nuclear waste and spent fuel repository.

After reviewing the extensive, indeed unprecedented, analysis the Department has undertaken, and in discharging the responsibilities made incumbent on the Secretary under the Act, I am recommending to the President that Yucca Mountain be developed as the Nation's first permanent, deep underground repository for high-level radioactive waste. A decision to develop Yucca Mountain will be a critical step forward in addressing our Nation's energy future, our national defense, our safety at home, and protection for our economy and environment.

What This Recommendation Means,
and What It Does Not Mean

Even after so many years of research, this Recommendation is a preliminary step. It does no more than start the formal safety evaluation process. Before a license is granted, much less before repository construction or waste emplacement may begin, many steps and many years still lie ahead. The DOE must submit an application for a construction license; defend it through formal review, including public hearings; and receive authorization from the NRC, which has the statutory responsibility to ensure that any repository built at Yucca Mountain meets stringent tests of health and safety. The NRC licensing process is expected to take a minimum of three years. Opposing viewpoints will have every opportunity to be heard. If the NRC grants this first license, it will only authorize initial construction. The DOE would have then have to seek and obtain a second operating license from the NRC before any wastes could be received. The process altogether is expected to take a minimum of eight years.

The DOE would also be subject to NRC oversight as a condition of the operating license. Construction, licensing, and operation of the repository would also be subject to ongoing Congressional oversight.

At some future point, the repository is expected to close. EPA and NRC regulations require monitoring after the DOE receives a license amendment authorizing the closure, which would be from 50 to about 300 years after waste emplacement begins, or possibly longer. The repository would also be designed, however, to be able to adapt to methods future generations might develop to manage high-level radioactive waste. Thus, even after completion of waste emplacement, the waste could be retrieved to take advantage of its economic value or usefulness to as yet undeveloped technologies.

Permanently closing the repository would require sealing all shafts, ramps, exploratory boreholes, and other underground openings connected to the surface. Such sealing would discourage human intrusion and prevent water from entering through these openings. DOE's site stewardship would include maintaining control of the area, monitoring and testing, and implementing security measures against vandalism and theft. In addition, a network of permanent monuments and markers would be erected around the site to alert future generations to the presence and nature of the buried waste. Detailed public records held in multiple places would identify the location and layout of the repository and the nature and potential hazard of the waste it contains. The Federal Government would maintain control of the site for the indefinite future. Active security systems would prevent deliberate or inadvertent human intrusion and any other human activity that could adversely affect the performance of the repository....

Nuclear Science and the National Interest

Our country depends in many ways on the benefits of nuclear science: in the generation of twenty percent of the Nation's electricity; in the operation of many of the Navy's most strategic vessels; in the maintenance of the Nation's nuclear weapons arsenal; and in numerous research and development projects, both medical and scientific. All these activities produce radioactive wastes that have been accumulating since the mid-1940s. They are currently scattered among 131 sites in 39 states, residing in temporary surface storage facilities and awaiting final disposal. In exchange for the many benefits of nuclear power, we assume the cost of managing its byproducts in a responsible, safe, and secure fashion. And there is a near-universal consensus that a deep geologic facility is the only scientifically credible, long-term solution to a problem that will only grow more difficult the longer it is ignored.

Energy Security

Roughly 20 percent of our country's electricity is generated from nuclear power. This means that, on average, each home, farm, factory, and business in America runs on nuclear fuel for a little less than five hours a day.

A balanced energy policy—one that makes use of multiple sources of energy, rather than becoming dependent entirely on generating electricity from a single source, such as natural gas—is important to economic growth. Our vulnerability to shortages and price spikes rises in direct proportion to our failure to maintain diverse sources of power. To assure that we will continue to have reliable and affordable sources of energy, we need to preserve our access to nuclear power.

Yet the Federal government's failure to meet its obligation to dispose of spent nuclear fuel under the NWPA—as it has been supposed to do starting in 1998—is placing our access to this source of energy in jeopardy. Nuclear power plants have been storing their spent fuel on site, but many are running out of space to do so. Unless a better solution is found, a growing number of these plants will not be able to find additional storage space and will be forced to shut down prematurely. Nor are we likely to see any new plants built.

Already we are facing a growing imbalance between our projected energy needs and our projected supplies. The loss of existing electric generating capacity that we will experience if nuclear plants start going off-line would significantly exacerbate this problem, leading to price spikes and increased electricity rates as relatively cheap power is taken off the market. A permanent repository for spent nuclear fuel is essential to our continuing to count on nuclear energy to help us meet our energy demands.

National Security

Powering the Navy Nuclear Fleet

A strong Navy is a vital part of national security. Many of the most strategically important vessels in our fleet, including submarines and aircraft carriers, are nuclear powered. They have played a major role in every significant military action in which the United States has been involved for some 40 years, including our current operations in Afghanistan. They are also essential to our nuclear deterrent. In short, our nuclear-powered Navy is indispensable to our status as a world power.

For the nuclear Navy to function, nuclear ships must be refueled periodically and the spent fuel removed. The spent fuel must go someplace. Currently, as part of a consent decree entered into between the State of Idaho and the Federal Government, this material goes to temporary surface storage facilities at the Idaho National Environmental and Engineering Laboratory. But this cannot continue indefinitely, and indeed the agreement specifies that the spent fuel must be removed. Failure to establish a permanent disposition pathway is not only irresponsible, but could also create serious future uncertainties potentially affecting the continued capability of our Naval operations.

Allowing the Nation to Decommission Its Surplus Nuclear Weapons and Support Nuclear Non-Proliferation Efforts

A decision now on the Yucca Mountain repository is also important in several ways to our efforts to prevent the proliferation of nuclear weapons. First, the end of the Cold War has brought the welcome challenge to our country of disposing of surplus weapons-grade plutonium as part of the process of decommissioning weapons we no longer need. Current plans call for turning the plutonium into "mixed-oxide" or "MOX" fuel. But creating MOX fuel as well as burning the fuel in a nuclear reactor will generate spent nuclear fuel, and other byproducts which themselves will require somewhere to go. A geological repository is critical to completing disposal of these materials. Such complete disposal is important if we are to expect other nations to decommission their own weapons, which they are unlikely to do unless persuaded that we are truly decommissioning our own.

A respository is important to non-proliferation for other reasons as well. Unauthorized removal of nuclear materials from a repository will be difficult even in the absence of strong institutional controls. Therefore, in countries that lack such controls, and even in our own, a safe repository is essential in preventing these materials from falling into the hands of rogue nations. By permanently disposing of nuclear weapons materials in a facility of this kind, the United States would encourage other nations to do the same.

Protecting the Environment

An underground repository at Yucca Mountain is important to our efforts to protect our environment and achieve sustainable growth in two ways. First, it will allow us to dispose of the radioactive waste that has been building up in our country for over fifty years in a safe and environmentally sound manner. Second, it will facilitate continued use and potential expansion of nuclear power, one of the few sources of electricity currently available to us that emits no carbon dioxide or other greenhouse gases.

As to the first point: While the Federal government has long promised that it would assume responsibility for nuclear waste, it has yet to start implementing an environmentally sound approach for disposing of this material. It is past time for us to do so. The production of nuclear weapons at the end of the Second World War and for many years thereafter has resulted in a legacy of high-level radioactive waste and spent fuel, currently located in Tennessee, Colorado, South Carolina, New Mexico, New York, Washington, and Idaho. Among these wastes, approximately 100,000,000 gallons of high-level liquid waste are stored in, and in some instances have leaked from, temporary holding tanks. In addition to this high-level radioactive waste, about 2,100 metric tons of solid, unreprocessed fuel from a plutonium-production reactor are stored at the Hanford Nuclear Reservation, with another 400 metric tons stored at other DOE sites.

In addition, under the NWPA, the Federal government is also responsible for disposing of spent commercial fuel, a program that was to have begun in 1998, four years ago. More than 161 million Americans, well more than half the population, reside within 75 miles of a major nuclear facility—and, thus, within 75 miles of that facility's aging and temporary capacity for storing this material. Moreover, because nuclear reactors require abundant water for cooling, on-site storage tends to be located near rivers, lakes, and seacoasts. Ten *closed* facilities, such as Big Rock Point, on the banks of Lake Michigan, also house spent fuel and incur significant annual costs without providing any ongoing benefit. Over the long-term, without active management and monitoring, degrading surface storage facilities may pose a risk to any of 20 major U.S. lakes and waterways, including the Mississippi River. Millions of Americans are served by municipal water systems with intakes along these waterways. In recent letters, Governors Bob Taft of Ohio and John Engler of Michigan raised concerns about the advisability of long-term storage of spent fuel in temporary systems so close to major bodies of water. The scientific consensus is that disposal of this material in a deep underground repository is not merely the safe answer and the right answer for protecting our environment but the *only* answer that has any degree of realism.

In addition, nuclear power is one of only a few sources of power available to us now in a potentially plentiful and economical manner that could drastically reduce air pollution and greenhouse gas emissions caused by the generation of electricity. It produces no controlled air pollutants, such as sulfur and particulates, or greenhouse gases. Therefore, it can help keep our air clean, avoid generation of ground-level ozone, and prevent acid rain. A repository of Yucca Mountain is indispensable to the maintenance and potential expansion of the use of this environmentally efficient source of energy....

Summary

In short, there are important reasons to move forward with a repository at Yucca Mountain. Doing so will advance our energy security by helping us to maintain diverse sources of energy supply. It will advance our national security by helping to provide operational certainty to our nuclear Navy and by facilitating the decommissioning of nuclear weapons and the secure disposition of nuclear materials. It will help us clean up our environment by allowing us to close the nuclear fuel cycle and giving us greater access to a form of energy that does not emit greenhouse gases. And it will help us in our efforts to secure ourselves against terrorist threats by allowing us to remove nuclear materials from scattered above-ground locations to a single, secure underground facility. Given the site's scientific and technical suitability, I find that compelling national interests counsel in favor of taking the next step toward siting a repository at Yucca Mountain.

NO

Gar Smith

A Gift to Terrorists?

America's atomic powerplants are burdened with growing stockpiles of spent fuel-rods and other radioactive wastes. "Temporary" fuel storage ponds at most reactors were filled long ago and, as aging reactors face the end of their operating (and revenue-generating) lives, the atomic power industry is running short of space, time and patience.

After years of opposition by antinuclear activists, environmentalists and the governors of all the affected states, the [George W.] Bush administration is prepared to start shipping 70,000 tons of radioactive wastes from nearly 100 nuclear powerplants nationwide to an "interim" storage site at Yucca Mountain, Nevada.

When the nuclear power business first got its start in the 1960s, the Department of Energy (DOE) promised to assume final responsibility for each and every spent nuclear fuel rod. The DOE was supposed to start picking up and parking Big Nuke's hot rods on January 31, 1998. It didn't happen.

Back in the 1960s, nuclear power advocates believed that they could generate electricity "too cheap to meter." The hope was that, by the time the powerplants needed to be shut down, future scientists would have discovered how to store radioactive waste safely for the next 24,000 years.

Forty years later, science still hasn't solved the problem.

With storage pools brim-full, US facilities have been forced to start packing used fuel rods above-ground in "dry cask" storage. The operators of the Maine Yankee nukeplant recently invested $60 million to build a new fuel-rod storage facility. These surface "parking lots" will store uranium-filled rods in two-story-tall casks, stacked in rows. Though fenced in and protected by armed guards, the casks will still be exposed to the open sky. By 2005, there may be as many as 50 such parking lots scattered about the country.

Hiroshima on Wheels

The White House's nuclear waste transport plan (dubbed "Mobile Chernobyl" by its critics) would send caravans of casks filled with High Level Waste (HLW) rolling down highways and rail lines near major cities in 43 states. Fifty-two million Americans live within a mile of the proposed routes.

From *Earth Island Journal*, Winter 2002–2003, pp. 41–44. Copyright © 2002 by Gar Smith. Reprinted by permission of the author.

Any casks that survived the trip would not be buried in the belly of Yucca Mountain, however. The facility is not expected to be open for business until 2010 at the earliest. Instead, the casks would be placed in another temporary above-ground parking lot—a federalized version of the dry-cask scenario.

Nearly 80,000 truck and 13,000 rail shipments would be required to ship used nuclear fuel rods and assorted rad-waste from decommissioned nuke plants. The shipments would continue day and night for 30–40 years.

The radiation aboard a single truck would be equal to 40 times the radiation released by the US A-bomb dropped on the Japanese city of Hiroshima. Each atomic cask traveling by rail would contain 240 Hiroshimas.

The Politics of Nuclear Waste

In 1986, the DOE began examining three potential sites that might be used as nuclear dumps. The sites were located in Texas, Nevada and Washington state.

But something strange happened in Congress. Legislation was crafted to eliminate the sites in Texas and Washington. Was it coincidental that the Speaker of the House at that time was Texas Representative Jim Wright and the House Majority Leader was Washington's Tom Foley? Robert Loux, the head of Nevada's Agency for Nuclear Projects, thinks not. "Congress acted on political, not scientific criteria in choosing this site," Loux charges.

Government geologists have since discovered that Yucca Mountain sits between two active earthquake faults, 12 miles from the epicenter of a 5.6 Richter scale quake that struck in 1992. A 4.4 quake rattled the region in June [2002].

Another drawback: Yucca Mountain is located atop a major Western aquifer. Millions of tiny fissures in the volcanic rock would allow water to drip onto the stored casks. The canisters will have to be retrofitted with titanium drip shields.

Government engineers claim these casks can last 270,000 years, but Loux's studies show the casks could corrode within as few as 500 years.

If any of the casks were to crack, the wastes would move inexorably toward the aquifer.

Does any of this concern the White House, whose resident-in-chief insists that his judgements will be made on the basis of "the best science, not politics"? Apparently not. On February 14, Bush agreed with his advisors' recommendation: "We've found nothing so far that would disqualify the site.... There are no show stoppers."

Highways to Hell

The government admits there could be as many as 900 accidents involving these nuclear shipments over 30 years. Department of Energy officials confide radioactive shipping accidents are "inevitable."

If a single truck were to spill its radioactive load, federal studies estimate, it would contaminate 42 square miles. Decontaminating a single square mile would take four years.

If the accident happened in a rural location, federal studies estimate cleanup costs could reach $620 million. If the accident occurred in an urban location, the entire city would be rendered uninhabitable. The decontamination costs would top $9.5 billion.

Truck accidents and train derailments are in the news nearly every day. The DOE, however, says there is little danger, as its casks are crash- and fireproof. The US Conference of Mayors is not reassured. On June 18, the mayors called on the DOE to halt its plans to ship waste to Yucca Mountain, noting that the casks "have never undergone full-scale physical testing to determine if they can withstand likely transportation accident and terrorism scenarios."

If the shipments are to go ahead, the mayors stated, Congress must first pass legislation requiring "adequate funds, training and equipment to protect the public health and safety in the event of an accident."

On July 18, 2001, a CSX railroad train caught fire in the Howard Street tunnel beneath the streets of downtown Baltimore. It was an hour before the fire departments were notified, and nearly three before the public was warned. The inferno raged for five days and reached temperatures of 1,500°F—hot enough to have melted the DOE's "impregnable" casks within a few hours.

According to studies conducted by the New York-based Radioactive Waste Management Associates, had that train been hauling HLW, 390,388 residents would have been exposed to the radioactive cloud. Tens of thousands might have died of cancer as a result. The cleanup costs would have approached $14 billion.

Despite calls for heightened security in the wake of 9/11, the Department of Transportation (DOT) and the Nuclear Regulatory Commission (NRC) are planning to relax safety regulations governing these nuclear shipments. The NRC concedes the new rules will reduce public health and safety.

Under the joint DOT/NRC plan, hundreds of radioactive isotopes would be exempted from regulatory controls. The plan would allow the industry to ship the wastes in cheaper, stripped down single-shell casks instead of the sturdier double-shell models currently required.

Agency officials explain the scheme to deregulate nuclear waste shipments was written before September 11. Nonetheless, NRC officials have refused to abandon plans to loosen security in the post 9/11 world. Their response is that these unforeseen new threats will be addressed "later."

The agency entrusted with safeguarding these rolling terror targets is the DOE's Transportation Security Division (TSD). In simulations run to assess the TSD's readiness to protect the cargo against terrorist attack, the Project On Government Oversight [www.pogo.org] reports, TSD defenders "were annihilated in ten seconds after an attack was started."

An internal DOE memo dated December 12, 1998 reported on the results of a computerized Joint Tactical Simulation of TSD's readiness. The results of the first test: three losses and no wins. The results of the second simulation: three losses and one win. At that point, all further simulations were cancelled.

DOE decided to purchase fleets of armored Humvees to help TSD's troopers patrol the shipments. That was before the Security Director at DOE's

Pantex nuclear weapons assembly plant in Texas pointed out the Humvees were motorized death traps and it would be "just as effective to buy Yugos."

The problem? Armor-piercing incendiary rounds could penetrate the Humvees, turning the passengers into toast. A Government Accounting Office investigation has revealed the Pentagon has released more than 100,000 of these deadly surplus rounds for sale on the open market.

The shipping casks could be equally vulnerable. According to the Nuclear Information and Resource Service [www.nirs.org] the White House has been informed "rocket launchers that are for retail sale... around the world are capable of penetrating a shipping cask, releasing deadly amounts of radioactivity." As NIRS spokesperson Kevin Kamps observes: "Providing security over a 30-year period for tens of thousands of moving targets is not realistic."

In July, the US Senate voted 60-39 to override Nevada's veto of the Yucca Mountain nuclear waste dump. This does not mean Yucca Mountain will ever open; instead, it sets the stage for years of courtroom activity, Nuclear Regulatory Commission (NRC) licensing proceedings, continued Congressional action, and an increased likelihood of large protests and blockades of highways and railways.

POSTSCRIPT

Should the United States Continue to Focus Plans for Permanent Nuclear Waste Disposal Exclusively at Yucca Mountain?

Abraham notes that the state of Nevada has the right to object to his recommendation. Not surprisingly, Nevada governor Kenny Guinn did exactly that on April 8, 2002. On May 8 the House of Representatives promptly voted to set aside the veto, and on July 9 the Senate voted to do the same. News reports said that this ends "years of political debate over nuclear waste disposal," but Nevada still has half a dozen lawsuits challenging the project pending. As of summer 2004, the U.S. Nuclear Regulatory Commission was still evaluating DOE's Yucca Mountain license application; see `http://www.nrc.gov/waste/hlw-disposal.html`.

Even those who favor using Yucca Mountain for high-level nuclear waste disposal admit that in time the site is bound to leak. The intensity of the radioactivity emitted by the waste will decline rapidly as short—half-life materials decay, and by 2300, when the site is expected to be sealed, that intensity will be less than 5 percent of the initial level. After that, however, radiation intensity will decline much more slowly. The nickel-alloy containers for the waste are expected to last at least 10,000 years, but they will not last forever. The Department of Energy's computer simulations predict that the radiation released to the environment will rise rapidly after about 100,000 years, with a peak annual dose after 400,000 years that is about double the natural background exposure. Many people are skeptical that the site can be protected for any significant fraction of such time periods. These are among the considerations that lead James Flynn et al., in "Overcoming Tunnel Vision," *Environment* (April 1997), to urge stopping work on the Yucca Mountain project and rethinking the entire nuclear waste disposal issue. On the other hand, Jonah Goldberg, in "Dead and Buried," *National Review* (April 8, 2002), contends that such considerations are irrelevant and that critics exaggerate the dangers of storing waste at Yucca Mountain. However, on July 9, 2004, a federal appeals court demanded that the Department of Energy come up with a plan to protect the public against radiation releases after the 10,000-year mark. Nevada officials declared that this made the project "effectively dead," but Energy Secretary Spencer Abraham said in a press release, "The Court dismissed all challenges to the site selection of Yucca Mountain. Our scientific basis for the Yucca Mountain Project is sound. The project will protect the public health and safety.... DOE will be working with the EPA and

Congress to determine appropriated steps to address [the 10,000 year compliance period]."

The nuclear waste disposal problem in the United States is real, and it must be dealt with. If it is not, America may face the same kinds of problems created by the former Soviet Union, which disposed of some nuclear waste simply by dumping it into the sea. For a recent summary of the nuclear waste problem and the disposal controversy, see Michael E. Long, "Half Life: The Lethal Legacy of America's Nuclear Waste," *National Geographic* (July 2002). Gary Taubes, in "Whose Nuclear Waste?" *Technology Review* (January/February 2002), argues that a whole new approach may be necessary. The need for care is underlined by experience with other nuclear waste disposal sites. Tom Carpenter and Clare Gilbert, in "Don't Breathe the Air," *Bulletin of the Atomic Scientists* (May/June 2004), describe the Hanford Site in Hanford, Washington, where wastes from nuclear weapons production were stored in underground tanks. Leaks from the tanks have contaminated groundwater, and an extensive cleanup program is under way. But cleanup workers are being exposed to both radioactive materials and toxic chemicals, and they are falling ill. In June 2004, the U.S. Senate voted to ease cleanup requirements; see http://us.cnn.com/2004/TECH/science/06/04/nuclearsludge.ap/.

ISSUE 20

Should the Military Be Exempt from Environmental Regulations?

YES: Benedict S. Cohen, from "Impact of Military Training on the Environment," Testimony before the Senate Committee on Environment and Public Works (April 2, 2003)

NO: Jamie Clark, from "Impact of Military Training on the Environment," Testimony before the Senate Committee on Environment and Public Works (April 2, 2003)

ISSUE SUMMARY

YES: Benedict S. Cohen argues that environmental regulations interfere with military training and other "readiness" activities, and that though the U.S. Department of Defense will continue "to provide exemplary stewardship of the lands and natural resources in our trust," those regulations must be revised to permit the military to do its job without interference.

NO: Jamie Clark argues that reducing the Department of Defense's environmental obligations is dangerous because both people and wildlife would be threatened with serious, irreversible, and unnecessary harm.

Most of us have heard of "scorched earth" wars, in which an army destroys forests and farms in order to deny the enemy their benefit. We have surely seen the images of a Europe laid waste by World War II. More recently, we may recall, the Gulf War saw oil deliberately released to flood desert sands and the waters of the Persian Gulf. Enough smoke poured from burning oil wells to threaten both local climate change and human health. See, for example, Randy Thomas, "Eco War," *Earth Island Journal* (Spring 1991), B. Ruben, "Gulf Smoke Screens," *Environmental Action* (July/August 1991), and Jeffrey L. Lange, David A. Schwartz, Bradley N. Doebbeling, Jack M. Heller, and Peter S. Thorne, "Exposures to the Kuwait Oil Fires and Their Association with Asthma and Bronchitis among Gulf War Veterans," *Environmental Health Perspectives* (November 2002). Weaponry can have environmental effects by destroying dams, by physically destroying plants and animals, by causing erosion, and by disseminating toxic materials; see Henryk Bem and Firyal Bou-Rabee, "Environmental and Health Consequences of Depleted Uranium Use in the 1991 Gulf War," *Environment International* (March 2004).

The environmental impact of war would seem impossible to deny. But even after the Gulf War, the U.S. Department of Defense tried to suppress satellite photos showing the extent of the damage; see Shirley Johnston, "Gagged on Smoke," *Earth Island Journal* (Summer 1991). And for many years, it insisted that nuclear war was survivable, until researchers made it clear that even a small nuclear war would produce a "nuclear winter" that would probably destroy civilization, if not the human species. See T. Rueter and T. Kalil, "Nuclear Strategy and Nuclear Winter," *World Politics* (July 1991), and Carl Sagan and Richard Turco, *A Path Where No Man Thought: Nuclear Winter and Its Implications* (Random House, 1990).

Preparations for war may also have serious environmental impacts. Puerto Rico's island of Vieques was long a bomb depot and bombing range for the U.S. Navy. Local residents protested vigorously and documented heavy-metal contamination of the local ecosystem. After the Navy left the island, it "has continued to deny that it has been anything but an excellent environmental steward in Vieques." See Shane DuBow and Scott S. Warren, "Vieques on the Verge," *Smithsonian* (January 2004).

In 2002, the U.S. Congress, through the Readiness and Range Preservation Initiative, granted the Department of Defense a temporary exemption to the Migratory Bird Treaty Act that allowed the "incidental taking" of endangered birds during bombing and other training on military lands. In 2003, the Department of Defense asked Congress for additional exemptions from environmental regulations, specifically the Clean Air Act, Marine Mammal Protection Act, Endangered Species Act, Migratory Bird Treaty Act, and federal toxic waste laws. Paul Mayberry, deputy undersecretary of defense for readiness, said the exemptions were justified because many environmental restrictions were putting the nation's military readiness at stake. See "Pentagon Seeks Clarity in Environmental Laws Affecting Ranges," Agency Group 09, FDCH Regulatory Intelligence Database (March 21, 2003). The United States Senate Committee on Environment and Public Works held a hearing on the "Impact of Military Training on the Environment" on April 2, 2003. In the following selections, Benedict S. Cohen, deputy general counsel for environment and installations, Department of Defense, argues in his testimony before the Committee that environmental regulations interfere with military training and other "readiness" activities, and that though the Department of Defense will continue "to provide exemplary stewardship of the lands and natural resources in our trust" those regulations must be revised to permit the military to do its job without interference. Jamie Clark, senior vice president for conservation programs, National Wildlife Federation, argues that reducing the Department of Defense's environmental obligations is dangerous for two reasons. First, both people and wildlife will be threatened with serious, irreversible, and unnecessary harm. Second, other federal agencies and industry sectors with important missions, using the same logic as used here by the Department of Defense, would demand similar exemptions from environmental laws.

Benedict S. Cohen

 **YES**

Impact of Military Training on the Environment

Mr. Chairman and distinguished members of this Committee, I appreciate the opportunity to discuss with you the very important issue of sustaining our test and training capabilities, and the legislative proposal that the Administration has put forward in support of that objective. In these remarks I would like particularly to address some of the comments and criticisms offered concerning these legislative proposals.

Addressing Encroachment

We have only recently begun to realize that a broad array of encroachment pressures at our operational ranges are increasingly constraining our ability to conduct the testing and training that we must do to maintain our technological superiority and combat readiness. Given World events today, we know that our forces and our weaponry must be more diverse and flexible than ever before. Unfortunately, this comes at the same time that our ranges are under escalating demands to sustain the diverse operations required today, and that will be increasingly required in the future.

This current predicament has come about as a cumulative result of a slow but steady process involving many factors. Because external pressures are increasing, the adverse impacts to readiness are growing. Yet future testing and training needs will only further exacerbate these issues, as the speed and range of our weaponry and the number of training scenarios increase in response to real-world situations our forces will face when deployed. We must therefore begin to address these issues in a much more comprehensive and systematic fashion and understand that they will not be resolved overnight, but will require a sustained effort.

Environmental Stewardship

Before I address our comprehensive strategy, let me first emphasize our position concerning environmental stewardship. Congress has set aside 25 million acres of land—some 1.1% of the total land area in the United States.

From Testimony before the Senate Committee on Environment and Public Works, April 2, 2003. Notes omitted.

These lands were entrusted to the Department of Defense (DoD) to use efficiently and to care for properly. In executing these responsibilities we are committed to more than just compliance with the applicable laws and regulations. We are committed to protecting, preserving, and, when required, restoring, and enhancing the quality of the environment.

- We are investing in pollution prevention technologies to minimize or reduce pollution in the first place. Cleanup is far more costly than prevention.
- We are managing endangered and threatened species, and all of our natural resources, through integrated natural resource planning
- We are cleaning up contamination from past practices on our installations and are building a whole new program to address unexploded ordnance on our closed, transferring, and transferred ranges.

The American people have entrusted these 25 million acres to our care. Yet, in many cases, these lands that were once "in the middle of nowhere" are now surrounded by homes, industrial parks, retail malls, and interstate highways.

On a daily basis our installation and range managers are confronted with a myriad of challenges—urban sprawl, noise, air quality, air space, frequency spectrum, endangered species, marine mammals, and unexploded ordnance. Incompatible development outside our fence-lines is changing military flight paths for approaches and take-offs to patterns that are not militarily realistic—results that lead to negative training and potential harm to our pilots. With over 300 threatened and endangered species on DoD lands, nearly every major military installation and range has one or more endangered species, and for many species, these DoD lands are often the last refuge. Critical habitat designations for an ever increasing number of threatened or endangered species limit our access to and use of thousands of acres at many of our training and test ranges. The long-term prognosis is for this problem to intensify as new species are continually added to the threatened and endangered list.

Much too often these many encroachment challenges bring about unintended consequences to our readiness mission. This issue of encroachment is not going away. Nor is our responsibility to "train as we fight."

2003 READINESS AND RANGE PRESERVATION INITIATIVE (RRPI)

Overview

DoD's primary mission is maintaining our Nation's military readiness, today and into the future. DoD is also fully committed to high-quality environmental stewardship and the protection of natural resources on its lands. However, expanding restrictions on training and test ranges are limiting realistic preparations for combat and therefore our ability to maintain the readiness of America's military forces.

Last year, the Administration submitted to Congress an eight-provision legislative package, the Readiness and Range Preservation Initiative (RRPI). Congress enacted three of those provisions as part of the National Defense Authorization Act for Fiscal Year 2003. Two of the enacted provisions allow us to cooperate more effectively with local and State governments, as well as private entities, to plan for growth surrounding our training ranges by allowing us to work toward preserving habitat for imperiled species and assuring development and land uses that are compatible with our training and testing activities on our installations.

Under the third provision, Congress provided the Department a regulatory exemption under the Migratory Bird Treaty Act for the incidental taking of migratory birds during military readiness activities. We are grateful to Congress for these provisions, and especially for addressing the serious readiness concerns raised by recent judicial expansion of the prohibitions under the Migratory Bird Treaty Act. I am pleased to inform this Committee that as a direct result of your legislation, Air Force B-1 and B-52 bombers, forward deployed to Anderson Air Force Base, Guam, are performing dry run training exercises over the Navy's Bombing Range at Farallon de Medinilla in the Commonwealth of the Northern Mariana Islands.

Last year, Congress also began consideration of the other five elements of our Readiness and Range Preservation Initiative. These five proposals remain essential to range sustainment and are as important this year as they were last year—maybe more so. The five provisions submitted this year reaffirm the principle that military lands, marine areas, and airspace exist to ensure military preparedness, while ensuring that the Department of Defense remains fully committed to its stewardship responsibilities. These five remaining provisions:

- Authorize use of Integrated Natural Resource Management Plans in appropriate circumstances as a substitute for critical habitat designation;
- Reform obsolete and unscientific elements of the Marine Mammal Protection Act, such as the definition of "harassment," and add a national security exemption to that statute;
- Modestly extend the allowable time for military readiness activities like bed-down of new weapons systems to comply with Clean Air Act; and
- Limit regulation of munitions on operational ranges under the Comprehensive Environmental Response, Compensation, and Liability Act (CERCLA) and Resource Conservation and Recovery Act (RCRA), if and only if those munitions and their associated constituents remain there, and only while the range remains operational.

Before discussing the specific elements of our proposal, I would like to address some overarching issues. A consistent theme in criticisms of our proposal is that it would bestow a sweeping or blanket exemption for the Defense Department from the Nation's environmental laws. No element of this allegation is accurate.

First, our initiative would apply only to military readiness activities, not to closed ranges or ranges that close in the future, and not to "the routine operation of installation operating support functions, such as administrative offices, military exchanges, commissaries, water treatment facilities, storage, schools, housing, motor pools . . . nor the operation of industrial activities, or the construction or demolition of such facilities." Our initiative thus is not applicable to the Defense Department activities that have traditionally been of greatest concern to state and federal regulators. It does address only uniquely military activities—what DoD does that is unlike any other governmental or private activity. DoD is, and will remain, subject to precisely the same regulatory requirements as the private sector when we perform the same types of activities as the private sector. We seek alternative forms of regulation only for the things we do that have no private-sector analogue: military readiness activities.

Moreover, our initiative largely affects environmental regulations that don't apply to the private sector or that disproportionately impact DoD:

- Endangered Species Act "critical habitat" designation has limited regulatory consequences on private lands, but can have crippling legal consequences for military bases.
- Under the Marine Mammal Protection Act, the private sector's Incidental Take Reduction Plans give commercial fisheries the flexibility to take significant numbers of marine mammal each year, but are unavailable to DoD—whose critical defense activities are being halted despite far fewer marine mammal deaths or injuries a year.
- The Clean Air Act's "conformity" requirement applies only to federal agencies, not the private sector.

Our proposals therefore are of the same nature as the relief Congress afforded us last year under the Migratory Bird Treaty Act, which environmental groups are unable to enforce against private parties but, as a result of a 2000 circuit court decision were able and willing to enforce, in wartime, against vital military readiness activities of the Department of Defense.

Nor does our initiative "exempt" even our readiness activities from the environmental laws; rather, it clarifies and confirms existing regulatory policies that recognize the unique nature of our activities. It codifies and extends EPA's existing Military Munitions Rule; confirms the prior Administration's policy on Integrated Natural Resource Management Plans and critical habitat; codifies the prior Administration's policy on "harassment" under the Marine Mammal Protection Act; ratifies longstanding state and federal policy concerning regulation under RCRA and CERCLA of our operational ranges; and gives states and DoD temporary flexibility under the Clean Air Act. Our proposals are, again, of the same nature as the relief Congress provided us under the Migratory Bird Treaty Act last year, which codified the prior Administration's position on DoD's obligations under the Migratory Bird Treaty Act.

Ironically, the alternative proposed by many of our critics—invocation of existing statutory emergency authority—would fully exempt DoD from the waived statutory requirements for however long the exemption lasted, a more far-reaching solution than the alternative forms of regulation we propose.

Accordingly, our proposals are neither sweeping nor exemptive; to the contrary, it is our critics who urge us to rely on wholesale, repeated use of emergency exemptions for routine, ongoing readiness activities that could easily be accommodated by minor clarifications and changes to existing law.

Existing emergency authorities

As noted above, many of our critics state that existing exemptions in the environmental laws and the consultative process in 10 USC 2014 render the Defense Department's initiative unnecessary.

Although existing exemptions are a valuable hedge against unexpected future emergencies, they cannot provide the legal basis for the Nation's everyday military readiness activities.

- The Marine Mammal Protection Act, like the Migratory Bird Treaty Act the Congress amended last year, has no national security exemption.
- 10 USC 2014, which allows a delay of at most five days in regulatory actions significantly affecting military readiness, is a valuable insurance policy for certain circumstances, but allows insufficient time to resolve disputes of any complexity. The Marine Corps' negotiations with the Fish and Wildlife Service over excluding portions of Camp Pendleton from designation as critical habitat took months. More to the point, Section 2014 merely codifies the inherent ability of cabinet members to consult with each other and appeal to the President. Since it does not address the underlying statutes giving rise to the dispute, it does nothing for readiness in circumstances where the underlying statute itself— not an agency's exercise of discretion—is the source of the readiness problem. This is particularly relevant to our RRPI proposal because none of the five amendments we propose have been occasioned by the actions of state or federal regulators. Four of the five proposed amendments (RCRA, CERCLA, MMPA, and ESA), like the MBTA amendment Congress passed last year, were occasioned by private litigants seeking to overturn federal regulatory policy and compel federal regulators to impose crippling restrictions on our readiness activities. The fifth, our Clean Air Act amendment, was proposed because DoD and EPA concluded that the Act's "general conformity" provision unnecessarily restricted the flexibility of DoD, state, and federal regulators to accommodate military readiness activities into applicable air pollution control schemes. Section 2014, therefore, although useful in some circumstances, would be of no use in addressing the critical readiness issues that our five RRPI initiatives address.
- Most of the environmental statutes with emergency exemptions clearly envisage that they will be used in rare circumstances, as a last resort, and only for brief periods.
- Under these statutes, the decision to grant an exemption is vested in the President, under the highest possible standard: "the paramount interest of the United States," a standard understood to involve exceptionally grave threats to national survival. The exemptions are also

usually limited to renewable periods of a year (or in some cases as much as three years for certain requirements).

- The ESA's section 7(j) exemption process, which differs significantly from typical emergency exemptions, allows the Secretary of Defense to direct the Endangered Species Committee to exempt agency actions in the interest of national security. However, the Endangered Species Committee process has given rise to procedural litigation in the past, potentially limiting its usefulness—especially in exigent circumstances. In addition, because it applies only to agency actions rather than to ranges themselves, any exemption secured by the Department would be of limited duration and benefit: because military testing and training evolve continuously, such an exemption would lose its usefulness over time as the nature of DoD actions on the range evolved.
- The exemption authorities do not work well in addressing those degradations in readiness that result from the cumulative, incremental effects of many different regulatory requirements and actions over time (as opposed to a single major action).
- Moreover, readiness is maintained by thousands of discrete test and training activities at hundreds of locations. Many of these are being adversely affected by environmental provisions. Maintaining military readiness through use of emergency exemptions would therefore involve issuing and renewing scores or even hundreds of Presidential certifications annually.
- And although a discrete activity (e.g., a particular carrier battle group exercise) might only rarely rise to the extraordinary level of a "paramount national interest," it is clearly intolerable to allow all activities that do not individually rise to that level to be compromised or ended by overregulation.
- Finally, to allow continued unchecked degradation of readiness until an external event like Pearl Harbor or September 11 caused the President to invoke the exemption would mean that our military forces would go into battle having received degraded training, with weapons that had received degraded testing and evaluation. Only the testing and training that occurred after the emergency exemption was granted would be fully realistic and effective.

The Defense Department believes that it is unacceptable as a matter of public policy for indispensable readiness activities to require repeated invocation of emergency authority—particularly when narrow clarifications of the underlying regulatory statutes would enable both essential readiness activities and the protection of the environment to continue. Congress would never tolerate a situation in which another activity vital to the Nation, like the practice of medicine, was only permitted to go forward through the repeated use of emergency exemptions.

That having been said, I should make clear that the Department of Defense is in no way philosophically opposed to the use of national security waivers or exemptions where necessary. We believe that every environmental statute should have a well-crafted exemption, as an insurance policy, though we continue to hope that we will seldom be required to have recourse to them. . . .

Specific Proposals

This year's proposals do include some clarifications and modifications based on events since last year. Of the five, the Endangered Species Act (ESA) and Clean Air Act provisions are unchanged. Let me address the changed provisions first.

RCRA and CERCLA

The legislation would codify and confirm the longstanding regulatory policy of EPA and every state concerning regulation of munitions use on operational ranges under RCRA and CERCLA. It would confirm that military munitions are subject to EPA's 1997 Military Munitions Rule while on range, and that cleanup of operational ranges is not required so long as material stays on the range. If such material moves off range, it still must be addressed promptly under existing environmental laws. Moreover, if munitions constituents cause an imminent and substantial endangerment on range, EPA will retain its current authority to address it on range under CERCLA section 106. (Our legislation explicitly reaffirms EPA's section 106 authority.) The legislation similarly does not modify the overlapping protections of the Safe Drinking Water Act, NEPA, and the ESA against environmentally harmful activities at operational military bases. The legislation has no effect whatsoever on DoD's cleanup obligations under RCRA or CERCLA at Formerly Used Defense Sites, closed ranges, ranges that close in the future, or waste management practices involving munitions even on operational ranges (such as so-called OB/OD activities).

The core of our concern is to protect against litigation the longstanding, uniform regulatory policy that (1) use of munitions for testing and training on an operational range is not a waste management activity or the trigger for cleanup requirements, and (2) that the appropriate trigger for DoD to address the environmental consequences of such routine test and training uses involving discharge of munitions is (a) when the range closes, (b) when munitions or their elements migrate or threaten to migrate off-range, or (c) when munitions or their elements create an imminent and substantial endangerment on-range. . . .

This legislation is needed because of RCRA's broad definition of "solid waste," and because states possess broad authority to adopt more stringent RCRA regulations than EPA (enforceable both by the states and by environmental plaintiffs). EPA therefore has quite limited ability to afford DoD regulatory relief under RCRA. Similarly, the broad statutory definition of "release" under CERCLA may also limit EPA's ability to afford DoD regulatory relief. And the President's site-specific, annually renewable waiver (under a paramount national interest standard in RCRA and a national security standard in CERCLA) is inapt for the reasons discussed above. . . .

Marine Mammal Protection Act

Although I realize this Committee is not centrally concerned with the Marine Mammal Protection Act (MMPA), I would like to take a moment to discuss it for purposes of completeness. This year's MMPA proposal includes some new provisions. This year's proposal, like last year's, would amend the term "harassment" in the MMPA, which currently focuses on the mere "potential" to injure or disturb marine mammals.

Our initiative adopts verbatim a reform proposal developed during the prior Administration by the Commerce, Interior, and Defense Departments and applies it to military readiness activities. That proposal espoused a recommendation by the National Research Council (NRC) that the currently overbroad definition of "harassment" of marine mammals—which includes "annoyance" or "potential to disturb"—be focused on biologically significant effects. As recently as 1999, the National Marine Fisheries Service (NMFS) asserted that under the sweeping language of the existing statutory definition harassment "is presumed to occur when marine mammals react to the generated sounds or visual cues"—in other words, whenever a marine mammal notices and reacts to an activity, no matter how transient or benign the reaction. As the NRC study found, "If [this] interpretation of the law for level B harassment (detectable changes in behavior) were applied to shipping as strenuously as it is applied to scientific and naval activities, the result would be crippling regulation of nearly every motorized vessel operating in U.S. waters."

Under the prior Administration, NMFS subsequently began applying the NRC's more scientific, effects-based definition. But environmental groups have challenged this regulatory construction as inconsistent with the statute. As you may know, the Navy and the National Oceanic and Atmospheric Administration suffered an important setback last year involving a vital anti-submarine warfare sensor—SURTASS LFA, a towed array emitting low-frequency sonar that is critical in detecting ultra-quiet diesel-electric submarines while they are still at a safe distance from our vessels. In the SURTASS LFA litigation environmental groups successfully challenged the new policy as inconsistent with the sweeping statutory standard, putting at risk NMFS' regulatory policy, clearly substantiating the need to clarify the existing statutory definition of harassment that we identified in our legislative package last year. . . .

The last change we are proposing, a national security exemption process, also derives from feedback the Defense Department received from environmental advocates last year after we submitted our proposal, as I discussed above. Although DoD continues to believe that predicating essential military training, testing, and operations on repeated invocations of emergency authority is unacceptable as a matter of public policy, we do believe that every environmental statute should have such authority as an insurance policy. The comments we received last year highlighted the fact that the MMPA does not currently contain such emergency authority, so this year's submission does include a waiver mechanism. Like the Endangered Species Act, our proposal would allow the Secretary of Defense, after conferring with the Secretaries of Commerce or Interior, as appropriate, to waive MMPA provisions for actions or categories of actions when required by national security. This provision is not a substitute for the other clarifications we have proposed to the MMPA, but rather a failsafe mechanism in the event of emergency.

The only substantive changes are those described above. The reason that the text is so much more extensive than last year's version is that last year's version was drafted as a freestanding part of title 10—the Defense Department title—rather than an amendment to the text of the MMPA itself. This year, because we were making several changes, we concluded that as a drafting matter we should include our changes in the MMPA itself. That necessitated a lot more language, largely just reciting existing MMPA language that we are not otherwise modifying.

The environmental impacts of our proposed reforms would be minimal. Although our initiative would exclude transient, biologically insignificant effects from regulation, the MMPA would remain in full effect for biologically significant effects—not only death or injury but also disruption of significant activities. The Defense Department could neither harm marine mammals nor disrupt their biologically significant activities without obtaining authorization from FWS or NMFS, as appropriate.

Nor does our initiative depart from the precautionary premise of the MMPA. The Precautionary Principle holds that regulators should proceed conservatively in the face of scientific uncertainty over environmental effects. But our initiative embodies a conservative, science-based approach validated by the National Research Council. By defining as "harassment" any readiness activities that "injure or have the significant potential to injure," or "disturb *or are* likely to disturb," our initiative includes a margin of safety fully consistent with the Precautionary Principle. The alternative is the existing grossly overbroad, unscientific definition of harassment, which sweeps in any activity having the "potential to disturb." As the National Research Council found, such sweeping overbreadth is unscientific and not mandated by the Precautionary Principle. . . .

The Defense Department already exercises extraordinary care in its maritime programs: all DoD activities worldwide result in fewer than 10 deaths or injuries annually (as opposed to 4800 deaths annually from commercial fishing activities). And DoD currently funds much of the most significant research on marine mammals, and will continue this research in future.

Although the environmental effects of our MMPA reforms will be negligible, their readiness implications are profound. Application of the current hair-trigger definition of "harassment" has profoundly affected both vital R&D efforts and training. Navy operations are expeditionary in nature, which means world events often require planning exercises on short notice. To date, the Navy has been able to avoid the delay and burden of applying for a take permit only by curtailing and/or dumbing down training and research/testing. For six years, the Navy has been working on research to develop a suite of new sensors and tactics (the Littoral Advanced Warfare Development Program, or LWAD) to reduce the threat to the fleet posed by ultraquiet diesel submarines operating in the littorals and shallow seas like the Persian Gulf, the Straits of Hormuz, the South China Sea, and the Taiwan Strait. These submarines are widely distributed in the world's navies, including "Axis of Evil" countries such as Iran and North Korea and potentially hostile great powers. In the 6 years that the program has operated, over 75% of the tests have been impacted by environmental considerations. In the last 3 years, 9 of 10 tests have been affected. One was cancelled entirely, and 17 different projects have been scaled back.

Endangered Species Act

Our Endangered Species Act provision is unchanged from last year. The legislation would confirm the prior Administration's decision that an Integrated Natural Resources Management Plan (INRMP) may in appropriate circumstances obviate the need to designate critical habitat on military installations. These plans for conserving natural resources on military property, required by the Sikes Act, are developed in cooperation with state wildlife agencies, the U.S. Fish and Wildlife Service, and the public. In most cases they offer comparable or better protection for the species because they consider the base's environment holistically, rather than using a species-by-species analysis. The prior Administration's decision that INRMPs may adequately provide for appropriate endangered species habitat management is being challenged in court by environmental groups, who cite Ninth Circuit caselaw suggesting that other habitat management programs provided an insufficient basis for the Fish and Wildlife Service to avoid designating Critical Habitat. These groups claim that no INRMP, no matter how protective, can ever substitute for critical habitat designation. This legislation would confirm and insulate the Fish and Wildlife Service's policy from such challenges.

Both the prior and current Administrations have affirmed the use of INRMPs as a basis for possible exclusion from critical habitat. Such plans are required to provide for fish and wildlife management, land management, forest management, and fish and wildlife-oriented recreation; fish and wildlife habitat enhancement; wetland protection, enhancement, and restoration; establishment of specific natural resource management goals, objectives, and timeframes; and enforcement of natural resource laws and regulations. And unlike the process for designation of critical habitat, INRMPs assure a role for state regulators. Furthermore, INRMPs must be reviewed by the parties on a regular basis, but not less than every five years, providing a continuing opportunity for FWS input.

By contrast, in 1999, the Fish and Wildlife Service stated in a Notice of Proposed Rulemaking that "we have long believed that, in most circumstances, the designation of 'official' critical habitat is of little additional value for most listed species, yet it consumes large amounts of conservation resources. [W]e have long believed that separate protection of critical habitat is duplicative for most species."

Our provision does not automatically eliminate critical habitat designation, precisely because under the Sikes Act, the statute giving rise to INRMPs, the Fish & Wildlife Service is given approval authority over those elements of the INRMP under its jurisdiction. This authority guarantees the Fish & Wildlife Service the authority to make a case-by-case determination concerning the adequacy of our INRMPs as a substitute for critical habitat designation.

And if the Fish & Wildlife Service does not approve the INRMP, our provision will not apply to protect the base from critical habitat designation.

Our legislation explicitly requires that the Defense Department continue to consult with the Fish and Wildlife Service and the National Marine Fisheries Service under Section 7 of the Endangered Species Act (ESA); the other provisions of the ESA, as well as other environmental statutes such as the National Environmental Policy Act, would continue to apply, as well.

The Defense Department's proposal has vital implications for readiness. Absent this policy, courts, based on complaints filed by environmental litigants, compelled the Fish and Wildlife Service to re-evaluate "not prudent" findings for many critical habitat determinations, and as a result FWS proposed to designate over 50% of the 12,000-acre Marine Corps Air Station (MCAS) Miramar and over 56% of the 125,000-acre Marine Corps Base (MCB) Camp Pendleton. Prior to adoption of this policy, 72% of Fort Lewis and 40% of the Chocolate Mountains Aerial Gunnery Range were designated as critical habitat for various species, and analogous habitat restrictions were imposed on 33% of Fort Hood. These are vital installations.

Unlike Sikes Act INRMPs, critical habitat designation can impose rigid limitations on military use of bases, denying commanders the flexibility to manage their lands for the benefit of both readiness and endangered species.

Clean Air Act General Conformity Amendment

Our Clean Air Act amendment is unchanged since last year. The legislation would provide more flexibility for the Defense Department in ensuring that emissions from its military training and testing are consistent with State Implementation Plans under the Clean Air Act by allowing DoD and the states a slightly longer period to accommodate or offset emissions from military readiness activities.

The Clean Air Act's "general conformity" requirement, applicable only to federal agencies, has repeatedly threatened deployment of new weapons systems and base closure/realignment despite the fact that relatively minor levels of emissions were involved. . . .

Conclusion

In closing Mr. Chairman, let me emphasize that modern warfare is a "come as you are" affair. There is no time to get ready. We must be prepared to defend our country wherever and whenever necessary. While we want to train as we fight, in reality our soldiers, sailors, airmen and Marines fight as they train. The consequences for them, and therefore for all of us, could not be more momentous.

DoD is committed to sustaining U.S. test and training capabilities in a manner that fully satisfies that military readiness mission while also continuing to provide exemplary stewardship of the lands and natural resources in our trust. . . .

NO

Jamie Clark

Impact of Military Training on the Environment

. . . Prior to arriving at the National Wildlife Federation in 2001, I served for 13 years at the U.S. Fish and Wildlife Service, with the last 4 years as the Director of the agency. Prior to that, I served as Fish and Wildlife Administrator for the Department of the Army, Natural and Cultural Resources Program Manager for the National Guard Bureau, and Research Biologist for U.S. Army Medical Research Institute. I am the daughter of a U.S. Army Colonel, and lived on or near military bases throughout my entire childhood.

Based on this experience, I am very familiar with the Defense Department's long history of leadership in wildlife conservation. On many occasions during my tenures at FWS and the Defense Department, DOD rolled up its sleeves and worked with wildlife agency experts to find a way to comply with environmental laws and conserve imperiled wildlife while achieving military preparedness objectives.

The Administration now proposes in its Readiness and Range Preservation Initiative that Congress scale back DOD's responsibilities to conserve wildlife and to protect people from the hazardous pollution that DOD generates. This proposal is both unjustified and dangerous. It is unjustified because DOD's longstanding approach of working through compliance issues on an installation-by-installation basis works. As DOD itself has acknowledged, our armed forces are as prepared today as they ever have been in their history, and this has been achieved without broad exemptions from environmental laws.

The DOD proposal is dangerous because, if Congress were to broadly exempt DOD from its environmental protection responsibilities, both people and wildlife would be threatened with serious, irreversible and unnecessary harm. Moreover, other federal agencies and industry sectors with important missions, using the same logic as used here by DOD, would line up for their own exemptions from environmental laws.

My expertise is in the Endangered Species Act (ESA), so I would like to focus my testimony on why exempting the Defense Department from key provisions of the ESA would be a serious mistake. I will rely on my fellow witnesses to explain why the proposed exemptions from other environmental and public health and safety laws is similarly unwise.

From Testimony before the Senate Committee on Environment and Public Works, April 2, 2003.

Concerns with the ESA Exemption

The Defense Department's proposed ESA exemption suffers from three basic flaws: it would severely weaken this nation's efforts to conserve imperiled species and the ecosystems on which all of us depend; it is unnecessary for maintaining military readiness; and it ignores the Defense Department's own record of success in balancing readiness and conservation objectives under existing law.

1. Section 2017 Removes a Key Species Conservation Tool

Section 2017 of the Administration's Readiness and Range Preservation Initiative would preclude designations of critical habitat on any lands owned or controlled by DOD if DOD has prepared an Integrated Natural Resources Management Plan (INRMP) pursuant to the Sikes Act and has provided "special management consideration or protection" of listed species pursuant to Section 3(5)(A) of the ESA.

This proposal would effectively eliminate critical habitat designations on DOD lands, thereby removing an essential tool for protecting and recovering species listed under the ESA. Of the various ESA protections, the critical habitat provision is the only one that specifically calls for protection of habitat needed for recovery of listed species. It is a fundamental tenet of biology that habitat must be protected if we ever hope to achieve the recovery of imperiled fish, wildlife and plant species.

Section 2017 would replace this crucial habitat protection with management plans developed pursuant to the Sikes Act. The Sikes Act does not require the protection of listed species or their habitats; it simply directs DOD to prepare INRMPs that protect wildlife "to the extent appropriate." Moreover, the Sikes Act provides no guaranteed funding for INRMPs and the annual appropriations process is highly uncertain. Even the best-laid management plans can go awry when the anticipated funding fails to come through. Yet, under Section 2017, even poorly designed INRMPs that allow destruction of essential habitat and put fish, wildlife or plant species at serious risk of extinction would be substituted for critical habitat protections.

Section 2017 contains one minor limitation on the substitution of INRMPs for critical habitat designations: such a substitution is allowed only where the INRMP provides "special management consideration or protection" within the meaning of Section 3(5)(A) of the ESA. Unfortunately, this limitation does nothing to ensure that INRMPs truly conserve listed species.

The term "special management consideration or protection" was never intended to provide a biological threshold that land managers must achieve in order to satisfy the ESA. The term is found in Section 3(5) of the ESA, which sets forth a two-part definition of critical habitat. Section 3(5)(A) states that critical habitat includes areas occupied by a listed species that are "essential for the conservation of the species" and "which may require special management consideration or protection." Section 3(5)(B) states that critical habitat also includes areas not currently occupied by a listed species that are simply "essential for the conservation of the species."

As this language makes clear, an ESA §3(5) finding by the U.S. Fish and Wildlife Service or National Marine Fisheries Service (Services) that a parcel of land "may require special management consideration or protection" is not the same as finding that it is already receiving adequate protection. Such a finding simply highlights the importance of a parcel of land to a species, and it should lead to designation of that land as critical habitat. See *Center for Biological Diversity v. Norton*, 240 F. Supp. 2d 1090 (D. Ariz. 2003) (rejecting, as contrary to plain meaning of ESA, defendant's interpretation of "special management consideration or protection" as providing a basis for substituting a U.S. Forest Service management plan for critical habitat protection). By allowing DOD to substitute INRMPs for critical habitat designations whenever it unilaterally makes a finding of "special management consideration or protection," Section 2017 significantly weakens the ESA.

Section 2017 is also problematic because it would eliminate many of the ESA Section 7 consultations that have stimulated DOD to "look before it leaps" into a potentially harmful training exercise. As a result of Section 7 consultations, DOD and the Services have routinely developed what is known as "workarounds," strategies for avoiding or minimizing harm to listed species and their habitats while still providing a rigorous training regimen.

Section 2017 purports to retain Section 7 consultations. However, the duty to consult only arises when a proposed federal action would potentially jeopardize a listed species or adversely modify or destroy its critical habitat. By removing critical habitat designations on lands owned or controlled by DOD, Section 2017 would eliminate one of the two possible justifications for initiating a consultation, reducing the likelihood that consultations will take place. This would mean that DOD and the Services would pay less attention to species concerns and would be less effective in conserving imperiled species and maintaining the sustainability of the land.

The reductions in species protection proposed by DOD would have major implications for our nation's rich natural heritage. DOD manages approximately 25 million acres of land on more than 425 major military installations. These lands are home to at least 300 federally listed species. Without the refuge provided by these bases, many of these species would slide rapidly toward extinction. These installations have played a crucial role in species conservation and must continue to do so.

2. The ESA Exemption Is Not Necessary to Maintain Military Readiness

The ESA already has the flexibility needed for the Defense Department to balance military readiness and species conservation objectives. Three key provisions provide this flexibility. First, under the consultation provision of Section 7(a)(2) of the Act, DOD is provided with the opportunity to develop solutions in tandem with the Services to avoid unnecessary harm to listed species from military activities. Typically, the Services conclude, after informal consultation, that the proposed action will not adversely affect a listed species or its designated critical habitat or, after formal consultation, that it will not likely jeopardize a listed species or destroy or adversely modify its critical habitat. See, e.g., U.S. Army Environmental Center, Installation Summaries from the FY 2001 Survey of Threatened and Endangered Species on Army Lands (August 2002) at 9 (noting successful conclusion of 282 informal consultations and 36 formal consultations, with no "jeopardy" biological opinions). In both informal and formal consultations, the Services either will recommend that the action go forward without changes, or it will work with DOD to design "work arounds" for avoiding and minimizing harm to the species and its habitat. In either case, DOD accomplishes its readiness objectives while achieving ESA compliance.

Second, under Section 4(b)(2) of the ESA, the Services are authorized to exclude any area from critical habitat designation if they determine that the benefits of exclusion outweigh the benefits of specifying the area. (An exception is made for when the Services find that failure to designate an area as critical habitat will result in the extinction of a species—a finding that the Services have never made.) In making this decision, the Services must consider "the economic impact, and any other relevant impact" of the critical habitat designation. DOD has recently availed itself of this provision to convince the U.S. Fish and Wildlife Service to exclude virtually all of the habitat at Camp Pendleton—habitat deemed critical to five listed species in proposed rulemakings—from final critical habitat designations. Thus, for situations where the Section 7(a)(2) consultation procedures place undue burdens on readiness activities, DOD already has a tool for working with the Services on excluding land from critical habitat designation. Attached to my testimony is a factsheet that shows how the Services have worked cooperatively with DOD on these exclusions, and another factsheet showing the importance of maintaining the Services' role in evaluating proposed exclusions.

Third, under Section 7(j) of the ESA an exemption "shall" be granted for an activity if the Secretary of Defense finds the exemption is necessary for reasons of national security. To this date, DOD has never sought an exemption under Section 7(j)—highlighting the fact that other provisions of the ESA have provided DOD with all the flexibility it needs to reconcile training needs with species conservation objectives.

Where there are site-specific conflicts between training needs and species conservation needs, the ESA provides these three mechanisms for resolving them in a manner that allows DOD to achieve its readiness objectives. Granting DOD a nationwide ESA exemption, which would apply in many places where no irreconcilable conflicts between training needs and conservation needs have arisen, would be harmful to imperiled species and totally unnecessary to achieve readiness objectives.

a. DOD Has Misstated the Law Regarding Its Ability to Continue with a Cooperative, Case-by-Case Approach to Critical Habitat Designations

DOD has stated that the ESA exemption is necessary because a recent court ruling in Arizona would prevent DOD from taking the cooperative, case-by-case approach to critical habitat designations that was developed when I served as Director of the Fish and Wildlife Service. This description of the court ruling is inaccurate—the ruling clearly allows DOD to continue the cooperative, case-by-case approach if it wishes.

The court ruling at issue is entitled *Center for Biological Diversity v. Norton*, 240 F. Supp. 2d 1090 (D. Ariz. 2003). In this case, FWS excluded San Carlos Apache tribal lands from a critical habitat designation pursuant to ESA §4(b)(2) because the tribal land management plan was adequate and the benefits of exclusion outweighed the benefits of inclusion. The federal district court upheld the exclusion as within FWS's broad authority under ESA §4(b)(2). At the same time, the court held that lands could not legitimately be excluded from a critical habitat designation on the basis of the "special management" language in ESA §3(5).

Under the court's reasoning, FWS continues to have the broad flexibility to exclude DOD lands from a critical habitat designation on the basis of a satisfactory INRMP and the benefits to military training that the exclusion would provide. The ruling simply clarifies that such exclusions must be carried out pursuant to ESA §4(b)(2) rather than ESA §3(5). Thus, DOD's assertion that the Center for Biological Diversity ruling prevents it from working with FWS to secure exclusions of DOD lands from critical habitat designations is inaccurate.

b. DOD's Anecdotes Do Not Demonstrate That the ESA Has Reduced Readiness

The DOD has offered a series of misleading anecdotes describing difficulties it has encountered in balancing military readiness and conservation objectives. Before Congress moves forward with any exemption legislation, the appropriate Congressional committees should get a more complete picture of what is really happening at DOD installations.

Some of DOD's anecdotes are simply unpersuasive on their face, such as DOD's repeated assertion that environmental laws have prevented the armed services from learning how to dig foxholes and that troops abroad have been put at greater risk as a result. There is simply no evidence that environmental laws have ever prevented foxhole digging. Moreover, given its vast and varied landholdings and the many management options available, the Defense Department certainly can find places on which troops can learn to dig foxholes without encountering endangered species or other environmental issues.

Other anecdotes have simply disregarded the truth. For example, DOD and its allies have repeatedly argued that more than 50 percent of Camp Pendleton may not be available for training due to critical habitat designations. In fact, only five species have been proposed for critical habitat designations at Camp Pendleton. In each of these five instances, DOD raised concerns about impacts to military readiness, and in each instance, FWS worked closely with DOD to craft a solution. FWS ultimately excluded virtually all of the habitats for the five listed species on Camp Pendleton from critical habitat designations—even though FWS had earlier found that these habitats were essential to the conservation of the species. As a result of FWS's exclusion decisions, less than one percent of the training land at Camp Pendleton, and less than 4 percent of all of Camp Pendleton, is designated critical habitat. (Most of the critical habitat designated at Camp Pendleton is non-training land leased to San Onofre State Park, agricultural operations, and others. DOD's repeated suggestion that more than 50 percent of Camp Pendleton is at risk of being rendered off-limits to training due to critical habitat is simply inaccurate.

DOD also has argued that training opportunities and expansion plans at Fort Irwin have been thwarted by the desert tortoise. Yet just two weeks ago this official line was contradicted by the reality on the ground. In an article dated March 21, 2003, Fort Irwin spokesman Army Maj. Michael Lawhorn told the Barstow Desert Dispatch that he is unaware of any environmental regulations that interfere with troops' ability to train there. He also said there isn't any environmental law that hinders the expansion. . . .

These examples of misleading anecdotes highlight the need for Congress to look behind the reasons that are being put forward by DOD as the basis for weakening environmental laws. DOD uses the anecdotes in an attempt to demonstrate that conflicts between military readiness and species conservation objectives are irreconcilable. However, solutions to these conflicts are within reach if DOD is willing to invest sufficient time and energy into finding them. DOD has vast acres of land on which to train and vast stores of creativity and expertise among its land managers. With careful inventorying and planning, DOD can find a proper balance.

Has DOD made the necessary effort to inventory and plan for its training needs? In June 2002, the General Accounting Office issued a report entitled "Military Training: DOD Lacks a Comprehensive Plan to Manage Encroachment on Training Ranges," suggesting that the answer is no. The GAO found:

- DOD has not fully defined its training range requirements and lacks information on training resources available to the Services to meet those requirements, and that problems at individual installations may therefore be overstated.
- The Armed Services have never assessed the overall impacts of encroachment on training.
- DOD's readiness reports show high levels of training readiness for most units. In those few instances of when units reported lower training readiness, DOD officials rarely cited lack of adequate training ranges, areas or airspace as the cause.
- DOD officials themselves admit that population growth around military installations is responsible for past and present encroachment problems.
- The Armed Services' own readiness data do not show that environmental laws have significantly affected training readiness.

Ten months after the issuance of the GAO report, DOD still has not produced evidence that environmental laws are at fault for any of the minor gaps in readiness that may exist. EPA Administrator Whitman confirmed this much at a recent hearing. At a February 26, 2003, Senate Environment and Public Works Committee hearing on EPA's budget, EPA Administrator Whitman stated that she was "not aware of any particular area where environmental protection regulations are preventing the desired training."

To this date, DOD has not provided Congress with the most basic facts about the impacts of ESA critical habitat requirements on its readiness activities. Out of DOD's 25 million acres of training land, how many acres are designated critical habitat? At which installations? Which species? In what ways have the critical habitat designations limited readiness activities? What efforts did DOD make to alert FWS to these problems and to negotiate resolutions? Without answers to these most basic questions, Congress cannot fairly conclude that the ESA is at fault for any readiness gaps or that a sweeping ESA exemption is warranted.

3. DOD Has Worked Successfully with the Services to Balance Readiness and Species Conservation Objectives

The third reason why enacting DOD's proposed ESA changes would be a mistake is because the current approach—developing solutions at the local level, rather than relying on broad, national exemptions—has worked. My experience at both FWS and DOD has shown me that solutions developed at the local level are sometimes difficult to arrive at, but they are almost always more intelligent and long-lasting than one-size-fits-all solutions developed at the national level.

Allow me to provide a few brief examples. At the Marine Corps Base at Camp Lejeune in North Carolina, every colony tree of the endangered red-cockaded woodpecker is marked on a map, and Marines are trained to operate their vehicles as if those mapped locations are land mines. Here is the lesson that Major General David M. Mize, the Commanding General at Camp Lejeune, has drawn from this experience:

> "Returning to the old myth that military training and conservation are mutually exclusive; this notion has been repeatedly and demonstrably debunked. In the overwhelming majority of cases, with a good plan along with common sense and flexibility, military training and the conservation and recovery of endangered species can very successfully coexist."
>
> "Military installations in the southeast are contributing to red-cockaded woodpecker recovery while sustaining our primary mission of national military readiness."
>
> "I can say with confidence that the efforts of our natural resource managers and the training community have produced an environment in which endangered species management and military training are no longer considered mutually exclusive, but are compatible."

These sentiments, which I share, were relayed by Major General Mize just eight weeks ago at a National Defense University symposium sponsored by the U.S. Army Forces Command (FORSCOM) and others. At that symposium, representatives of Camp Lejeune Marine Corps Base, Eglin Air Force Base, Fort Bragg Army Base, Fort Stewart Army Base, Camp Blanding Training Center in Florida, the U.S. Army Environmental Center, and other Defense facilities—some of the most heavily utilized training bases in the country—heralded the success that Defense Department installations have had in furthering endangered species conservation while maintaining military readiness.

On the Mokapu Peninsula of Marine Corps Base Hawaii, the growth of non-native plants, which can decrease the reproductive success of endangered waterbirds, is controlled through annual "mud-ops" maneuvers by Marine Corps Assault Vehicles. Just before the onset of nesting season, these 26 ton vehicles are deployed in plow-like maneuvers that break the thick mats of invasive plants, improving nesting and feeding opportunities while also giving drivers valuable practice in unusual terrain. . . .

These success stories highlight a major trend that I believe has been missed by those promoting the DOD exemptions. In recent years, DOD has increasingly recognized the importance of sustainability because it meets several importance objectives at once. Sustainable use of the land helps DOD achieve not only compliance with environmental laws, but also long-term military readiness and cost-effectiveness goals. For example, by operating tanks so that they avoid the threatened desert tortoise, DOD prevents erosion, a problem that is extremely difficult and costly to remedy. If DOD abandons its commitment to environmental compliance, it will incur greater long-term costs for environmental remediation and will sacrifice land health and military readiness.

A November 2002 policy guidance issued by the then-Secretary of the Navy to the Chief of Naval Operations and the Commandant of the Marine Corps suggests that certain members of DOD's leadership are indeed willing to abandon the sustainability goal. The policy guidance on its face seems fairly innocuous—it purports to centralize at the Pentagon all decisionmaking on proposed critical habitat designations and other ESA actions. However, the Navy Secretary's cover memo makes clear that its purpose is also to discourage any negotiation of solutions to species conservation challenges by Marines or Navy personnel in the field, lest these locally-developed "win-win" solutions undercut DOD's arguments on Capitol Hill that the ESA is broken. According to paragraph 2 of the cover memo, "concessions ... could run counter to the legislative relief that we are continuing to pursue with Congress."

Similar sentiments were voiced by Deputy Defense Secretary Paul Wolfowitz in his March 7, 2003, memo to the chiefs of the Army, Navy and Air Force. Deputy Secretary Wolfowitz argued that "it is time for us to give greater consideration to requesting exemptions" from environmental laws and pleaded for specific examples of instances in which environmental regulations hamper training. The implicit message is that efforts at the installation level to resolve conflicts between conservation and training objectives should be suspended, and that such conflicts instead should be reported to the Pentagon, where environmental protections will simply be overridden.

These messages to military personnel in the field mark a very unfortunate abdication of DOD's leadership in wildlife conservation. To maintain its leadership role as steward of this nation's endangered wildlife, DOD must encourage its personnel to continue developing innovative solutions and not thwart those efforts.

Conclusion

With the Iraq war ongoing and terrorism threats always present, no one can dismiss the importance of military readiness. However, there is no justification for the Defense Department to retreat from its environmental stewardship commitments at home. As base commanders have been telling us, protecting endangered species and other important natural resources is compatible with maintaining military readiness.

Surveys show that the American people today want environmental protection from the federal government, including the Defense Department, as much as ever. According to an April 2002 Zogby Poll, 85% of registered voters believe that the Defense Department should be required to follow America's environmental and public health laws and not be exempt. Americans believe that no one, including the Defense Department, should be above the law.

Congress should reject the proposed environmental exemptions in the Administration's defense authorization package. This proposal, along with the parallel proposal in the Administration's FY04 budget request that Congress cut spending on DOD's environmental programs by $400 million, are a step in the wrong direction.

DOD has a long and impressive record of balancing readiness activities with wildlife conservation. The high quality of wildlife habitats at many DOD installations provides tangible evidence of DOD's positive contribution to the nation's conservation goals. At a time when environmental challenges are growing, DOD should be challenged to move forward with this successful model and not to sacrifice any of the progress that has been made. . . .

POSTSCRIPT

Should the Military Be Exempt from Environmental Regulations?

After the April 2003 Senate committee hearing, this issue received some press attention as environmentalists and congressional Democrats prepared to oppose the Bush administration. See "War on the Environment," *The Ecologist* (May 2003), and John Stanton, "Activists, Democrats Brace for Defense Environment Showdown," *CongressDaily AM* (May 13, 2003). The Senate voted for the exemptions, and early in 2004, the issue came before the House of Representatives. Dan Miller, first assistant attorney general, Natural Resources and Environmental Section, Colorado Department of Law, testified against them before the House Committee on Energy and Commerce Subcommittee on Energy and Air Quality, saying that.

"Even read in the narrowest possible fashion, the [proposed exemptions] would hamstring state and EPA cleanup authorities at over 24 million acres of 'operational ranges,' an area the size of Maryland, Massachusetts, New Jersey, Hawaii, Connecticut and Rhode Island combined. As a practical matter, environmental regulators would likely be precluded from using RCRA [Resource Conservation and Recovery Act], CERCLA [Comprehensive Environmental Response, Compensation, and Liability Act], and related state authorities to require any investigation or cleanup of groundwater contamination on these ranges, even if the contamination had migrated off-range, polluted drinking or irrigation water supplies, and even if it posed an imminent and substantial endangerment to human health. And it is likely that DOD's amendments would be construed more broadly to exempt even more contamination from state and EPA oversight. . . . If we have learned anything in the past thirty years of environmental regulation, it is that relying on federal agencies to 'voluntarily' address environmental contamination is often fruitless. One need look no further than the approximately 130 DOD facilities on the Superfund National Priorities List, or DOD's poor record of compliance with state and federal environmental laws to see that independent, legally enforceable state oversight of federal agencies is required to achieve effective results."

In May, the House Committees on Armed Services Readiness and Energy and Commerce announced that they were not about to consider the exemption. See "Hefley: No Plans to Exempt Military From Enviro Laws," *CongressDaily AM* (May 6, 2004).

But the issue is not about to go away. The Government Accounting Office (GAO) prepared a background paper, "Military Training: DOD Approach to Managing Encroachment on Training Ranges Still Evolving," GAO-03-621T (April 2, 2003), delivered by Barry W. Holman, director,

Defense Infrastructure Issues, as testimony before the Senate Committee on Environment and Public Works. It discussed eight encroachment issues, including urban growth around military bases, air and noise pollution, unexploded ordinance and other munitions, endangered species habitat, and protected marine resources. Since urban growth is not likely to cease and the number of endangered species and protected marine resources is sure to increase, encroachment is not about to diminish. The Department of Defense, says the GAO, must better document the impact of the encroachment on training and costs; it has not yet produced required reports to Congress. So far, "workarounds" have been enough to deal with the problem, but that may not remain sufficient. E. G. Willard, Tom Zimmerman, and Eric Bee, "Environmental Law and National Security: Can Existing Exemptions in Environmental Laws Preserve DOD Training and Operational Prerogatives without New Legislation?" *Air Force Law Review* (2004), conclude that existing exemptions are not enough to support military readiness and say that more are needed. "The bottom line is that we must be able to train the way we fight, and we must be able to operate to defend the country and its interests." Paul D. Thacker, "Are Environmental Exemptions for the U.S. Military Justified?" *Environmental Science & Technology* (October 15, 2004), noted that "many critics of the administration say that the campaign is more about undermining environmental laws than protecting military readiness."

A new approach to Congress was imminent in April 2005, when Bruce Geiselman, "Exemption vs. Clarification," *Waste News* (April 25, 2005), wrote that instead of seeking exemptions from environmental regulations, the Department of Defense is now seeking "clarification" of how such regulations apply to military activities. The issue gained fresh urgency in May 2005, when the Department of Defense announced its list of military bases proposed to be closed as a cost-saving measure, and the press reported that 34 military bases shut down since 1988 remain on the EPA's Superfund list.

ISSUE 21

Is Additional Federal Oversight Needed for the Construction of LNG Import Facilities?

YES: Edward J. Markey, from "LNG Import Terminal and Deepwater Port Siting: Federal and State Roles," Testimony before House Committee on Government Reform, Subcommittee on Energy Policy, Natural Resources, and Regulatory Affairs (June 22, 2004)

NO: Donald F. Santa, Jr., from "LNG Import Terminal and Deepwater Port Siting: Federal and State Roles," Testimony before House Committee on Government Reform, Subcommittee on Energy Policy, Natural Resources, and Regulatory Affairs (June 22, 2004)

ISSUE SUMMARY

YES: Edward J. Markey argues that the risks—including those associated with terrorist attack—associated with LNG (liquefied natural gas) tankers and terminals are so great that additional federal regulation is essential in order to protect the public.

NO: Donald F. Santa, Jr., argues that meeting demand for energy requires public policies that "do not unreasonably limit resource and infrastructure development." The permitting process for LNG import facilities should be governed by existing Federal Energy Regulatory Commission procedures without additional regulatory impediments.

The environmental movement has, since its dawn in the 1960s, added several interesting terms to the English language. NIMBY means "not in my back yard." NIMTOF means "not in my term of office." BANANA means "build absolutely nothing anywhere near anything." The sentiments behind these acronyms can be seen in connection with many of the issues in this book. They help drive objections to landfills, nuclear power plants, nuclear and toxic waste dumps, and many more projects that pose real or potential risks to human beings and their property.

Liquefied natural gas (LNG) provokes the same feelings. Natural gas used to be considered an unwanted by-product of oil wells. Disposing of it meant "flaring it off"; oil well towers were routinely topped by monstrous plumes of flame and smoke. After World War II, pipelines were built to carry the gas to factories, power plants, and urban areas where it could be used for its energy content. Later, the technology was developed to compress and liquefy the natural gas so it could be transported in tankers across oceans where pipelines could not go. Of course, the tankers had to arrive someplace where their gas could be transferred to tanks on trucks and trains or converted from liquid to gaseous form and put into a pipeline. These LNG terminals are essential if the gas is to be used.

But natural gas is highly flammable, and accidents do happen. Many places where terminals have been proposed have objected strenuously to the associated risks. In Maine, the LNG industry has for several years sought to convince communities to accept terminals (along with local economic benefits). So far, all communities but one have rejected the proposals; in the summer of 2004, the Pleasant Point Passamaquoddy reservation (near Eastport) voted to permit a terminal on its land, precisely because of the economic benefits. The terminal could open by 2008. However, debate has been and continues to be vigorous; neighboring communities in Maine and Canada have lobbied against the terminal. The Pleasant Point Passamaquoddy's Sipayik Environmental Department has released a report stressing the terminal's safety and the effectiveness of existing safety measures (http://www.wabanaki.com/SED/GOALS/LNG_SED_Letter_to_community.htm). Critics stress the hazards (http://www.savepassamaquoddybay.org/assets.html). And another terminal has been proposed for the region, in St. John, New Brunswick.

When Doug Ose (R-CA) opened the July 5, 2004, meeting of the Subcommittee on Energy Policy, Natural Resources, and Regulatory Affairs of the House Committee on Government Reform, he noted that "the United States, especially California, is relying more and more on natural gas. It is the fuel-of-choice for the electric power generation because it is reliable and is much cleaner than other fossil fuels. Natural gas is also used by individual citizens, and by industry, agriculture and transportation as a raw material. As a critical resource used throughout the economy, shortages in natural gas have a more profound impact than most other commodities. [But] North American natural gas fields are depleting at an increasing rate. Even if some new domestic natural gas comes onto the market, most experts believe that we will need even more. Pipeline imports from Canada make up about 15 percent of total U.S. consumption but, there too, experts anticipate diminishing sources. . . . I believe that increasing U.S. importation of liquefied natural gas (LNG) should be a component of the solution—either by on-shore or off-shore facilities."

In the following selections, Representative Edward J. Markey (D-MA) argues that the risks—including those associated with terrorist attack—associated with LNG tankers and terminals are so great that additional federal regulation is essential in order to protect the public. Existing regulations are not sufficient. Donald F. Santa, Jr., president of the Interstate Natural Gas Association of America, argues that meeting demand for energy requires public policies that "do not unreasonably limit resource and infrastructure development." A comprehensive energy policy must include removing barriers both to pipeline and storage infrastructure development and to enhancing gas supply by the importation of LNG. The permitting process should be governed by existing Federal Energy Regulatory Commission procedures without additional regulatory impediments.

Edward J. Markey  **YES**

Testimony Before the Subcommittee on Energy Policy, Natural Resources and Regulatory Affairs

. . . LNG is an important component of the energy supply of New England, and . . . it has great potential to help the nation meet its growing need for natural gas. As Federal Reserve Chairman Alan Greenspan noted in his testimony before the Energy and Commerce Committee, one notable difference between the oil and natural gas markets in the United States is that our nation is able to obtain access to global supplies of oil via tanker. In contrast, virtually all of our natural gas supply comes from either U.S. or Canadian resources delivered via pipeline. Only a small portion of our supply comes in via tanker in the form of LNG. Increasing LNG imports is therefore one important way to help address America's increasing demand for natural gas. Obtaining access to the global natural gas supply through LNG imports is also one way of helping to reduce the current volatility in the U.S. natural gas marketplace.

The question then is where is the most appropriate place for these facilities to be sited? I would suggest to the Subcommittee that this is an issue that the Congress already considered nearly 25 years ago, based in large part about public safety concerns surrounding the siting of the Distrigas facility in a densely populated urban area, and the inherent difficulties in trying to address the consequences of an accident or an act of sabotage at this type of facility. At that time, the Congress enacted a law, which I authored, which tried to learn from the Everett experience by directing the Secretary of Transportation to consider the need for remote siting as part of the rules applicable to all new LNG importation terminals. The Secretary of Transportation, unfortunately, has chosen to largely ignore this law and has failed to comply with Congress' intent regarding what factors the Department needs to take into account in writing rules for the siting of new LNG facilities. This failure had little consequence for more than 25 years, as no new LNG importation terminals were being built. Today, however, with dozens of LNG terminals being proposed around the country, this failure can no longer be tolerated.

From Testimony before the House Committee on Government Reform, Subcommittee on Energy Policy, Natural Resources, and Regulatory Affairs, June 22, 2004.

Key Issues

As I see it, there currently are four critical issues that need to be addressed at the federal level.

First, we need to have a much better scientific and technical assessment of the consequences of a terrorist attack against an LNG tanker or LNG terminal. Such a hazard assessment is needed to better inform federal siting decisions with respect to any new LNG terminals around the nation. It is also needed to better inform state and local emergency planning and response activities with respect to existing LNG facilities.

Second, we need help from both the federal government and the facility operator to defray the costs that local governments incur in securing LNG or other critical infrastructure facilities from a terrorist attack. While Distrigas provides some funding for this purpose today, and has taken other actions to facilitate the efforts of local law enforcement to secure the facility, I believe that federal support is needed to help ensure that local firefighters are given realistic training to deal with the types of large fires or explosions that could occur, that local police departments have the resources needed to help provide security during times of elevated Homeland Security alert status, and during LNG shipments.

Third, we need to get the Transportation Department to upgrade its LNG siting regulations to comply with the Congressional intent that all future LNG terminals be remotely sited, and demand that the Department stop merely incorporating the National Fire Protection Agency Standards into its siting rules.

Fourth, we need the Coast Guard to undertake a more thorough analysis of the safety of LNG tankers, including the issues of brittle fracture and insulation flammability. . . .

Consequences of an Attack

On page 15 of the memoirs of Richard Clark, the White House's former antiterrorism czar, and a man who served in the Clinton Administration, the first Bush Administration, and the Reagan Administration there is a disturbing passage that describes one of the discussions he had on 9/11 with Admiral James Loy, then the Commandant of the Coast Guard, as follows:

> "Jim, you have a Captain in the Port in every harbor, right." He nodded. "Can they close the harbors? I don't want anything coming in and blowing up, like the LNG in Boston." After the Millennium Terrorist Alert we had learned that al Qaeda operatives had been infiltrating Boston by coming in on liquid natural gas tankers from Algeria. We had also learned that had one of the giant tankers blown up in the harbor, it would have wiped out downtown Boston.

> "I have that authority." Loy turned and pointed at another admiral. "And I have just exercised it."

The fact that al Qaeda terrorists had come into Boston on LNG tankers was extremely disturbing to those of us who live near the Distrigas LNG facility, and it heightens the importance of ensuring that this facility, and others like it, are fully protected against terrorist attack. It also underscores the need for us to better understand the hazardous presented by such an attack. In recent months, numerous press reports have raised concerns about nature and adequacy of some of the hazard studies that were performed for the Distrigas facility shortly after the September 11th attacks.

In the fall of 2001, the Department of Energy commissioned a study by Quest Consultants, Inc. regarding public safety issues relating to the transportation of LNG to the Distrigas facility and the storage of LNG at the facility. Secretary of Transportation Mineta wrote me about the study on October 26, 2001, noting that:

> "Quest Consultants, Inc., has been hired by DOE [the Department of Energy] to perform studies related to security on vessels transporting LNG and on the onshore LNG storage tanks."

On page 10 Secretary Mineta indicated that:

> "Quest Consultants, an engineering firm, has been asked by DOE to perform a study to analyze the threat that could result from a five-meter diameter hole in an LNG tank on a vessel. Quest has performed some initial calculations to quantify the gas dispersion and fire scenarios that could follow a large release from the LNG storage tanks."

Also on page 10, Secretary Mineta further stated that in addition to actions undertaken by the Department of Transportation to enhance security at the Distrigas facility, it was his understanding that:

> "To improve security measures, DOE will work directly with the local law enforcement officials and Distrigas. MEMA [Massachusetts Emergency Management Agency] will review the studies performed by Quest and develop a plan of action. RSPA [the Department of Transportation's Research and Special Projects Administration] will be involved in the review of the onshore plant protection security features."

My office was subsequently provided with a copy of the Quest study. This Quest study, along with a study prepared for the facility operator by Lloyd's Register of Shipping, which my office was also provided, has been used by the federal government and the facility operator to reassure the Commonwealth Massachusetts about the potential danger of a fire and explosion at or near the Distrigas facility, thereby allowing the facility to reopen.

Last fall, several press reports called the accuracy of these studies into question. For example, the Quest study focused on accidents in Boston's Outer Harbor, when the most troubling public safety threats could occur in the Inner Harbor. The methodology of the study has also been called into question by numerous experts. Even the author backed away from the study's findings and conclusions. According to an October 19, 2003 article in the *Mobile Register* quotes John Cornwell, the lead scientists on the Quest study of LNG fires, as stating:

> "Some of the modeling we did for DOE—in hindsight, we should have done a more complete paper. . . . I've learned you never write anything you don't want public. We violated our own rules on that score."

The *Register* article goes on to report that Mr. Cornwell did the Quest study on short notice and that he . . . believed that it would be employed in-house by federal agencies as one of several tools used to examine LNG fire scenarios. However, according to the *Register* article:

> "In Boston, the Quest study—which has never been published in scientific journals—was apparently used by the DOE to suggest that a terrorist attack on an LNG tanker would result in only limited damage immediately around the ship. In stark contrast, published scientific studies have suggested that an LNG fire could have disastrous consequences for densely populated neighborhoods around Boston Harbor."

An article in the *Boston Herald* further suggests that the Quest study also was used by the Coast Guard to justify the resumption of LNG shipments in the months after the September 11th attacks. . . .

I wrote to the Department of Energy, the Department of Transportation, and the FERC about this study. In response, DOE acknowledged that it had commissioned the study, and reported that it had been used by DOE officials in a presentation to an interagency working group formed to assist Massachusetts following the September 11th attacks. FERC indicated that it had cited the Quest Study in the Environmental Impact Statements for four LNG terminals (The Trunkline LNG Expansion Project, the Elba Island Expansion Project, the Hackberry LNG Project, and the Freeport LNG project). DOT reported that it had used the Quest study "as a hazard assessment model that was applied specifically to the Distrigas facility" and that "the results were used to justify enhanced security procedures for vessels transporting LNG and the onshore LNG storage tanks.

All three agencies seem to have tacitly admitted the shortcomings of the Quest study in deciding to support additional LNG safety studies.

The FERC commissioned a study by ABS Consulting, which was recently released to widespread criticism from both the industry and independent experts. The ABS Study found the earlier Quest study to have several flaws, and did not recommend that it be used to analyze the consequences of a terrorist attack on an LNG tanker or terminal. While the FERC put the ABS study out for public comment, it has also indicated that it regards the ABS Study to be a final study and does not plan to request a formal peer review of this study or update it to take account of the comments that have been submitted. Both industry and expert commentary submitted to the FERC about the ABS Consulting study has been largely critical, nothing several flaws in its methodology and urging that it be peer reviewed before it is used. Despite this recommendation, FERC appears to have no plans to request a peer review of the ABS study, but has nonetheless cited the study in the Environmental Impact Statement for the Freeport LNG project. . . .

I understand that the DOE has commissioned a study by the Sandia Laboratory, which is expected to be available later in the year. While I don't know what is in the Sandia Study, I can only hope that it is more thorough than the previous government-funded LNG hazard studies. I would suggest to the Subcommittee that if the EPA issued an environmental regulation based on studies with as many flaws or shortcomings at the Quest and ABS studies, the regulated industry would be in an uproar and we would be hearing complaints about "junk science" being used to justify new regulations. Here, when we are talking about a matter that directly affects public safety; Congress also needs to demand that the science be done right, that it be methodologically sound, and that it be subjected to peer review. . . .

LNG Carrier Vessel Vulnerabilities

A second issue that I would call to the Subcommittee's attention is the potential for a terrorist attack on an LNG carrier vessel to result in failure of the cargo containment systems. Earlier this year, my office received a copy of a letter that Professor Jerry Havens of the University of Arkansas had sent Secretary Ridge regarding potential LNG tanker vulnerabilities. The Department's response suggested that the concerns posed by Professor Havens regarding: 1) the susceptibility of the foam insulation used on LNG carrier vessels to fire; 2) the possibility of rupture of the LNG containment system; and, 3) the potential for vapor pressure in the ship's LNG tanks to be elevated to levels beyond the capacity of the relief valves are either unfounded or are already being adequately addressed.

I have written the Department to request further information about the Department's basis for reaching such conclusions, based on contradictory evidence which is readily available from the public record. Here are my concerns:

First, the Department alleges that "foam polystyrene insulation, cited by Professor Havens, is not used on LNG carriers precisely because it's susceptible to melting and deformation in a fire."

This statement appears to be inaccurate. The Finnish LNG vessel manufacturer, Kvaerner Masa-Yards, reports in a sales brochure that, "the majority of the world's present LNG fleet, including those on order, incorporate the [the company's] Kvaerner Moss LNG tank design." This document goes on to state that "The design of the cargo tank insulation is based on panels made of expanded *polystyrene*." [Emphasis added]

A quick look at the Kvaerner Masa-Yard website confirms that polystyrene is still being used by the company for its LNG carrier vessels (see http://www.masayards.fi/publications/pdf/LNG.pdf). This publication describes the use of "inserts of very soft polystyrene for flexibility and fiberglass fibre reinforcement to absorb forces which are built up during the cooling down of the cargo tank.". . .

The Department told me that "the insulation on LNG carriers is a complex assembly of many layers" and that "each layer is tested for fire resistance, and its ability to stop the spread of a fire, before it can be used on LNG carriers in U.S. waters." I have several questions about this statement, . . . including:

1. Who in the federal government tests the insulation on LNG carriers for fire resistance?
2. Who is responsible for determining whether this insulation is acceptable for use on LNG carrier vessels operating in US waters?
3. What are the standards used by the federal government for determining whether or not the insulating materials used on LNG carrier vessels are acceptable?
4. What hazard analysis has been done to examine what would happen in the event that a fire on an LNG carrier vessel ignited the insulation or otherwise compromised it?
5. Are older ships required to be retrofitted with new insulation if they use insulating materials, like polystyrene, which have now been determined to be highly flammable? If not, why not? If so, how does the federal government verify that this has occurred?
6. In light of the post-9/11 threat, is there any plan by the Department, or by the Coast Guard, to review the safety standards applicable to LNG carriers (including fire safety standards) to determine whether they need to be upgraded to better address the threat of sabotage or terrorist attack?

In its letter, the Department stated that "the relief valve capacity of LNG carriers is designed based upon exposure to fire." This statement appears to assume that the insulation will continue to function properly. My concern is that if the insulation should fail as the result of a fire, the relief valves would not be capable of handling the increased vapor pressure that would result, since they would not allow for a sufficient flow through the valves. Professor Havens . . ., has suggested that if this were to be the case, the vessels, which are designed for only a few pounds overpressure, would be endangered.

The Department further suggests that concerns about the brittle fracture problem have been anticipated by U.S. regulations, which "require the use of a special crack-arresting steel in strategic locations throughout the vessel's hull." However, she goes on to acknowledge that "both the U.S. and international standards for LNG carriers were developed with the potential consequences posed by conventional maritime risks such as groundings, collisions, and equipment failures in mind." The Department then goes on to say that in recognition of the "new risks now possible in our post 9/11 world, the United States and the international community have responded by implementing additional operational security measures" under U.S. law and international maritime codes. My question is this: How does adoption of additional operational security measures suffice to address an issue—brittle fracture—that seems to go to the fundamental design of an LNG tanker? Might not terrorist threats require the use of additional measures to address the problem of brittle fracture of the ship's hull resulting from an LNG spill? . . .

Need for a New DOT Rulemaking

Let me now turn to . . . DOT's failure to properly exercise its authorities over LNG siting. Under a provision of the Pipeline Safety Act 1979, the Secretary of Transportation is supposed to ensure that the siting of all new LNG terminals is subject to standards which consider: 1) the kind and use of the facility; 2) existing and projected population and demographic characteristics of the location; 3) existing and proposed land use near the location; 4) natural physical aspects of the location; 5) medical, law enforcement, and fire prevention capabilities near the location that can cope with a risk caused by the facility; and 6) the need to encourage remote siting (see 49 U.S.C. 60103).

I am concerned about the nature and adequacy of the Transportation Department's efforts to carry out this authority. In the Committee report accompanying the House Energy and Commerce Committee's version of what became the Pipeline Safety Act of 1979 (H.Rept. 96-201, Part 1), the Committee noted:

> "One area of particular concern to the committee has been the failure to adopt comprehensive Federal standards regarding the sting, design, operation, and maintenance of liquefied natural gas facilities." In 1972, the industry consensus standards developed by the National Fire Protection Association were incorporated into the federal gas pipeline safety regulations, supposedly as an interim measure pending the development of comprehensive standards. Despite widespread concern over the adequacy of these interim standards and the growing importance of LNG as an energy source, the promised comprehensive standards have never been adopted. H.R. 51 addresses this problem by identifying the criteria to be considered by the Secretary in developing standards and setting firm deadlines for proposing and adopting them."

However, if you read the DOT regulations at 40 CFR Part 193, for example, you will find that the DOT's regulations still continue to largely incorporate by reference the National Fire Protection Association (NFPA) standards—specifically, NFPA Standard 59A.

Deputy Chief Joseph Flemming of Boston Fire Department, in his May 25, 2004 comments on the ABS Consulting Report, has raised some very serious concerns about the wisdom of continuing to rely on the NFPA standards. . . . [He also] notes that the NFPA Committee that made up these standards is largely comprised of representatives of the LNG industry or energy industry consultants, and that public officials—including firefighters who may have to deal with an LNG fire, are not routinely brought into discussion about what the appropriate standards should be. A quick check of the NFPA website reveals that the NFPA LNG Committee has representatives from BP Amoco, Distrigas, ExxonMobil, Weaver's Cove Energy, Keyspan, the American Gas Association, the American Petroleum Institute, the American Concrete Institute, and the Steel Plate Fabricators Association. . . .

Shortly after enactment of the 1979 Act, changes in the natural gas market place resulting from the decontrol of natural gas wellhead prices lead to the withdrawal of proposals for new LNG terminals and the shut down of all but the Everett, Massachusetts terminal. In a period when no new LNG terminals were being built, and existing ones were being shuttered, it is perhaps understandable that DOT did not take action to replace the NFPA standards with standards of its own. However, given the current resurgence of interest in LNG and the flood of new proposals to build LNG terminals, I think that DOT needs to revisit this matter now and consider revising its standards. I would also note that FERC has the legal authority to impose additional standards for LNG terminals. If DOT fails to Act, perhaps it is time for FERC to do so.

Conclusion

. . . Looking to the future, LNG is likely to become an increasing part of our energy mix. Given that fact, Congress needs to ensure that the federal government takes further steps to ensure that any future LNG terminals are sited in locations that prevent them from becoming an attractive terrorist target. Adhering to the Congressional directive that the Secretary consider the need for remote siting, looking at offshore siting alternatives, and updating the LNG siting rules so that they reflect sound science and decisions by federal agencies—as opposed to industry self regulatory bodies—is desperately needed. Finally, a more thorough examination of the potential consequences of a terrorist attack on an LNG tanker needs to be done. Perhaps the Sandia study will address this issue, but based on my experiences with the previous Quest and ABS Consulting studies, I think that the Congress needs to step up oversight in this area and demand that the studies that are being funded by the federal government are scientifically sound and subjected to a full peer review. . . .

Testimony Before the Subcommittee on Energy Policy, Natural Resources and Regulatory Affairs

. . . Over the past year, LNG [liquefied natural gas] has captured the attention of the energy industry and energy policy makers. Still, the reality is that LNG is not a new product in the U.S. energy market. LNG has been utilized in various applications in this country since the Second World War. Many of our pipelines and distribution companies, for example, use LNG as a method for storing natural gas. In the 1970s, as a result of supply shortages in the U.S. interstate market, the nation developed and constructed a number of LNG importation terminals in order to supplement domestic supply with natural gas from other parts of the world. LNG's role in the domestic natural gas market was short-lived, however, once wellhead decontrol and the removal of other artificial market barriers ended the supply shortage. Imported LNG quickly became too expensive to compete against much more affordable natural gas supplies from the U.S. and Canada. Three of the four terminals that were built in the 1970s were, to a large extent, mothballed until several years ago.

Why are we again focused on LNG? It now is widely recognized that North America is experiencing a fundamental shift in the supply and demand equation for natural gas. For many years, this country had a significant excess of natural gas deliverability (what was commonly referred to as the "natural gas bubble"). This kept prices low and contributed to a shift to greater use of natural gas for electric power generation, home heating and industrial processes. Demand growth gradually eliminated this excess deliverability. Supplies now are tight and prices are considerably higher—on a sustained basis—than in previous years.

Therefore, we now must develop new natural gas supply options from multiple sources to keep pace with the still growing demand for this clean-burning fuel. INGAA agrees with the assessment that we are not running out of natural gas; rather, we are running out of places where we are permitted to explore for and produce it. Abundant natural gas resources do still exist in North America and worldwide, and can supply the market with natural gas at

From Testimony before the House Committee on Government Reform, Subcommittee on Energy Policy, Natural Resources, and Regulatory Affairs. Notes omitted.

reasonable prices, provided that public policies do not unreasonably limit resource and infrastructure development.

While it is the focus of today's hearing, LNG should not be mistaken for a "silver bullet" that alone will solve the Nation's natural gas supply problem. Our current natural gas supply challenges will not be solved *only* by expanding production in the Rocky Mountain region or the Outer Continental Shelf, or *only* by building an Alaska natural gas pipeline, or *only* by importing more LNG. In order to meet anticipated demand, we must avail ourselves of *all of these options,* and more.

An important corollary to this supply message is the critical role that pipeline and storage infrastructure play in ensuring that natural gas supply can satisfy market demand. As part of a comprehensive energy policy, removing barriers to pipeline and storage infrastructure development must go hand-in-hand with efforts to enhance gas supply.

The Existing LNG Regulatory Framework

The Federal Energy Regulatory Commission (FERC) and the U.S. Coast Guard, respectively, have the authority for the approval and siting of on-shore and off-shore LNG import terminals. Both agencies have done an excellent job in streamlining the approval process for these facilities.

The Coast Guard has demonstrated its willingness, . . . to consider off-shore terminal siting proposals expeditiously. . . .

FERC's authority to approve and site on-shore LNG terminals is pursuant to section 3(a) of the Natural Gas Act (NGA). While this statutory provision does not expressly refer to the authorization and siting of facilities for importing natural gas, the courts have made clear that this function is an integral part of authorizing natural gas imports and, therefore, is within the scope of the authority conferred by section 3(a). This was addressed by the U.S. Court of Appeals for the D.C. Circuit in the 1974 *Distrigas* decision. The court said, in part:

> . . . while imports of natural gas are a useful source of supply, their potentially detrimental effect of domestic commerce can be avoided and the interests of consumers protected only if . . . the Commission exercises with respect to them the same detailed regulatory authority that it exercises with respect to interstate commerce in natural gas. In short, we find it fully within the Commission's power, so long as that power is responsibly exercised, to impose on imports of natural gas *the equivalent of Section 7 certificate requirements both as to facilities and . . . as to sales within and without the state of importation* (emphasis added). Indeed, we think that Section 3 supplies the Commission not only with the power necessary to prevent gaps in regulation, but also with the flexibility in exercising that power.

Section 7 of the NGA empowers FERC to issue certificates of public convenience and necessity authorizing the construction and operation of interstate natural gas pipelines and storage facilities. The U.S. Department of Energy and FERC have consistently applied the *Distrigas* case's construction of section 3 of the NGA in administering this part of the law.

Mr. Chairman, without going into the extensive case law, let me state that, whenever FERC's authority under either section 3 or section 7 of the NGA has come into conflict with state law, courts have consistently held in favor of federal primacy in matters of interstate and foreign commerce. The Commerce Clause of the U.S. Constitution provides the foundation for these decisions.

While FERC has exclusive jurisdiction under the NGA over the threshold decision on whether an LNG facility or interstate pipeline can be constructed, other state and federal agencies still play a substantive role in permitting this natural gas infrastructure. There are a myriad of other state and federal permits that must be obtained before a project sponsor may begin constructing its facility. FERC's application process requires that a project sponsor list all other permits that must be obtained. And FERC's orders authorizing these facilities routinely are conditioned upon the sponsor obtaining these other authorizations.

As part of discharging its responsibilities under the National Environmental Policy Act (NEPA), FERC makes all other federal, state and local permitting agencies "participating agencies" for purposes of the comprehensive NEPA process. Apart from the NEPA process and these independent sources of authority over pipeline permitting, state agencies can, and do, participate in FERC's proceedings as intervenors in order to represent the interests of their citizens.

The industry's experience in the context of interstate natural gas pipelines has been that FERC devotes significant resources to working cooperatively with these other agencies. Furthermore, the pipeline industry's experience has been that these other sources of authority over pipeline permitting, which often are federal authorities delegated to the states, provide state agencies with considerable leverage.

Industry Concerns

Safety and Security

While regulatory certainty and permit streamlining are important to constructing new LNG terminal capacity, the most significant immediate challenge facing the industry is public perception regarding safety and security. Fear of the unknown appears to be the greatest hurdle, followed closely by the various misconceptions about LNG. Such misconceptions are difficult to overcome. All of us—industry, regulators, the Executive Branch and the Congress—have a role to play in educating the public, so that we can make informed decisions about constructing needed energy infrastructure.

Fortunately, better information is on the way. In May, FERC released a report prepared by a contractor that addressed the consequences of potential LNG spill scenarios. While the Center for LNG believes that this report needs further refinement, it still is an important step in developing a public record that will support a balanced, fact-based consideration of the safety issues associated with LNG. Within the next several weeks, the Department of Energy's Sandia National Laboratory is scheduled to complete an LNG safety and security analysis that should supplement the FERC report by addressing probability of an LNG incident. Finally, Det Norske Veritas, a private risk analysis firm, soon will be completing its own study. We hope that these studies will put to rest many of the misconceptions that have characterized some of the recent public discussion of LNG safety and security issues.

Are there risks associated with LNG? Of course there are. Still, just as with any activity, this must be placed in perspective. LNG has a long and outstanding safety record. The robust worldwide trade in LNG that takes place every day is proof that LNG can be handled safely and securely. And here in the United States, FERC and the Coast Guard, working with the Department of Transportation's Office of Pipeline Safety, can mitigate risk to an even greater extent through their safety/security regulations and enforcement. We need your help, and your leadership, in getting that message out to the public.

Approval and Siting Authority

Another set of challenges facing the industry concerns jurisdictional disputes over LNG siting authority and the potential for protracted proceedings before multiple permitting agencies. The focal point for the jurisdictional issue is the dispute between FERC and the California Public Utility Commission (CPUC) regarding the authority to site an LNG terminal in the State of California.

The jurisdictional issue has been fully adjudicated by FERC and is now ripe for judicial review. FERC has gotten it right on both the law and the policy. As already noted, the courts have interpreted the NGA to provide FERC with the authority to site an LNG import facility and to attach the necessary conditions to its determination. The facts of the California case do not include anything that we believe would cause a reviewing court to reach a conclusion at odds with the *Distrigas* decision. FERC also is on firm ground as a matter of policy. To an even greater extent than with interstate commerce, the regulation of foreign commerce clearly is a function for the federal government. The siting of facilities directly associated with foreign commerce is an obvious extension of such regulation. If this regulation were left to the states, LNG facilities almost certainly would be subject to inconsistent regulation and likely would not be constructed if they were subject to traditional public utility regulation or other burdens. The nation as a whole would suffer if the ability to enhance the capacity to import this critical source of supplemental natural gas supply were frustrated. FERC jurisdiction is important to ensuring that the larger, national public interest is served, rather than just local, parochial interests.

Some have asked whether the Congress should amend section 3 of the NGA to clarify jurisdictional boundaries. We believe that, in exercising exclusive jurisdiction over the siting of LNG import facilities, FERC is acting within the bounds of the authority already conferred by the Congress under section 3 of the NGA. Still, to the extent that such an amendment would "clear the air" and permit worthy LNG projects to proceed without what may be perceived to be a cloud over jurisdiction, such an amendment may be advisable.

Beyond this threshold jurisdictional question, we also want to draw the Subcommittee's attention to the ability of federal, state and local regulators to erect impediments to the efficient, timely construction of natural gas infrastructure already authorized by FERC. While the NGA provides FERC with the exclusive authority for determining whether such projects should be constructed, other agencies increasingly are using the jurisdictional hook provided by other laws to second guess aspects of the decisions that FERC has made following the thorough review conducted under the NGA.

As noted earlier, other state and federal agencies have an integral role to play in permitting decisions related to interstate pipeline and LNG facility construction. Our point is that fairness and administrative efficiency would be served best if these other agencies coordinate the timing of their reviews with the FERC process. The already inclusive FERC NEPA process provides a vehicle for this to occur. In that way, all of the interested federal, state and local government agencies can come together under one concurrent and comprehensive review, so that all parties have equal standing and balanced decisions can be made.

In discussing regulatory impediments to LNG import facilities, we have referred frequently to the experience with interstate pipelines. We have done so for several reasons. First, the experience with interstate pipelines provides a window on what LNG facilities likely will experience as they attempt to reach the finish line of the regulatory gauntlet that must be run before ground can be broken. Second, adequate pipeline capacity is critical to bringing new natural gas supplies to consumers, whether it be LNG or North American supply. Third, specifically with respect to LNG, import facilities must be able to interconnect with the transmission pipeline network in order for the natural gas supply to reach customers. This point is demonstrated by Dominion Resources' recent announcement of plans to increase the capacity of its Cove Point LNG terminal from 1 billion cubic feet per day ("Bcf/day") to 1.8 Bcf/day, which is dependent upon FERC approval of two associated pipelines that will move the increased supply from the terminal and into the market.

Economic Impacts

What happens if the United States is unable to construct the natural gas infrastructure that we need? Quite simply, delays in pipeline and LNG terminal construction will reduce the amount of natural gas available to consumers and thereby increase the price that they must pay. This likely will cause further job losses in industrial sectors that depend on affordable supplies of natural gas, such as chemical and fertilizer manufacturing. Because an increasing amount of electricity is generated by natural gas, electricity prices will be higher for virtually all consumers.

The INGAA Foundation, Inc. now is completing an economic analysis that quantifies some of the consumer costs associated with delays in constructing new pipeline and LNG import capacity. The preliminary results are startling. The study estimates that a two-year delay in building natural gas infrastructure (both pipelines and LNG terminals) would cost U.S. natural gas consumers in excess of $200 billion by 2020. . . . California, alone, would experience increased natural gas costs of almost $30 billion over that period. And, of course, should the end result be that certain facilities are never constructed, the economic effect would be even more severe. This INGAA Foundation study is scheduled to be published in mid-July. . . .

The bottom line is that natural gas infrastructure delays and cancellations have consequences. Every consumer will pay higher prices for natural gas, electricity and the goods produced using natural gas if we do not act to ensure that adequate LNG and pipeline capacity are constructed in time to keep supplies affordable.

Legislative Proposals

Several important provisions in H.R. 6, the pending comprehensive energy legislation, would remove impediments to building LNG and pipeline infrastructure. . . .

These provisions represent areas where changes in the statutory framework for U.S. energy policy can make a real contribution to ensuring that there is adequate LNG import and pipeline infrastructure to serve the energy needs of the nation's consumers and its economy. We continue to urge the Congress to pass this legislation.

We also wish to comment on H.R. 4413, a bill recently introduced by Representative Lee Terry that would establish clear authority for LNG terminal approval, siting, and regulation. The bill would clarify exclusive FERC authority for on-shore terminal siting decisions, and require other federal and state agencies involved in permitting to work within the FERC process and make final decisions within one year of the original application. . . . Both the Center for LNG and INGAA strongly support this legislation, and believe that it should be the model for future discussions in Congress on removing impediments to new LNG import capacity.

Conclusion

In conclusion, let me emphasize the importance of public policies that foster a positive environment for natural gas infrastructure construction and investment. These large and capital-intensive projects will be constructed only if there is a rational process for reviewing and siting these facilities. Delays and detours are costly, both to project sponsors and ultimately to consumers, and in some cases the cumulative effect can be fatal to a project. We believe that the FERC provides an appropriate and inclusive forum for authorizing on-shore LNG import terminals and that FERC has done an admirable job in discharging its responsibilities. If anything, FERC's authority in these matters should be enhanced by Congress, to send a clear message as to the national importance of building natural gas infrastructure on a timely, responsible basis. . . .

POSTSCRIPT

Is Additional Federal Oversight Needed for the Construction of LNG Import Facilities?

Jerry Havens, in "Ready to Blow?" *Bulletin of the Atomic Scientists* (July/August 2003), says that the potential of LNG tankers and terminals for catastrophic explosions and fires has been understood since the 1970s, when those continental U.S. terminals now in use were approved and built. Today, an additional, major danger is that terrorists could strike at tankers or terminals. "For nearly 50 years now, all discussions of risk and probability in LNG transport have focused on how to account for human errors. The new reality is that we must now consider malicious acts as well." Havens reiterated his concerns in "LNG: Safety in Science," *Bulletin of the Atomic Scientists* (January/February 2004). For a lengthier overview of LNG, its safety record, existing legislation, industry proposals for new terminals, and risk concerns, see Paul W. Parfomak and Aaron M. Flynn, "Liquefied Natural Gas (LNG) Import Terminals: Siting, Safety and Regulation," Congressional Research Service (Order Code RL32205) (January 28, 2004) (available at http://www.wildcalifornia.org/cgi-files/0/pdfs/1078177225_LNG_Ignites_Controversy_CRS_Report_to_Congress_LNG_Jan_04.pdf).

Unfortunately, there are few (if any) locations that are so far from cities, suburbs, and protected natural environments that the risks can be ignored. Yet the need is growing. Testifying before the House Committee on Government Reform, Subcommittee on Energy Policy, Natural Resources, and Regulatory Affairs, hearing on "LNG Import Terminal and Deepwater Port Siting: Federal and State Roles" (June 22, 2004), David K. Garman, acting under secretary for energy, science, and environment of the U.S. Department of Energy, noted that according to the Energy Information Administration's *Annual Energy Outlook 2004* (AEO2004), demand for imported LNG will increase 54 percent between 2007 and 2010, which will require four new LNG terminals just on the Atlantic and Gulf Coasts. Additional terminals will be needed elsewhere. The debate is being played out on a local scale in communities such as Fall River, Massachusetts, where an LNG terminal has been proposed; see http://www.greenfutures.org/projects/LNG/. The local Coalition for Responsible Siting of LNG (http://www.nolng.org/) contends that the risks are unacceptable, while an opposing group, Friends of LNG, argues that the risks are outweighed by the economic benefits.

The ABS report criticized by Markey was released May 13, 2004 (see http://ferc.gov/industries/gas/indus-act/lng-model.pdf).

Among other things, it noted that the data available for properly assessing risk have serious shortcomings. Available models of LNG release do not take account of the actual structure of tankers, or of wave action and currents, and there are no experimental data. Further research is essential.

The Sandia report referred to in the readings was originally scheduled for release early in 2004. But in January 2004, in response to critics who thought the study reviewed too few previous safety studies, the Department of Energy asked Sandia to increase the scope of the report. The expanded report, released in December 2004 as "Guidance on Risk Analysis and Safety Implications of a Large Liquefied Natural Gas (LNG) Spill Over Water" (see http://www.fossil. energy.gov/programs/oilgas/storage/lng/sandia_lng_1204.pdf), said that the greatest risks are associated with intentional releases (as from terrorist attacks) but that appropriate security measures can keep these risks to acceptable levels. If they fail, the resulting fires may cause "major injuries and significant damage to structures" up to a third of a mile from the source and cause second-degree burns on people over a mile away.

On February 15, 2005, the Energy Subcommittee of the U.S. Senate's Committee on Energy and Natural Resources held a meeting on "The Future of Liquefied Natural Gas: Siting and Safety," at which Sandia's Mike Hightower summarized the conclusions of the Sandia study; see http:// energy.senate.gov/public/index.cfm?FuseAction=Hearings. Hearing&Hearing_ID=1384.

Contributors to This Volume

EDITORS

THOMAS A. EASTON is a professor of life sciences at Thomas College in Waterville, Maine, where he has been teaching since 1983. He received a B.A. in biology from Colby College in 1966 and a Ph.D. in theoretical biology from the University of Chicago in 1971. He has also taught at Unity College, Husson College, and the University of Maine. He is a prolific writer, and his articles on scientific and futuristic issues have appeared in the scholarly journals *Experimental Neurology* and *American Scientist,* as well as in such popular magazines as *Astronomy, Consumer Reports,* and *Robotics Age.* He is also the science columnist for the online magazine *Tomorrowsf* (http://www.tomorrowsf.com). His publications include *Focus on Human Biology,* 2d ed., coauthored with Carl E. Rischer (HarperCollins, 1995) and *Careers in Science,* 3rd ed. (National Textbook, 1996). Dr. Easton is also a well-known writer and critic of science fiction.

STAFF

Larry Loeppke   Managing Editor
Jill Peter   Senior Developmental Editor
Nichole Altman   Developmental Editor
Lori Church   Permissions Coordinator
Beth Kundert   Production Manager
Jane Mohr   Project Manager
Kari Voss   Lead Typesetter
Luke David   eContent Coordinator
Charles Vitelli   Cover Designer

AUTHORS

SPENCER ABRAHAM was sworn in as the tenth secretary of energy on January 20, 2001. Before that he had been a Senator (R-Michigan) and cochairman of the National Republican Congressional Committee.

JANET N. ABRAMOVITZ is a senior researcher at the Worldwatch Institute.

STEPHEN ANSOLABEHERE a is professor of political science at the Massa¦chusetts Institute of Technology, where he studies elections, democracy, and the mass media. He is coauthor, with Shanto Iyengar, of *The Media Game* (Macmillan, 1993) and of *Going Negative: How Political Advertising Alienates and Polarizes the American Electorate* (Free Press, 1996).

RONALD BAILEY is a science correspondent for *Reason* magazine. A member of the Society of Environmental Journalists, his articles have appeared in many popular publications, including the *Wall Street Journal, The Public Interest,* and *National Review*. He has produced several series and documentaries for PBS television and *ABC News,* and he was the Warren T. Brookes Fellow in Environmental Journalism at the Competitive Enterprise Institute in 1993. He is the editor of *Earth Report 2000: Revisiting the True State of the Planet* (McGraw-Hill, 1999) and the author of *Global Warming and Other Eco-Myths: How the Environmental Movement Uses False Science to Scare Us to Death* (Prima, 2002).

KATHERINE BALPATAKY is a Canadian freelance journalist and an executive editor of *Canadian Wildlife, WILD,* and *Biosphere* magazines. A graduate in journalism from Ryerson University, Toronto, Ontario, she is currently pursuing a master's in environmental studies at York University in Toronto.

ROBERTO BERTOLLINI is director of the division of health determinants at the WHO Regional Office for Europe, European Centre for Environment and Health, Rome, Italy.

LESTER R. BROWN is the president of the Earth Policy Institute and the founder of the Worldwatch Institute. His most recent book is *Plan B: Rescuing a Planet Under Stress and a Civilization in Trouble* (W. W. Norton, 2003).

ROBERT D. BULLARD is Ware Professor of Sociology and director of the Environmental Justice Resource Center at Clark Atlanta University. He is the author of numerous books, including *Dumping in Dixie: Race, Class and Environmental Quality* (Westview Press, 1990, 1994, 2000), *People of Color Environmental Groups Directory 2000* (Charles Stewart Mott Foundation, 2000), *Unequal Protection: Environmental Justice and Communities of Color* (Sierra Club Books, 1994), and *Confronting Environmental Racism: Voices From the Grassroots* (South End Press, 1993).

GREGORY CONKO is a policy analyst and director of Food Safety Policy at the Competitive Enterprise Institute. He specializes in issues of food and pharmaceutical drug safety regulation and on the general treatment of health risks in public policy. Conko's interests include the safety of genetically engineered foods and the application of the precautionary principle to domestic and international environmental and safety regulations.

GIULIO A. De LEO is an associate professor of applied ecology and environmental impact assessment in the Dipartimento di Scienze Ambientali at the Universit degli Studi di Parma in Parma, Italy.

DAVID FRIEDMAN is a writer, an international consultant, and a fellow in the MIT Japan program.

MARINO GATTO is a professor of applied ecology in the Dipartimento di Elettronica e Informazione at Politecnico di Milano in Milan, Italy. His main research interests include ecological models and the management of renewable resources. Gato is associate editor of *Theoretical Population Biology*.

MICHAEL GOUGH, a biologist and expert on risk assessment and environmental policy, has participated in science policy issues at the Congressional Office of Technology Assessment, in Washington think tanks, and on various advisory panels. He most recently edited *Politicizing Science: The Alchemy of PolicyMaking* (Hoover Institution Press, 2003).

ALEXANDER GOUREVITCH was an *American Prospect* writing fellow for 2002-2003.

KARL GROSSMAN is a professor of journalism at the State University of New York/College at Old Westbury. He has specialized in investigative and environmental journalism for over 35 years and is the author of *Cover Up: What You Are NOT Supposed to Know About Nuclear Power* (Permanent Press, 1994). He also hosts documentaries and interview shows for New Yorkñbased EnviroVideo.

BRIAN HALWEIL is a staff researcher at the Worldwatch Institute.

ROBERT H. HARRIS is a principal with ENVIRON International Corporation. He has over 25 years of experience in the area of environmental health and toxic chemicals, with particular emphasis on water and air pollution and hazardous waste issues. He is recognized nationally as an expert con¦sultant on the treatment and disposal of municipal solid and hazardous waste, as well as on air, soil, and groundwater contamination.

DIANE KATZ is the director of science, environment, and technology policy at the Mackinac Center for Public Policy in Michigan.

DAVID N. LABAND is a professor of economics and policy at the Forest Policy Center in the School of Forestry and Wildlife Sciences at Auburn University in Alabama. He is the author, with George McClintock, of *The Transfer Society: Economic Expenditures on Transfer Activity* (National Book Network, 2001). Labandís research interests include forest economics, causes and consequences of environmental policy, and land use planning.

DWIGHT R. LEE is the Ramsey Professor of Economics and Private Enterprise in the Terry College of Business at the University of Georgia. He received his Ph.D. from the University of California at San Diego in 1972 and his research has covered a variety of areas, including personal finance, public finance, the economics of political decision making, and the economics of the environment and natural resources. Lee has published over 100 articles and commentaries in academic journals, magazines, and newspapers. He is coauthor, with Richard B. McKenzie, of *Getting Rich in America: Eight Simple Rules for Building a Fortune and a Satisfying Life* (HarperBusiness, 2000).

MARK LYNAS is the author of *High Tide: The Truth About Our Climate Crisis* (Picador, 2004).

MARCO MARTUZZI is on the staff of the WHO Regional Office for Europe, European Centre for Environment and Health, Rome, Italy.

GEORGE MARSHALL is an environmentalist who works with Rising Tide, a grassroots group dedicated to taking local action and building a movement against climate change.

ANNE PLATT McGINN is a senior researcher at the Worldwatch Institute and the author of "Why Poison Ourselves? A Precautionary Approach to Syn¦thetic Chemicals," Worldwatch Paper 153 (November 2000).

HENRY I. MILLER is a research fellow at the Hoover Institution at Stanford University and the author of *Policy Controversy in Biotechnology: An Insider's View* (Academic Press, 1997). His research focuses on public policy toward science and technology, especially pharmaceutical development and the new biotechnology. His work often emphasizes the excessive costs of government regulation and models for regulatory reform.

STEPHEN MOORE is a director of the Cato Institute and a contributing editor to *National Review.* Moore is also the president of the Club for Growth.

SEAN PAIGE is the Warren Brookes Fellow at the Competitive Enterprise Institute, a free-market policy group in Washington, D.C.

DINAH M. PAYNE is a professor of management at the University of New Orleans in Louisiana. She teaches business law, business ethics, and international management.

HENRY PAYNE is a freelance writer and an editorial cartoonist for the *Detroit News.*

DAVID PIMENTEL, a professor at Cornell University in Ithaca, New York, holds a joint appointment in the Department of Entomology and the Section of Ecology and Systematics. He has served as consultant to the Executive Office of the President, Office of Science and Technology, and as chairman of various panels, boards (including the Environmental Studies Board), and committees at the National Academy of Sciences, the United States Department of Energy, and the United States Congress.

CECILY A. RAIBORNE is a professor of accounting at Loyola University in New Orleans, Louisiana. She teaches financial and managerial accounting.

JEREMY RIFKIN is the president of the Foundation on Economic Trends in Washington, D.C., and author of *The Hydrogen Economy: The Creation of the World Wide Energy Web and the Redistribution of Power on Earth* (Tarcher/Putnam, 2002).

MARGOT ROOSEVELT is a national correspondent for *Time* magazine. Since moving to Los Angeles, California, in 1994, she has specialized in social issues, covering immigration, education, crime, trade, energy, and environmental stories, including controversies over genetically modified food.

CHARLES W. SCHMIDT is a freelance science writer based in Portland, Maine. He received an M.S. in public health from the University of Massachusetts at Amherst, with a concentration in toxicology. He has written for *Environmental Health Perspectives, Technology Review, New Scientist, Popular Science,* the *Washington Post, Child* magazine, *Environmental Science and Technology,* and more.

FRED SINGER is the president of the Science & Environmental Policy Project and distinguished research professor at George Mason University.

GAR SMITH is a former editor of *Earth Island Journal* and the current editor of the weekly online ecozine *The Edge.* He also co-founded Environmentalists Against War.

MIKE TILCHIN is a vice president of CH2M HILL.

BRIAN TOKAR is an associate faculty member at Goddard College in Plainfield, Vermont. A regular correspondent for Z magazine, he has been an activist for over 20 years in the peace, antinuclear, environmental, and green politics movements. He is the author of *The Green Alternative: Creating an Ecological Future,* 2d. ed. (R & E Miles, 1987).

MICHELE L. TRANKINA is a professor of biological sciences at St. Maryís University and an adjunct associate professor of physiology at the University of Texas Health Science Center, both in San Antonio, Texas.

JAY VANDEVEN is a principal with ENVIRON International Corporation. He has 16 years of experience in the assessment and remediation of soil and groundwater contamination, contaminant fate and transport, environmental cost allocation, and environmental insurance claims.

ROBERT R. WARNER is a professor of marine ecology at the University of California at Santa Barbara. He has served on the Science Advisory Panel to the Marine Reserves Working Group for the Channel Islands National Marine Sanctuary, and as the research chair on the Sanctuary Advisory Council.

HOWARD YOUTH is a researcher and writer on wildlife conservation issues.

Index

Abraham, Spencer, recommendation to store nuclear waste at Yucca Mountain by, Nevada, 331–41

Abramovitz, Janet N., on valuing goods/services of ecosystems, 41–52

Abuja Declaration (2000), 288

acid rain, 103, 106

Acid Rain Program, 98

agriculture: benefits of genetically-modified (transgenic) plants to, 234–48; disadvantages of genetically-modified plants to (transgenic), 249–59; use of DDT in, 287, 292–93, 294. *See also* food production and supply; pesticides

air pollution, 113; market solutions for controlling, 96–101; perils of "free market capitalism" and credits for, 102–8

Alaska Wilderness League, 138

animals: exotic (non-native), 74; in, and near marine reserves, 265–67

Arctic National Wildlife Refuge (ANWR), proposed oil drilling in, 128–40

Attaran, Amir, 296–97

Audubon Society, National, 78, 128; land ownership and land use by, 129–30

automobiles: fuel standards for, 135–36, 162; gasoline remains best fuel for, 162–65; hydrogen fuel cells as emerging fuel for, 158–61

Bacillus thuringiensis (Bt), 238

Bailey, Ronald: on myths of environmentalism, 112–15; on problems of sustainable-development concept, 34–36

Baliunas, Sallie, 148

Balpataky, Katherine, on oil drilling in Arctic National Wildlife Refuge, 137–40

Barbie dolls, 306

bats as pollinators, 48–49

Bean v. Southwestern Waste Management, Inc., 84

bednets, controlling malaria by using, 287–88, 289, 291

bees as pollinators, 42, 47–50

Bertollini, Roberto, on precautionary principle applied to human health protection, 4–8

biodiversity: conservation programs for, 76–79; disappearing bird species, 72–79; establishing marine reserves to protect ocean, 264–68; exotic/non-native species as threat to, 74; habitat loss as threat to, 72–73; hot spots of, 75–76; hunting, pesticides, and global warming as threat to, 74–75; problems with cost-benefit analysis and pricing of, 53–61;

regulations and overprotection of, 68–71; seed saving and preservation of, 256–57

biological hot spots, 75–76

biotechnology: advantages of, for world agriculture, 234–48; disadvantages of, for world agriculture, 249–59; precautionary principle and, 13–15

birds: conservation programs for protecting, 75–79; ecological services provided by, 49; effects of DDT on, 293; multiple threats to biodiversity of, 72–75

Blair, Tony, 146

Brazil, setting fires to clear land in, 41

Brown, Lester R., 112; on environmental problems and human population, 216–23

Bullard, Robert D., on environmental justice and environmental racism, 83–86

Burtraw, Dallas, 98, 99, 100, 101

Bush, George H.W., presidential administration of, 102

Bush, George W., presidential administration of, 86, 162, 274–75, 297; on automobile fuels, 163, 164; "Clear Skies" initiative, 182–84, 191; economic incentives in environmental policy of, 96–97; energy policy, 186, 188, 205–6; on oil drilling in Arctic National Wildlife Refuge, 139–40; policy of, on Clean Air Act's New Source Review provisions, 170–91; on science and space program, 152, 153; on Superfund program, 325

businesses, level of sustainable-development efforts by, 28–31

business ethics, sustainable development and, 24–33

Byrd, Dan, 313, 314

CAFE (Corporate Average Fuel Economy) standards, 135, 163

cancer: DDT and breast, 294, 304; questions about supposed environmental causes of, 308–12

cap-and-trade schemes, 97–99

Caribou Commons Project (CCP), 138, 139

caribou herds, oil drilling and, 134–35, 137–40

Carson, Rachel, 112, 285, 293, 308

chemicals. *See* toxic chemicals

Cheney, Dick, 175, 205

Chernobyl nuclear accident, 201, 205

Chicago Climate Exchange (CCX), 98

China, 250; demand for grains in, 219–20; one-child policy in, 225–26, 229

Cinergy utility company case, 176–77

Ciriacy-Wantrup, Siegfried V., 55

Citizens Against Nuclear Trash (CANT), 85

Citizens Against Toxic Exposure, 85

civil rights law, environmental discrimination and, 84–85, 88, 89

Clean Air Act, 86, 96, 309; amendments to, 98, 102, 103, 105, 106; emissions trading and, 98–99, 102–6; J. Holmstead on needed revisions to, 186–91; E. Spitzer on enforcing New Source Review provisions of, 170–85

Clean Water Act, 96, 309

"Clear Skies" initiative, 182–84, 191

climate change, 75, 217, 220; deniers of, 148. *See also* global warming

climate regulation, role of forests in, 46–47

Clinton, Bill, presidential administration of, 180, 275; environmental racism regulations in, 86, 88, 88–91; on population regulation, 225

"closed" nuclear fuel cycle, 198

Cohen, Stanley, 146

Colborn, Theo, 310–11

commercial fishing, threat of marine reserves to, 269–75

Comprehensive Environmental Response, Compensation and Liability Act of 1980, 324. *See also* Superfund program

Conko, Gregory, on perils of precautionary principle, 9–20

conservation biology, 76

conservation programs, bird, 76–79

Conservation Reserve Program, U.S. Department of Agriculture, 76–77

conservation tillage, 254, 258

consumption, environmental problems and excessive, 216–17

contingent valuation methods, 54–58

cost-benefit analysis, 7; inadequacies of, in pricing biodiversity and ecological services, 53, 54–58

cropland, 236; declining fertility of, 116–18; use of marginalized, 239

crops: declining yields of, 116–18; genetically-modified, 13–15, 236, 237–41; herbicide-resistant, 253–54; pollinators and, 48–49

DDT (dichlorodiphenyl trichloroethane), 75, 79; alternatives to, for controlling mosquitoes and malaria, 282–90; continuing need for, to control malaria, 291–97; derivative of, as hormone mimic, 303

decision-making: cost-benefit versus multicriteria analysis in, 53–51; precautionary principle and, 4–8, 15–17

deforestation, 41–42, 114

De Leo, Giulio A., on pricing biodiversity and ecosystem services, 53–61

denial, psychological, regarding global warming, 145–49

Department of Energy (DOE), U.S., nuclear power and, 204, 333, 334, 335–36

developing countries: critique of sustainable development applied to, 34–36; environmental policies in, 25–26; genetic pollution by transgenic plants in, 254–55; malaria control and DDT use in, 282–97; renewable energy technology in, 160; threat of gene patents and seed contracts on seed saving in, 255–58

distributed-generation associations (DGAs), 159

Dry Tortugas, commercial fishing near, 269–75

Dumping in Dixie: Race, Class and Environmental Quality (Bullard), 85

Earth Summit, Rio de Janeiro, 24, 25, 145

ecological footprint, 61

ecological goods and services: inadequacies of cost-benefit analysis related to pricing of biodiversity and, 53–61; support for valuing, 41–52

economic growth, environmental protection and poverty reduction linked to, 34–36

economics: cost-benefit versus multicriteria analysis for pricing biodiversity and ecological services, 53–61; human population, over consumption, and bubble economy, 216–17, 220, 222; incentives for reducing pollution based on, 96–101; B. Lomborg on environmental protection and, 114–15; of nuclear power, 199–201; perils of "free market environmentalism," 102–8; valuing ecological services and, 41–52. *See also* markets and market solutions

ecotourism, 77

Ehrlich, Paul, 112, 225

eMergy concept, 61

emissions trading schemes, 97–101; critique of, 103–8

Endangered Species Act, 107

endocrine disrupters. *See* hormone mimics (endocrine disrupters)

Endocrine Disrupter Screening and Testing Advisory Committee (EDSTAC), 312–14

energy issues: forestalling global warming, 145–53; hydrogen as replacement for fossil fuels, 158–65; oil drilling in Arctic National Wildlife Refuge (ANWR), 128–40; pollution controls on existing power plants, 170–91; reviving nuclear power, 196–209

environment, use of transgenic plants and impact on, 240, 243–45

Environmental Defense Fund (EDF), 106

environmental goods and services. *See* ecological goods and services

environmental groups, proposed ownership of ANWR oil reserves by, 128–36

environmental impact assessments (EIA), 53, 59–60

environmentalism: critique of sustainable-development concept in, 34–36; debunking ideological myths of, 112–15; perils of "free market," 102–8 environmental justice: civil rights laws and, 83–86; emissions trading and, 100; environmental racism as hoax and, 87–91

environmental policy: in developing world, 25–26; precautionary principle and, 4–8

environmental problems: critique of Bjorn Lomborg's skepticism regarding severity of, 116–21; debunking myths about, 112–15

Environmental Protection Agency (EPA), 96, 309; Clean Air Act and, 170, 172, 173–76, 186; environmental justice movement and, 86, 87, 88–91; pollution credits auction by, 104; Superfund program of (*see Superfund program*)

environmental racism: environmental justice concept and civil rights laws applied to, 83–86; as hoax, 87–91; nuclear waste storage and, 207

E-Screen test, 311–13

estrogens in environment. *See* hormone mimics (endocrine disrupters)

ethics: precautionary principle and, 7; sustainable development and business, 24–33

European Commission (EC), precautionary principle and policies of, 12, 13–15

"existence" value, defined, 56

exotic (nonnative) species as threat to biodiversity, 74

farmland: biodiversity conservation programs on, 76–77; declining fertility of, 116–18; as ecosystem, 258. *See also* cropland

Federal Drug Administration (FDA), 16–17

fisheries: establishing marine reserves to protect global, 264–68; threats to commercial fishing from marine reserve protection of, 269–75

Florida: biodiversity conservation programs in, 77–78; marine reserves and commercial fishing in, 269–75

Food and Agricultural Organization (FAO), 116

food production and supply: benefits of genetically-modified plants for increased, 234–59; dangers of genetically-modified plants for, 249–59; declining, 116; establishing marine reserves to protect global fisheries, 264–74; human population growth and, 218–20, 228; precautionary principle applied to, 13–15

Food Quality Protection Act (FQPA), 312, 313, 314

forests: deliberately-set fires in, 41, 43–44; ecological services provided by, 44–47; intensively-managed, and biodiversity protection, 68–69; B. Lomborg on, 114; loss of species habitat in, 73

fossil fuels, 218; hydrogen as replacement for, 158–65

freedom versus precaution, 19–20

"free market environmentalism," perils of, 102–8

Friedman, David, on environmental racism as hoax, 87–91

fuel cells, hydrogen powered, 158–61, 162, 164

fuel cycles, nuclear, 198–99, 204

fuel-economy standards, 135–36, 162, 163

Fungicide, Insecticide, and Rodenticide Act, 309

gasoline as best fuel for automobiles still, 162–65

Gatto, Marino, on pricing biodiversity and ecosystem services, 53–61

genes, patents on, 255–56

gene-splicing. *See* biotechnology

genetically-modified (transgenic) plants, 13–15, 234–59

genetic engineering. *See* biotechnology

genetic immunization of plants, 239

genetic pollution, 254–55

genetic resistance: of insects, to insecticidal transgenic plants, 251–52, 253f; of mosquitoes to DDT, 282, 286, 293; of plants, to herbicides, 253–54

globalization, stance against, 35

Global Malaria Eradication Program, 1955, 285

global warming: adapting to, 114, 149; interview with Fred Singer on bad science regarding, 150–53; B. Lomborg and skepticism about, 114, 148; psychological denial regarding threats of, 145–49; as threat to species biodiversity, 75

GM (genetic modification) technology. *See* biotechnology

Goffman, Joseph, 97

Goldman, Lynn, 312–13

Goode, Stephen, interview with Fred Singer about global warming, 150–53

goods and services. *See* ecological goods and services

Gore, Al, 88, 112, 163

Gough, Michael, on pesticides as endocrine disrupters and effects on food production, 308–14

Gourevitch, Alexander, on malaria and use of DDT, 291–97

grasslands, loss of species habitat in, 73

greenhouse gases, 145, 164; emissions trading scheme for, 98

Greenpeace, 18, 112

Green Revolution, 238

Grossman, Karl, on reviving nuclear power, 205–9

habitat loss, threats to biodiversity due to, 72–73

Halweil, Brian, on transgenic plants and food production, 249–59

Hardin, Garrett, 69

Harris, Robert H., on Superfund hazardous waste program, 319–22

hazardous wastes, 309; environmental racism concept and, 85, 87; mine tailings as,

322, 323–26; nuclear waste storage, 197, 207, 331–45; Superfund program to cleanup, 309, 319–26

health, human: effects of hormones mimics (endocrine disrupters) on, 10–11, 18, 302–14; infectious disease and, 119–20; lead poisoning from mine tailings, 323–26; nuclear radiation and, 207; poverty and, 11–12, 90; precautionary principle and protection of, 4–8; Superfund program effectiveness and protection of, 319–26; threat of toxic chemicals to, 118, 119, 323–26; transgenic plants and, 242–43; use of DDT to control mosquitoes and malaria affecting, 282–97

health impact assessment (HIA), 8

hedonic pricing, 55

herbicide-resistant crops (HRCs), 253–54

Hewett, Andrew, 34

HIV virus and AIDS, 217, 223

Holmstead, Jeffrey, on need to revise New Source Review of Clean Air Act, 186–91

hormesis, 207

hormone mimics (endocrine disrupters): hazards of, for human health, 302–7; lack of evidence for hazards of, and increased food costs, 308–14; precautionary principle and, 10–11, 18

Hotelling, Harold, 55

household production functions, 55

hunting and trapping as biodiversity threat, 74–75

hydrogen as fuel, 221; advantages of, as replacement for fossil fuels, 158–61; limitations on, as replacement for gasoline, 162–65

hydrogen energy webs (HEWs), 159

Iceland, 221

implicatory denial, 146

Indonesia, setting fires to clear land in, 41, 43–44, 51–52

induced-mutation breeding, 14

infectious disease, 119–20; HIV and AIDS, 217, 223; malaria, 282–97

Interdisciplinary MIT Study, on future of nuclear power, 196–204

Intergovernmental Panel on Climate Change (IPCC), 145

International Atomic Energy Agency (IAEA), 204

International Chamber of Commerce Commitment to Sustainable Development, 32

Israel, water use in, 221

justice, environmental, and racism, 83–86

Kassi, Norma, 137–40

Katz, Diane, on replacement of fossil fuels with hydrogen, 162–65

Kyoto Protocol, 24, 98, 145, 153

Laband, David N., on regulation of biodiversity, 68–71

lead poisoning, 323–26

Lee, Dwight R., on oil drilling in Arctic National Wildlife Refuge (AnWR), 128–36

life expectancy, human, 113, 217, 227

"limits to growth" model, 35

Lindzen, Richard, 148

Lomborg, Bjorn, on skepticism about severity of environmental problems, 112–15, 148; D. Pimentel's critique of, 116–21

Lynas, Mark, on psychological denial of global warming, 145–49

McGinn, Anne Platt, on malaria, mosquitoes, and use of DDT, 282–90

McLachlan, John, 312, 313

malaria: alternatives to DDT for controlling, 282, 284–90; causes of, 283–84; DDT as best method of preventing, 291–97

Malthus, Thomas and Malthusianism applied to human population and environmental problems, 224–30

marine reserves: protecting global fisheries by establishing, 264–68; threats of, to commercial fishing, 269–75

markets and market solutions: gas tax to cut consumption, 163; perils of using, to protect environment, 102–8; pricing ecological services based on, 54–55; support for pollution-rights trading, 96–101

Marshall, George, on psychological denial of global warming, 145–49

Martuzzi, Marco, on precautionary principle applied to human health protection, 4–8

Medred, Craig, 132–33

methane, hydrogen fuels extracted from, 164

Miller, Henry I., on perils of precautionary principle, 9–20

mining sites, hazardous wastes at, 322, 323–26

Mittler, Daniel, 35, 36

monitored natural attenuation (MNA) at Superfund sites, 320–21

Monsanto Corporation, 249, 251, 253, 257

Moore, Stephen, on Malthusian concepts and human population growth, 224–30

Morris, Julian, 148

mosquitoes, preventing malaria by using DDT to control, 282–97

mountains, loss of species habitat in, 74

multicriteria analysis in decision-making, 53, 54, 58–59

Narain, Sunita, 35

National Ambient Air Quality Standards (NAAQS) program, 187–88

National Black Environmental Justice Network (NBEJN), 84

National Cancer Institute (NCI), 308

National Contingency Plan (NCP), 319–20

National Energy Policy, Bush administration, 186, 188
National Environmental Justice Advisory Council (NEJAC), 88, 90
National People of Color Environmental Leadership Summits, 83
National Priorities List (NPL), Superfund program, 321–22
Native Americans: proposed oil drilling on lands of Alaskan, 137–40; storing nuclear waste on lands of, 207
Nature Conservancy, 77; land ownership and use by, 131–32
New Source Review (NSR) provisions of Clean Air Act: J. Holmstead on needed revisions of, 189–91; E. Spitzer on need to maintain and enforce, 170–85
Nigara Mohawk case, 178–79
Norton, Gale, 139, 140
Nuclear Information and Resource Service (NIRS), 205, 207
nuclear power, 164; perils of, as energy source, 205–9; storing waste produced by, 331–45; support for expanding, as energy source, 196–204
Nuclear Regulatory Commission (NRC), 85, 206, 333
nuclear waste, 197; perils of transporting and storing, at Yucca Mountain, 207, 342–45; recommendation on management of, 202–3; recommendation to store, at Yucca Mountain, 202, 331–41
nuclear weapons nonproliferation, 203–4, 339
nutritional benefits of transgenic plants, 239–40

Occupational Safety and Health Administration (OSHA), 309
oil drilling in Arctic National Wildlife Refuge: opposition to, 137–40; support for different approach to, 128–36
"once-through" nuclear fuel cycle, 198–99, 204
O'Neill, Paul, 205
"open market" emissions trading, 100
Our Common Future (United Nations), 34
Ozone Transport Commission, 98–99

Paige, Sean, on marine reserves as threat to commercial fishing, 269–75
Passive bystander effect, 149
Payne, Dinah M., on sustainable development and business ethics, 24–33
Payne, Henry, on replacement of fossil fuels with hydrogen, 162–65
PCBs (polychlorinated biphenyls), 75; environmental racism and exposure to, 84–85
pesticides: effects on pollinators, 48; genetic resistance to, 251–52, 253f, 282, 286, 293; lack of evidence for hazards of, and food costs, 308–14; as threat to biodiversity, 75; as threat to human

health, 119, 302–7, 312, 314. See also DDT (dichlorodiphenyl trichloroethane)
pests: genetic resistance of, to pesticides and herbicides, 251–54, 282, 286, 293; mosquitoes as, and malaria, 282–97; resistance of transgenic plants to, 238; transgenic plants lethal to, 251–52
pharmaceuticals: to control malaria, 288; transgenic plants and production of, 241
Philippines, 51
phytoestrogens, 305
Pilcher, Oklahoma, hazardous mining wastes at, 323–26
Pimentel, David, 46; critique of environmental skeptic E. Lomborg by, 116–21
Planned Parenthood, 225, 227
plants: exotic (nonnative), 74; genetically-modified, 13–15, 234–59; in, and near marine reserves, 265–67; relationship of, to pollinators, 47–48
plastics, 11; as source of environmental hormone mimics, 305–7
plausibility versus provability, 18
plutonium, 203–4
policy: precautionary principle and making, 4–8. See also environmental policy
pollinators, decline in populations of, 42, 47–50
pollution: perils of to trading credits for emissions of, 102–8; support for emissions trading and market incentives for reducing, 96–101. See also air pollution; water pollution
pollution taxes, 107
population, human, 31; growth of, and escalating environmental problems, 121, 216–30; Malthusian theory, and coercive controls on growth of, 224–30
poverty: adverse effects on health caused by, 11–12, 90; agricultural biotechnology and, 257–58; economic growth and reduction of, 34–36
power plants: air pollution controls on, required by law, 170–85; licensing nuclear, 206; need for revision on laws regarding air pollution controls on, 186–91
precautionary principle, 151; perils of basing policy on, 9–20; protecting human health and role of, 4–8
precaution versus freedom, 19–20
Prevention of Significant Deterioration (PSD) program, 171, 187–88
Price-Anderson Act, 207–8
pricing of ecological services: inadequacies of cost-benefit analysis for, 53–61; support for valuation and, 41–52
private landownership: overprotection of biodiversity and regulations affecting, 68–71; perspectives of, and land use, 129–32

property rights, excessive regulation and threats to, 69–70
provability versus plausibility, 18
PVC (polyvinyl chloride), 85
puberty, estrogen mimics linked to precocious, 304–5
public attitudes toward nuclear power, 199
Public Employees for Environmental Responsibility, 100
public lands, free-market schemes and, 107–8
pyrethroids as alternative to DDT, 286–88, 296

racism. See environmental racism
Raiborn, Cecily A., on sustainable development and business ethics, 24–33
Rainey Wildlife Sanctuary, Louisiana, 129–31
Reagan, Ronald, administration of, 106, 228
Redefining Progress, 216
reference group, pricing using, 56–57
refugia, 252
Regional Clean Air Incentives Market, California, 98
regulations, 114–15; biodiversity protection and excessive, 68–71; environmental justice concept and EPA, 86, 87, 88–91; precautionary principle applied in, 12, 15–17; threats to commercial fishing from, 269–75
replacement cost technique, 54–55
reproductive system, effects of estrogen mimics on human, 302–5
resistance management plans, 252
Resource Recovery and Control Act, 309
Rifkin, Jeremy, on hydrogen as fuel, 158–61
risk, precautionary principle and, 9–10
risk assessment, 7, 16; as Superfund sites, 319–20
risk perception, 9–10
Roll Back Malaria program (1998), 288–90, 295–97
Roosevelt, Margot, on hazardous waste and Superfund program, 323–26
Royal Society of London et al., on transgenic plants and world agriculture, 234–48
Rural Advancement Foundation International (RAFI), 257

Safe Drinking Water Act, 309, 311–12
safety and security, nuclear power and, 197, 201–4, 205, 207–9, 338–40
Safiotti, Umberto, 308
St. James Citizens for Jobs and the Environment, 85
Sawyer, Steve, 34, 36
Schmidt, Alexander, 17
Schmidt, Charles W., on market for pollution and emissions trading, 96–101
security. See safety and security, nuclear power and
seed contracts, 256–57
Shiva, Vandana, 35
Silent Spring (Carson), 285, 293

Simon, Julian, 112, 113
simulated market, contingent value methods and, 57–58
Singer, Fred, on weakness of global warming arguments, 150–53
The Skeptical Environmentalist (Lomborg), 112–15; critique of, by D. Pimentel, 116–21
Smith, Gar, 34; on nuclear waste and threat of terrorism, 342–45
soil erosion, 46, 116–17
Soon, Willie, 148
Soto, Anna, E-Screen test by, 311–13
source control treatment at Superfund sites, 320
South Africa, malaria-control programs in, 286–87
Spitzer, Eliot, support for maintaining and enforcing air pollution regulations, 170–85
States of Denial (Cohen), 146
Stockholm Convention on Persistent Organic Pollutants (POPs) of 2001, 283, 293, 294
Stott, Philip, 148
sulfur dioxide (SO2), emissions trading credits for, 98, 103
Superfund Amendments and Reauthorization Act (SARA), 319
Superfund program, 85, 87, 309; effectiveness of, in protecting public health, 319–22; ineffectiveness of, in protecting public health, 323–26
sustainable development: business ethics linked with, 24–33; defined, 24–25; shortcomings of environmental ideology regarding, 34–36
swine flu vaccine, 16

Tar Creek, Oklahoma, mine tailings as hazardous wastes at, 323–26
taxes: on gas, 163; on pollution, 107
Tennessee Valley Authority case, 177
terrorism, 223; nuclear reactors, transport of nuclear waste, and threat of, 208–9, 342–45
Three Mile Island nuclear accident, 201, 205
Tilchin, Mike, on Superfund hazardous waste program, 319–22
timber production and biodiversity protection, 68–71
Tokar, Brian, on pollution credits and "free market environmentalism," 102–8
toxic chemicals: alternatives to DDT for controlling mosquitoes and malaria, 282–90; continuing need for DDT to control malaria, 291–97; environmental racism and exposure to, 84–85; nuclear waste as, 197, 330–45; possible health threats of environmental hormone mimics (endocrine disrupters), 10–11, 18, 300–314; Superfund program for cleanup of, 319–26; as threat to biodiversity, 75
Toxic Substances Control Act, 309
Trading Thin Air, 100–101

tragedy of the commons, 69
Trankina, Michele L., on health hazards of environmental estrogens, 302–7
transgenic plants, 13–15; advantages of, for food production, 234–48; disadvantages of, 249–59
travel cost method, 55

Uganda, attempts to control malaria in, 291
U.N. Conference on Trade and Development, 106
U.N. Millennium Development Goals, 219
U.N. Population Fund (UNFPA), 225, 226
United States: environmental leadership and, 222; grain reserves, 219–20
U.S. Agency for International Development (AID): malaria program, 295, 297; population regulation programs, 224–25
U.S. Congress, role in maintaining Clean Air Act, 182–84
The United States Experience with Economic Incentives for Protecting the Environment (EPA), 96
U.S. Flood Control Act of 1936
urban and rural differences on biodiversity protection, 70–71

vaccines, transgenic plants and production of, 241
Vandeven, Jay, on Superfund hazardous waste program, 319–22

Virginia Electric Power Company (VEPCO) case, 176–77

Wackernagel, Mathis, 216
Warner, Robert R., on marine reserves and fish production, 264–68
water pollution, 118, 119
Watkins, Harvey, commercial fishing activities of, 269–75
weather anomalies, climate change and, 75
wetlands, loss of species habitat in, 73
Wildavsky, Aaron, 11–12
Wisconsin Electric Power Company (WEPCO) ruling, 181–82, 188
World Business Council for Sustainable Development, 25
World Commission on Environment and Development (Brundtland Commission), 24–25
World Health Organization (WHO), 119, 285
World Trade Organization (WTO), 12, 36
World Wildlife Fund, 138
xenoestrogens, health hazards of environmental, 302–7

Youth, Howard, on disappearing bird species, 72–79
Yucca Mountain, Nevada: perils of storing nuclear waste at, 342–45; recommendation to store nuclear wastes at, 331–41